Design Patterns for
Embedded Systems in C

Design Patterns for Embedded Systems in C

An Embedded Software Engineering Toolkit

Bruce Powel Douglass, PhD

AMSTERDAM • BOSTON • HEIDELBERG • LONDON
NEW YORK • OXFORD • PARIS • SAN DIEGO
SAN FRANCISCO • SINGAPORE • SYDNEY • TOKYO
Newnes is an imprint of Elsevier

Newnes

Newnes is an imprint of Elsevier
Linacre House, Jordan Hill, Oxford OX2 8DP, UK
30 Corporate Drive, Suite 400, Burlington, MA 01803, USA

First edition 2011

Library of Congress Cataloging-in-Publication Data
Douglass, Bruce Powel.
Design patterns for embedded C: an embedded software engineering toolkit / Bruce Powel Douglass – 1st ed.
    p.   cm.
Includes index.
ISBN 978-1-85617-707-8
1. C (Computer program language)   2. Embedded computer systems–Programming.   3. Software patterns.   I. Title.
QA76.73.C15D685 2011
005.13'3–dc22

                                                                                    2010027721

A catalog record for this book is available from the Library of Congress

British Library Cataloguing in Publication Data
A catalogue record for this book is available from the British Library

For information on all Newnes publications
visit our website at books.elsevier.com

ISBN: 978-1-85617-707-8

Printed and bound in USA
13   12   11   10   9   8   7   6   5   4   3   2

Dedication

This book is dedicated to my elder son Scott, who graduated with an almost-4.0 average in physics and is going on to graduate school — nicely done Scott!

Contents

Preface

The predominate language for the development of embedded systems is clearly C. Other languages certainly have their allure, but over 80% of all embedded systems are developed in this classic language. Many of the advances in the industry assume the use of object-oriented languages, web clients, and technologies that are either unavailable in the bare-metal development environment of embedded systems, or are too memory or resource intensive to be effectively deployed.

Design patterns are one of these profitable areas of advancement. A design pattern is a generalized solution to a recurring problem. Design patterns allow several advantages. First, they allow users to reason about solutions in a more abstract way and to codify their important properties. Since all design patterns are about optimization of some set of design criteria at the expense of others, different design patterns might address the same operational context but provide different benefits and costs. By reifying design patterns as a basic concept, we can reason about the best ways to optimize our systems and the technologies and approaches to accomplish that.

Secondly, design patterns allow us to reuse solutions that have proven themselves to be effective in other, similar circumstances. This is certainly a larger scope of reuse than reusing lines of code or individual functions. Because design patterns can be analyzed for their performance and optimization properties, we can select the best patterns for our particular problems.

Additionally, design patterns give us a larger set of reusable building blocks with which to describe our system. If you say the system uses a "symmetric deployment pattern with RMS multitasking scheduling and a triple modular redundancy approach" that sums up a significant portion of the architectural optimization decisions about your system. Design patterns make our vocabulary larger and more expressive.

Lastly, design patterns provide a unit of reference. If you Google "design patterns" you get over 3,000,000 hits. If you search the Amazon book department online, you get a list of 793 books. There is a great deal of work that goes on in the area of defining and using design patterns, so we are exposed to an enormous wealth of reusable and proven solutions from which we can select, mix, and bring to bear appropriate solutions.

This book brings the power of design patterns to the embedded C developer. Where appropriate, we use an object-based implementation policy with code examples to see how the patterns are implemented and used. The patterns are divided up into different areas of relevance to the embedded developer, including:

- Hardware access
- Concurrency
- State machine implementation
- Safety and reliability

In each area, different patterns are provided that address common concerns.

Audience

The book is oriented toward the practicing professional software developer and the computer science major, in the junior or senior year. It focuses on practical experience by solving posed problems from the context of the examples. The book assumes a reasonable proficiency in the C programming language. UML is used to graphically represent the structure and behavior of the patterns, but code examples are given to clearly show the design and implementation concepts.

Goals

The goal of this book is to provide a toolbox for the embedded C developer and to train him or her on the appropriate use of those tools. An agile development workflow, called the Harmony Process™ (developed by the author) is discussed in Chapter 2 to provide a process framework in which patterns have a definite role. However, the use of patterns transcends any individual process, no matter how coolly named.

By the time you've reached the end of the book, you will hopefully have the expertise to tackle the real design issues that you face in your life as a professional developer of embedded systems.

Bruce Powel Douglass, Ph.D.

Acknowledgements

I want to thank my wife Sarah for being supportive even though I work too much, and my great kids – Scott, Blake, and Tamera – for being awesome even though I work too much.

About the Author

Bruce Powel Douglass was raised by wolves in the Oregon wilderness. He taught himself to read at age 3 and learned calculus before age 12. He dropped out of school when he was 14 and traveled around the U.S. for a few years before entering the University of Oregon as a mathematics major. He eventually received his M.S. in exercise physiology from the University of Oregon and his Ph.D. in neurophysiology from the USD Medical School, where he developed a branch of mathematics called autocorrelative factor analysis for studying information processing in multicellular biological neural systems.

Bruce has worked as a software developer in real-time systems for over 30 years and is a well-known speaker, author, and consultant in the area of real-time embedded systems and systems engineering. He is on the Advisory Board of the *Embedded Systems* conference and has taught courses in software estimation and scheduling, project management, object-oriented analysis and design, communications protocols, finite state machines, design patterns, and safety-critical systems design. He develops and teaches courses and consults in real-time object-oriented analysis and design and project management and has done so for many years. He has authored articles for a many journals and periodicals, especially in the real-time domain.

He is the Chief Evangelist[1] for IBM Rational, a leading producer of tools for real-time systems development, including the widely-used Rhapsody modeling tool. Bruce worked with other UML partners on the specification of the UML. He is a former co-chair of the Object Management Group's Real-Time Analysis and Design Working Group. He is the author of several other books on software, including

- *Doing Hard Time: Developing Real-Time Systems with UML, Objects, Frameworks, and Patterns* (Addison-Wesley, 1999)
- *Real-Time Design Patterns: Robust Scalable Architecture for Real-Time Systems* (Addison-Wesley, 2002)
- *Real-Time UML Third Edition: Advances in the UML for Real-Time Systems* (Addison-Wesley, 2004)

---

[1] Being a Chief Evangelist is much like being a Chief Scientist, except for the burning bushes and stone tablets.

- *Real-Time UML Workshop for Embedded Systems* (Elsevier, 2007)
- *Real-Time Agility* (Addison-Wesley, 2009)

and several others, including a short textbook on table tennis.

Bruce enjoys classical music and has played classical guitar professionally. He has competed in several sports, including table tennis, bicycle racing, running, triathlons, and full-contact Tae Kwon Do, although he currently only fights inanimate objects that don't hit back.

What Is Embedded Programming?

What you will learn

- Basics of embedded systems

- OO versus structured programming

- Implementing classes, inheritance, and state machines in C

1.1 What's Special About Embedded Systems?

This book focuses solely on embedded systems development. In doing so, it is drawing a distinction between "embedded" systems and "others." Before we get into the depth of the discussion, we need to understand what this difference is so that we can appreciate what it forebodes with respect to the patterns and technologies we will use to develop embedded systems.

I define an embedded system as "a computerized system dedicated to performing a specific set of real-world functions, rather than to providing a generalized computing environment." Clearly, this is a broad categorization that includes tiny 8-bit computers embedded in cardiac pacemakers, linked 32-bit computers controlling avionics, communications, fire control for aircraft, and wide-area networks composed of hundreds of powerful computer systems for battlefield management in C^4ISR (Command, Control, Communications, Computers, Intelligence, Surveillance, and Reconnaissance) systems. Many embedded systems have no disks, human interface, and barely any memory but the scope of the embedded systems market is far broader than such simple devices.

Embedded systems are everywhere:

- In the medical field, embedded systems include implantable devices (e.g., cardiac pacemakers, defibrillators, and insulin pumps), monitoring equipment (e.g., ECG/EKG monitors, blood gas monitors, blood pressure monitors, EMG monitors), imaging systems (e.g., CT, SPECT, PET, TEM, and x-ray imagers), and therapy delivery devices (e.g., patient ventilator, drug vaporizers, and infusion pumps).
- In the telecom market, there are devices ranging from cell phones, switching systems, routers, modems, and satellites.
- In automotive environments, embedded systems optimize engine combustion, manage power delivery in transmissions, monitor sensor data, control anti-lock braking, provide security, and offer infotainment services such as CD and DVD players, and GPS routing (in some locations, they can offer radar and laser detection and even active radar and laser countermeasures).
- In the office, embedded systems manage phones, printers, copies, fax machines, lights, digital projectors, security systems, and fire detection and suppression systems.
- In the home, examples include ovens, televisions, radios, washing machines, and even some vacuum cleaners.

Embedded systems already control, augment, monitor, and manage virtually every high-tech device we have from televisions to trains to factory automation, and their use is on the rise.

An important subset of embedded systems are real-time systems. Many people have the mistaken impression that "real time" means "real fast" but that is not true. A real-time system is one in which timeliness constraints must be satisfied for system correctness. A common, if simplistic, categorization of real-time systems is into two groups. "Hard" real-time systems are ones in which timeliness constraints are modeled as *deadlines*, points in time by which the execution of specific actions are required to be complete. "Soft" real-time systems are those that are not "hard"[1]; that is, some other (usually stochastic) measure than deadlines is used to determine timeliness. This may include average throughput, average execution time, maximum burst length, or some other measure. All systems may be modeled as hard real-time systems, but this often results in "over-designing" the system to be faster or have more available resources than is necessary, raising the recurring cost (approximately "manufacturing cost") of the system.

---

[1] Although, as the saying goes, "Hard real-time systems are hard. Soft real-time systems are harder."

Even though all systems may be modeled as hard real-time systems, in actual fact, most are not. If the response is occasionally delayed or even an entire input event is missed, most systems will continue to function properly. The primary reason for modeling real-time systems as "hard" is because it eases the assurance of the system's timeliness through mathematical analysis.

1.1.1 Embedded Design Constraints

From the inside, one of the most striking characteristics of embedded systems is severity of their constraints. Unlike writing software for a general-purpose computer, an embedded system is usually shipped already integrated with all the hardware it needs. The hardware platform is not usually user-extensible, so resources such as memory, power, cooling, or computing power contribute to the per-unit cost (known as *recurring cost*). To maintain profitability, there is almost always tremendous pressure on the developer to minimize the use of such hardware resources. This means that embedded systems often require additional optimization efforts far beyond that required for desktop applications.

Beyond the need to minimize hardware, performance concerns are often critical to the success of a system. There are many aspects to performance, and different systems value these aspects differently. In some systems, throughput is a critical criterion. *Throughput* is normally measured in terms of the number of transactions, samples, connections, or messages that can be processed per unit time. In other systems, handling each request as quickly as possible is more important, a quality known as *responsiveness*, usually captured as a worst case execution time. Other systems value *predictability* of performance over maximum throughput or responsiveness. Predictability is usually measured as occurring within a range or as being drawn from a probability density function.

Reliability, robustness, and safety are other kinds of constraints levied on embedded systems. The reliability of a system is a (stochastic) measure of the likelihood that the system will deliver the correct functionality. Robustness refers to the ability of a system to deliver services properly when its preconditions (such as operating conditions or input data rates) are violated. Safety denotes the level of risk of a system, that is, the likelihood that using the system will result in an accident or loss. These concerns often require additional hardware and software measures to maintain the operation of the system within acceptable limits. For example, most embedded systems have a power on self-test (POST) as well as a periodic or continuous built-in test (BIT).

Collectively, these constraints on the system are known as the qualities of services (QoS) provided by the system. In addition to the various QoS constraints, to reduce recurring cost, it is common to create custom hardware that requires specialized device driver software.

1.1.2 The Embedded Tool Chain

Most embedded systems are created on a different system (the "host") than they execute on (the "target"). This has a number of effects on the developer and the tools used in the creation of embedded systems. The most obvious such tool is the cross-compiler. This is a compiler that runs on the host but creates executable code that runs on a different computer and operating environment. Many real-time operating systems (RTOSs) provide their own proprietary compilers or provide customizations for open-source compilers such as GCC (Gnu Compiler Collection)[2].

A linker is a program that combines a set of executable codes together into an executable for a target. Some operating systems don't require an explicit link step because the OS loader dynamically links the program elements as it loads into memory. This is true of Unix-style embedded operating systems but most embedded OSs require an explicit linking step. The linker often relocates the program as well, meaning that the start address is specified during the link step and assembly language jump instructions must be updated to reflect the actual starting base address.

A loader is a tool that loads the object image output from the linking step into the memory of the target environment. This may be done via a serial or network link or by burning the software image into nonvolatile memory such as Flash or EPROM. As an alternative to loading the software image on a target platform, many developers use simulators for the target that execute on their host development systems. It is common, for example, to use Z80 or 8051 simulators running on Windows to begin to run, debug, and test your software even before target boards are available.

So far, the tools mentioned in the tool chain have been limited to getting software onto the system. Beyond that, we need to ensure that the software works properly. The next set of tools are debuggers – tools that give us a great measure over the execution of the software, including the ability to step into (execute line by line) or step over (execute in entirety) functions, set breakpoints, and to examine and modify variables. These debuggers may work over standard serial and network links or over JTAG[3] ports.

Modern-day Integrated Development Environments (IDEs) link together most or all of the tools in the embedded tool chain to facilitate and automate the development process. The latest industry trend is to host the IDEs in the Eclipse platform[4] because of the power of the integrated environment and the availability of third-party and open-source plug-ins. The Jazz Foundation

---

[2] See *www.gnu.org* for releases and supported targets.

[3] Joint Test Architecture Group, an electronic and software standard interface used by many developers for testing on embedded targets. See *www.freelabs.com/%26whitis/electronics/jtag/* or *www.embedded.com/story/OEG2002 1028S0049* for more information.

[4] *www.eclipse.org*

takes the Eclipse platform further to integrate management, measurement, reporting, and other tools to better support collaborative software development and delivery.

1.1.3 OS, RTOS, or Bareback?

An *operating system* (OS) provides a set of system and platform execution services for the application developer, especially around the management and use of target system resources. These resources include memory, concurrency units (processes, tasks, or threads), event queues, interrupts, hardware, and application programs. Most OSs do not provide any guarantees about timeliness, and desktop OSs may invoke unpredictable delays due to internal processing, memory management, or garbage collection at unforeseeable times. This unpredictability, and the fact that most desktop OSs are huge (compared to available memory), makes them unsuitable for real-time and embedded environments that are constrained in both time and resources.

A *real-time operating system* (RTOS) is a multitasking operating system intended for real-time and embedded applications. RTOSs are generally written to provide services with good efficiency and performance, but usually the predictability of the performance is more important than the maximum throughput. In addition, RTOSs are smaller and often less capable than desktop OSs. RTOSs don't guarantee real-time performance but they provide an application environment so that appropriate developed applications can achieve real-time performance.

RTOSs run applications and tasks using one of three basic design schemas. Event-driven systems handle events as they arise and schedule tasks to handle the processing. Most such systems use task *priority* as a quantitative means by which to determine which task will run if multiple tasks are ready to run. Task priorities are most often static (i.e., specified at design time as such with *rate-monotonic scheduling*) but some are dynamic, varying the task priorities to account for current operating conditions (such as *earliest deadline first scheduling*). The other two approaches to task scheduling implement a "fairness doctrine" either by giving all tasks a periodic time slice in which to run (time-base schemas, such as *round robin scheduling*) or by running the task set cyclically (sequence-based schemas, such as *cyclic executive scheduling*).

Twenty years ago, most development houses specializing in real-time systems wrote their own proprietary RTOSs but the RTOS market has undergone a consolidation, enabling such houses to spend a greater percentage of their development efforts on their in-house specialties by buying a commercial RTOS instead.

Some of the challenges of using RTOSs include the fact that while there are some big players, such as Wind River (maker of VxWorks™) and Green Hills Software (maker of Integrity™), there are literally hundreds of RTOSs in use today, and each has its own Application Program Interface (API). Further, although RTOSs have a smaller footprint and timeliness impact than their desktop brethren, that is not the same as having *no* impact. Many embedded systems run on small 8-bit processors with only a few tens of kilobytes of memory – too little for a standard RTOS. Many RTOSs support a scalable set of services (based on the Microkernel Architecture Pattern[5]) but the degree of fidelity of that scalability may not meet the needs of a tightly-constrained application.

Some systems having too few resources to support both an RTOS and the actual application software opt to go *bareback* – that is, function without a commercial or proprietary RTOS at all. This means that the application code itself must replicate any RTOS services or functionality needed by the application. With the drop in hardware costs, systems formerly done with tiny 8-bit processors have moved up to larger 16- and 32-bit processors. This is needed to add the additional complexity of behavior required in modern embedded devices and is enabled by the lower recurring costs of parts and manufacturing.

For very simple embedded systems, there may be no explicit operating system functionality at all. The application may simply be a set of interrupt handlers that communicate via a shared resource scheme such as queuing or shared memory. Alternatively, a simple task loop implementation of a cyclic executive might suffice. Somewhat more complex systems may introduce operating system features such as memory or task management as they need them.

From one point of view, an operating system is an integrated set of design patterns that provides certain kinds of services and resources to applications. Chapter 3 in this book discusses patterns for accessing hardware resources, such as timers and memory while Chapter 4 focuses on ways of managing concurrency in your embedded system.

1.1.4 Embedded Middleware

Middleware is software that connects software components together in some way. Middleware as a term dates back to the 1968 NATO Software Engineering Conference[6]. Like operating systems, there is commercial support for middleware, but in small systems it may be developed as a part of the application software. The most common middlewares for real-time and embedded systems include Common Object Request Broker Architecture (CORBA) and its many variants, and the Data Distribution Service (DDS). Both these middleware architectures are based on standards published by the Object Management Group (OMG)[7]. In the case of CORBA, there are many specialized variant standards from which to choose including Real-Time CORBA, Embedded CORBA, and Minimum CORBA.

As you might guess, middleware is appropriate for larger-scale embedded applications – those consisting of multiple software components or applications that may be distributed across

---

[5] See my book *Real-Time Design Patterns: Robust Scalable Architecture for Real-Time Systems* (Addison-Wesley, 2002).

[6] *http://homepages.cs.ncl.ac.uk/brian.randell/NATO/nato1968.PDF*

[7] *www.omg.org*

multiple processors and networks. Particularly when the components are to be developed by different organizations, the system will be extended by different organizations, or when the system is particularly long-lived, using standard middleware provides significant benefit to the developer

These middleware architectures, again like operating systems, are integrated collections of design patterns, such as Proxy, Data Bus, and Broker Pattern.

1.1.5 Codevelopment with Hardware

Many embedded projects involve the development of electronic and mechanical hardware simultaneously with the software. This introduces special challenges for the software developer who must develop software solely on the basis of predesign specifications of how the hardware is intended to work. All too often, those specifications are nothing more than notions of functionality, making it impossible to create the correct software until hardware is eventually delivered. Any slips in the hardware schedule directly induce slips in the software schedule, for which the software developers "take the heat."

I am reminded of working against a hardware specification for a telemetry system in the 1980s. In this system, the hardware was supposed to pulse encode bits – eight pulses for a "0" bit and 15 pulses for a "1" bit, with a specific time period between pulses and a longer specific time between bits. I wrote the software predicated on that specification but much to my surprise[8], when the actual hardware arrived I could only intermittently communicate with the device. Despite intensive debugging, I couldn't figure out where the software was failing – until I got out the oscilloscope, that is. With the oscilloscope I could see that when the software put a "0" bit out, the coil pulsed anywhere from six to 12 times. For a "1" bit, it pulsed anywhere from 10 to 18 times (note that the ranges for "0" and "1" bits actually overlap). The solution was to write predictor-corrector algorithm that estimated probabilities of bit error and corrected on-the-fly based on message CRC, the bit values, and their certainties. The software was certainly more complex than I had expected it to be! Such is the life of the embedded programmer.

Although much of the hardware-software integration pain can be eliminated with good specifications and early delivery of prototype hardware, much of it cannot. One of the truths of software development is that reengineering is a necessary consequence of codevelopment.

Agile methods address this concern in a couple of different ways[9]. First, with agile methods, software is constantly being developed throughout the software lifecycle. Each day, software is being written, debugged, unit tested, and delivered. And each day, different software units are integrated together and tested as a cohesive whole in a workflow known as *continuous*

---

[8] I was naïve back then. ;-)
[9] See my book *Real-Time Agility* (Addison-Wesley, 2009) for a thorough discussion of the topic.

integration. This allows software defects to be found as early as possible. This integration can (and should) include early hardware as well.

1.1.6 Debugging and Testing

The Prussian general von Clausewitz wrote "Everything in war is very simple, but the simplest thing is very difficult[10]." He might as well have said it about software!

The simplest thing that is very difficult about software isn't writing the software – it's writing the *right software with the right functionality.* The state of the art in developing defect-free software is an agile practice known as *test-driven development* (TDD). In TDD, the unit tests for a piece of software are written simultaneously with, or even slightly before, the software it will verify. All too commonly, unit testing is skipped entirely or performed far too late to have any beneficial effect on the software.

There are many different kinds of unit tests that can be applied to software in general and embedded software in particular. These include:

- Functional – tests the behavior or functionality of a system or system element
- Quality of Service – tests the "performance" of a system or system element, often to measure the performance of the system or element against its performance requirements
- Precondition tests – tests that the behavior of the system or system element is correct in the case that the preconditional invariants are met and in the case that the preconditional invariants are violated
- Range – tests values within a data range
- Statistical – tests values within a range by selecting them stochastically from a probability density function (PDF)
- Boundary – tests values just at the edges of, just inside, and just outside a data range
- Coverage – tests that all execution paths are executed during a test suite
- Stress – tests data that exceeds the expected bandwidth of a system or system element
- Volume – also known as "load testing" – tests the system with large amounts of data that meet or exceed its design load
- Fault Seeding – tests in which a fault is intentionally introduced to the system to ensure the system handles it properly
- Regression tests – normally a subset of previously passed tests to ensure that modification to a system did not introduce errors into previously correctly functioning systems

In a desktop system, unit testing software with test cases that explore all of these concerns is rare (and difficult) enough. When you embedded that software on a less-accessible target

---

[10] von Clausewitz, C., 1976. *On War*. Howard, M., Paret, P. (Eds, tr.), Princeton University Press; published originally in 1832 as *Vom Kriege*.

platform it becomes even more difficult. A number of different strategies can be employed to execute test cases, such as

- "printf" testing – tests the system by writing to a file or to stdout
- "Test buddies" – writing test fixtures that embed the test cases in their own functionality
- Testing on host – performing most of the tests on the host platforms using a host native complier and a critical subset on the target platform using a cross compiler
- Simulating on host – simulating the target platform on the host with cross-compiled software and retesting a critical subset on the target with the same object code
- Commercial software testing tools – using software testing tools, such as TestRT™, LDRA™, or VectorCAST™
- Commercial hardware-software integrated tools – this includes tools such as logic analyzers, in-circuit emulators, JTAG-compliant testing tools, and ROM emulators

Of the various kinds of tests, performance tests are usually the most difficult to adequately perform. This is because most of the common methods for executing test cases involve running addition software on the same target platform, which affects the timing and performance of the software under test. Where possible, it is clearly best to use hardware-assisted debugging with in-circuit emulators or ROM emulators for performance testing.

1.2 OO or Structured – It's Your Choice

Structured programming is a disciplined form of software development that emphasizes two separate and distinct aspects.

On one hand, functions or procedures form the foundation of behavioral programming: a procedure is a collection of primitive actions with a single entry point that performs a coherence behavioral goal; a function is simply a procedure that returns a value. Procedures can be broken up into call trees by one procedure invoking another, permitting algorithmic decomposition. Procedures are usually synchronously invoked (i.e., called) but by adding additional means, asynchronous invocations can be invoked as well.

The other side of structured programming is the notion of data structuring. All third-generation computer languages have the notion of building complex data structures from more primitive elements, ultimately basing them on the basic types provided by the computer language. These may be homogeneous collections, such as arrays, or heterogeneous ones, as with C structs.

Object-oriented programming is based on an orthogonal paradigm. Rather than have two separate taxonomies, object-oriented programming has a single one based on the notion of a class. A *class* combines together both data (stored in *attributes*) and procedures (known as *operations*) that operate on that data, because they tend to be inherently tightly bound anyway.

An *object* is an instance of a class. This makes an object equivalent to, but more powerful than, a variable in a structured language because the object provides both the values held in its attributes and the operations that manipulate them.

A *structured language* is a language that directly supports structured programming. In the context of this book, C is clearly a structured language. C is by far the most prevalent language for creating embedded systems and C is an inherently "structured" language. It has all the earmarks – functions, variables, and so forth. However, an interesting question arises: can object-oriented programming be done in a structured language, such as C? And even if it *can* be done, *should* it be done?

I am reminded of a similar discussion back when the most common programming languages were assembly languages. Was it possible to write structured programs in assembly code? After all, assembly languages do not provide structured language features intrinsically. As it happens, the answer is clearly, "Yes!" It is not only possible but profitable to write structured assembly language programs. It requires some discipline on the part of the programmer and decisions about the mapping of assembly language instructions to structured concepts (such as passing parameters on the stack), but it is easy enough to do.

The same is true of object-oriented programming in structured languages such as C. In this book, we will take a primarily object-based perspective in the programming (object-oriented without subclassing) because I believe there are benefits to doing so. However, the creation of object-based or object-oriented programs in C is pretty straightforward. Let's briefly discuss the important aspects of object-oriented programming and how to implement those concepts in C.

1.2.1 Classes

A class is really nothing more than a C struct, but what is special about it is that it contains two different kinds of features: data (attributes) and behavioral (operations).

The simplest way to implement classes is simply to use the file as the encapsulating boundary; public variables and functions can be made visible in the header file, while the implementation file holds the function bodies and private variables and functions. Multiple files can be linked with «Usage» dependencies to support the call tree, such as in Figure 1-1. In this case, a "class" is represented as a pair of files and its implementation uses some features (variables or functions) of the Display class (file), in this case, the `displayMsg()` function.

A somewhat more flexible approach uses structs within the files to represent the classes. The operations of the class are defined as functions located within the same file as the struct. To make sure that the function has access to the correct object data, we need to pass in a me pointer.

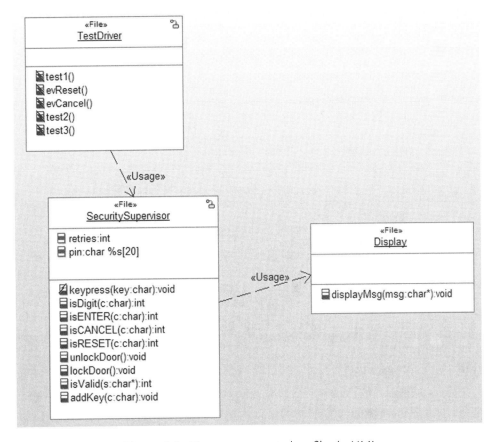

Figure 1.1: Classes represented as files in UML

This allows us to have multiple instances (objects) of the same class and make sure that the member functions work on the correct copies of the data. In addition, classes are endowed with "special" operations. A constructor creates an object of the class. An initializer (optionally) initializes the object and its attributes. A destructor destroys the class and releases the memory used.

Consider a simple class `Sensor` with data elements `value`, `updateFrequency`, and `filterFrequency` and operations, `getValue()`, `setValue(v: int)`, `setUpdateFreq(r:int)`, `getUpdateFreq()`, `setFilterFreq(ff: int)`, and `getFilterFreq()`. The header file would look something like Code Listing 1.

```
#ifndef Sensor_H
#define Sensor_H

/*## class Sensor */
typedef struct Sensor Sensor;
```

```
struct Sensor {
    int filterFrequency;
    int updateFrequency;
    int value;

};

int Sensor_getFilterFrequency(const Sensor* const me);

void Sensor_setFilterFrequency(Sensor* const me, int p_filterFrequency);

int Sensor_getUpdateFrequency(const Sensor* const me);

void Sensor_setUpdateFrequency(Sensor* const me, int p_updateFrequency);

int Sensor_getValue(const Sensor* const me);

Sensor * Sensor_Create(void);

void Sensor_Destroy(Sensor* const me);
```

Code Listing 1-1: Sensor Header File

The associated implement file is shown in Code Listing 2.

```
#include "Sensor.h"

void Sensor_Init(Sensor* const me) {
}

void Sensor_Cleanup(Sensor* const me) {
}

int Sensor_getFilterFrequency(const Sensor* const me) {
    return me->filterFrequency;
}

void Sensor_setFilterFrequency(Sensor* const me, int p_filterFrequency) {
    me->filterFrequency = p_filterFrequency;
}
int Sensor_getUpdateFrequency(const Sensor* const me) {
    return me->updateFrequency;
}

void Sensor_setUpdateFrequency(Sensor* const me, int p_updateFrequency) {
    me->updateFrequency = p_updateFrequency;
}

int Sensor_getValue(const Sensor* const me) {
    return me->value;
}

Sensor * Sensor_Create(void) {
    Sensor* me = (Sensor *) malloc(sizeof(Sensor));
```

```
        if(me!=NULL)
           {
                Sensor_Init(me);
           }
        return me;
    }
    void Sensor_Destroy(Sensor* const me) {
        if(me!=NULL)
           {
                Sensor_Cleanup(me);
           }
        free(me);
    }
```

Code Listing 2: Sensor Implementation File

In this case, the file serves as the encapsulation boundary, and stylistically two files (a header .h file and an implementation .c file) group together the elements within a single class. This approach supports *object-based* programming very well, but doesn't implement virtual functions in a fashion that can be easily overridden in subclasses.

An alternative approach that supports *object-oriented* programming is to embed function pointers within the struct itself. This will be discussed shortly in Section 1.2.3.

1.2.2 Objects

Objects are instances of classes, in the same way that a specific variable x (of type int) is an instance of its type (int). The same set of operators (such as arithmetic operators) apply to all variables of that type, but the specific values of one instance differ from another instance (say, y).

In standard C programming, complex algorithms can still be embedded in classes, but it is common that these classes are *singletons*, meaning that there is exactly one instance of the class in the application. A singleton Printer class, for example, may have variables such as currentPrinter and operations like print(), but the application would have only a single instance. There are still benefits to encapsulating the data used by a class with the operations that act on that data, even if there is only one instance ever running. In other cases, usually more data-centric (as opposed to algorithm- or service-centric) classes will have multiple instances.

To create an instance of a class, simply create an instance of the struct. Consider one possible main() that creates and uses two instances of the previous Sensor class, shown in Code Listing 3:

```
#include "Sensor.h"
#include <stdlib.h>
#include <stdio.h>
int main(int argc, char* argv[]) {
    Sensor * p_Sensor0, * p_Sensor1;
    p_Sensor0 = Sensor_Create();
    p_Sensor1 = Sensor_Create();

    /* do stuff with the sensors ere */
    p_Sensor0->value = 99;
    p_Sensor1->value = -1;
    printf("The current value from Sensor 0 is %d\n",
        Sensor_getValue(p_Sensor0));
    printf("The current value from Sensor 1 is %d\n",
        Sensor_getValue(p_Sensor1));
    /* done with sensors */

    Sensor_Destroy(p_Sensor0);
    Sensor_Destroy(p_Sensor1);
    return 0;
}
```

Code Listing 3: Sensor main()

1.2.3 Polymorphism and Virtual Functions

Polymorphism is a valuable feature of object-oriented languages. It allows for the same function name to represent one function in one context and another function in a different context. In practice, this means that when either the static or dynamic context of an element changes, the appropriate operation can be called.

Of course, the standard way to do this in C is to use conditional statements such as if or switch. The problem with the standard approach is that it quickly becomes unwieldy when many different contexts are available. Additionally, the approach requires that all possible contexts must be known when the original function is written, or at least the function must be modified to allow a new context. With polymorphic functions, prescience isn't required. When the new context is discovered, the polymorphic function may be created and added without requiring changes to the original function.

Consider a sensor in which the actual interface may differ. The acquireValue function in one context may require accessing a memory-mapped sensor, waiting for a while and then reading a different address to acquire the value. A different context may use a port-mapped sensor. We would definitely prefer that the client of the sensor doesn't know how the interface to the sensor works – requiring such knowledge makes the application far more fragile and difficult to maintain. With the standard C approach, we'd write code something like Code Listing 4 with the attributes whatKindOfInterface added to the Sensor struct.

```c
int acquireValue(Sensor *me) {
    int *r, *w; /* read and write addresses */
    int j;
    switch(me->whatKindOfInterface) {
    case MEMORYMAPPED:
        w = (int*)WRITEADDR; /* address to write to sensor */
        *w = WRITEMASK; /* sensor command to force a read */
        for (j=0;j<100;j++) { /* wait loop */ };
        r = (int *)READADDR; /* address of returned value */
        me->value = *r;
        break;
    case PORTMAPPED:
        me->value = inp(SENSORPORT);
        /* inp() is a compiler-specific port function */
        break;
    }; /* end switch */
    return me->value;
};
```

Code Listing 4: Polymorphism the hard way

1.2.4 Subclassing

Subclassing (known as *generalization* in the Unified Modeling Language [UML] or as *inheritance*) is a way of representing classes that are a "specialized kind of" another class. The basic rule is that the specialized class inherits the features of the more general class. That is, all of the features of the more general class (also known as the *base* or *parent class*) are also features of the specialized class (also known as the *derived* or *child class*), but the latter is allowed to both specialize and extend the more general class.

One of the key advantages of subclassing is the ability to reuse design and code. If I design a class for a specific context and want to reuse it later in a different context, I can just design and code the specializations and extensions without touching the base class. Further, in usage, I can write algorithms that manipulate an instance of a type, without regard to which subclass it might be, when that is appropriate and modify the behavior to be semantically correct with subclass when that is necessary.

By "specialize," I mean that the specialized class can redefine operations provided by the base class. Because a subclass is a specialized form of a base class, any operations that make sense for an instance of the base class should also make sense with an instance of the derived class. However, it often happens that the implementation of that operation must be slightly different. For example, if I write the code for queuing data from a sensor, it is likely to have operations such as insert(int value), int remove(), int getSize(), and so on. If I create a specialized kind of a queue that can store large amounts of data out to a mass storage device such as a flash drive, those operations still make sense but their implementation will be different. Note that specialization refers exclusively to changing the implementation of operations and not at all to the redefinition of data.

Specialization can be easily done with the switch-case statement approach mentioned in the previous section, but it is much more versatile if function pointers are used instead. Then the designer can simply write the specialized functions and create a new constructor that points to the new functions instead of the old. That is, subclasses can override the operations by inserting pointers to different functions that provide specialized behaviors. The downside is that function pointers are a bit tricky and pointers are the leading cause of programmer-induced defects in C programs.

By "extend," I mean that the child class will have new features, such as new attributes or operations. In the case of the data queue, this means that I can add new attributes such as the name of the data file used to store cached data to the flash drive as well as new operations such as flush() and load() operations for writing out the in-memory buffer to flash or reading data from the flash drive into the in-memory buffer, respectively. The UML class diagram in Figure 1-2 shows both the Queue and CachedQueue classes. To simplify the problem, we assume the queue is never misused (so we don't have to handle over and underflow conditions) and we want to store a maximum of 100 integers in the queue[11].

The CachedQueue is conceptually straightforward. If there is room in the insert buffer, then insert the data as with the normal queue. However, if you fill up the internal buffer, then the insert() operation will call flush() to write the insert buffer out to disk, then the internal buffer will be reset. For removing data, it is only slightly more complicated. If the internal buffer is empty, then remove() calls the load() function to read the oldest data off the disk.

Reading the UML Class diagram

The class diagram shown has a number of classic UML class diagram features. The boxes represent the classes while the lines represent relations between the classes. The class box is (optionally) divided into three segments. The upper-most segment holds the name of the class. The middle segment lists the attributes (data members), along with their types. The bottom-most segment shows the operations within the class.

Two different relations are shown in the figure. The line with the closed arrowhead is the generalization relation; the line points to the more base class. The other line is known as composition. The composition is relation that implies strong ownership, and the responsibility for creation and destruction of the instance of the class at the other end of the relation. The open arrowhead depicts navigation (the owner has a pointer to the part in this case, but not the other way around). The name near the arrowhead is the name of the pointer within the CachedQueue class. The number near the arrowhead is the multiplicity – the number of instances needed to play the role (in this case, a single instance is needed).

---

[11] If you're feeling a little ambitious, feel free to elaborate the example to handle under flow and overflow as well as use C preprocessor macros to be able to change the type and number of elements for the data being queued. Don't worry – I'll wait until you're done.

This book will show patterns and examples with class diagrams, even though the primary emphasis will be code representations. Appendix A has a brief overview of UML. If you'd like more detail, see any of a number of books on UML, such as my *Real-Time UML 3rd Edition: Advances in the UML for Real-Time Systems* (Addison-Wesley, 2004).

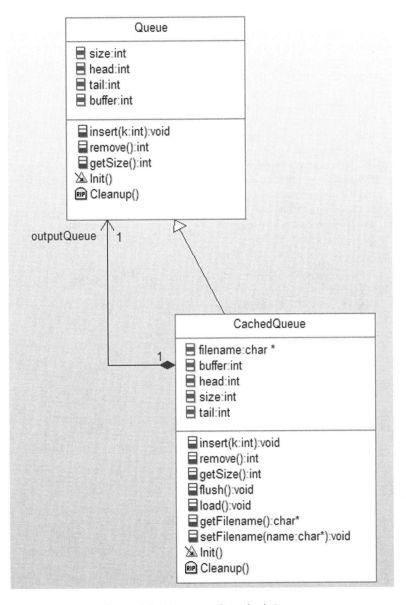

Figure 1.2: Queue and Cached Queue

There is certainly more than one way to implement subclassing in C. In the approach used here, to handle specialization, we will employ the member function pointers described in the previous section. For extension, we will simply embed the base class as a struct within the child class.

For example, let's look at the code for the queue example of Figure 1-2. Code Listing 5 is the header file for the Queue class, the structure of which should be pretty much as you expect.

```c
#ifndef QUEUE_H_
#define QUEUE_H_

#define QUEUE_SIZE 10

/* class Queue */
typedef struct Queue Queue;
struct Queue {
    int buffer[QUEUE_SIZE];     /* where the data things are */
    int head;
    int size;
    int tail;
    int (*isFull)(Queue* const me);
    int (*isEmpty)(Queue* const me);
    int (*getSize)(Queue* const me);
    void (*insert)(Queue* const me, int k);
    int (*remove)(Queue* const me);
};

/* Constructors and destructors:*/
void Queue_Init(Queue* const me,int (*isFullfunction)(Queue* const me),
        int (*isEmptyfunction)(Queue* const me),
        int (*getSizefunction)(Queue* const me),
        void (*insertfunction)(Queue* const me, int k),
        int (*removefunction)(Queue* const me) );

void Queue_Cleanup(Queue* const me);

/* Operations */
int Queue_isFull(Queue* const me);
int Queue_isEmpty(Queue* const me);
int Queue_getSize(Queue* const me);
void Queue_insert(Queue* const me, int k);
int Queue_remove(Queue* const me);

Queue * Queue_Create(void);
void Queue_Destroy(Queue* const me);

#endif /*QUEUE_H_*/
```

Code Listing 5: Queue Header

Code Listing 6 shows the (simple) implementation of the Queue operations and Code Listing 7 provides a simple little test program that shows elements inserted and removed into and from the queue.

```c
#include <stdio.h>
#include <stdlib.h>
#include "queue.h"

void Queue_Init(Queue* const me, int (*isFullfunction)(Queue* const me),
            int (*isEmptyfunction)(Queue* const me),
            int (*getSizefunction)(Queue* const me),
            void (*insertfunction)(Queue* const me, int k),
            int (*removefunction)(Queue* const me) ){
        /* initialize attributes */
    me->head = 0;
    me->size = 0;
    me->tail = 0;
    /* initialize member function pointers */
    me->isFull = isFullfunction;
    me->isEmpty = isEmptyfunction;
    me->getSize = getSizefunction;
    me->insert = insertfunction;
    me->remove = removefunction;
}
/* operation Cleanup() */
void Queue_Cleanup(Queue* const me) {
}

/* operation isFull() */
int Queue_isFull(Queue* const me){
    return (me->head+1) % QUEUE_SIZE == me->tail;
}

/* operation isEmpty() */
int Queue_isEmpty(Queue* const me){
    return (me->head == me->tail);
}
/* operation getSize() */
int Queue_getSize(Queue* const me) {
    return me->size;
}
/* operation insert(int) */
void Queue_insert(Queue* const me, int k) {
    if (!me->isFull(me)) {
        me->buffer[me->head] = k;
        me->head = (me->head+1) % QUEUE_SIZE;
        ++me->size;
    }
}

/* operation remove */
int Queue_remove(Queue* const me) {
    int value = -9999; /* sentinel value */
    if (!me->isEmpty(me)) {
        value = me->buffer[me->tail];
        me->tail = (me->tail+1) % QUEUE_SIZE;
        -me->size;
```

```
        }
        return value;
}

Queue * Queue_Create(void) {
    Queue* me = (Queue *) malloc(sizeof(Queue));
    if(me!=NULL)
        {
            Queue_Init(me, Queue_isFull, Queue_isEmpty, Queue_getSize,
            Queue_insert, Queue_remove);
        }
    return me;
}

void Queue_Destroy(Queue* const me) {
    if(me!=NULL)
        {
            Queue_Cleanup(me);
        }
    free(me);
}
```

Code Listing 6: Queue Implementation

```
#include <stdio.h>
#include <stdlib.h>
#include "queue.h"

int main(void) {
    int j,k, h, t;

    /* test normal queue */
    Queue * myQ;
    myQ = Queue_Create();
    k = 1000;

    for (j=0;j<QUEUE_SIZE;j++) {
        h = myQ->head;
        myQ->insert(myQ,k);
        printf("inserting %d at position %d, size =%d\n",k--,h, myQ->getSize(myQ));
    };
    printf("Inserted %d elements\n",myQ->getSize(myQ));
    for (j=0;j<QUEUE_SIZE;j++) {
        t = myQ->tail;
        k = myQ->remove(myQ);
        printf("REMOVING %d at position %d, size =%d\n",k,t, myQ->getSize(myQ));
    };

    printf("Last item removed = %d\n", k);
    printf("Current queue size %d\n", myQ->getSize(myQ));
puts("Queue test program");

    return EXIT_SUCCESS;
}
```

Code Listing 7: Queue Test Program

Now let us suppose that memory is very tight and so we need to reduce the size of the in-memory buffer and store most elements to disk. Our CachedQueue fits the bill.

Code Listing 8 shows the header file for the class. You can see some important aspects of the definition of the class. First, note that the base class is retained in the subclass as an attribute named queue. This brings in the operations and attributes from the base class within the subclass. Secondly, note that the operations defined for the base class are replicated as member function pointers in the subclass. In that way, they can be directly invoked on the subclass instance. Each such virtual function can then either deal entirely with the request or deal with some aspect of it and delegate the rest to the original function defined on the Queue class (still held within its virtual functions). Lastly, note that the aggregate outputQueue is also an attribute of the subclass. This allows the CachedQueue to manage two in-memory buffers. The "normal" one (represented by the base class) is where new data gets inserted. The other one (represented by the aggregated outputQueue) is where the data comes from for a remove.

```c
#ifndef CACHEDQUEUE_H_
#define CACHEDQUEUE_H_

#include "queue.h"

typedef struct CachedQueue CachedQueue;
struct CachedQueue {
    Queue* queue;          /* base class */
    /* new attributes */
    char filename[80];

    int numberElementsOnDisk;
    /* aggregation in subclass */
    Queue* outputQueue;

    /* inherited virtual functions */
    int (*isFull)(CachedQueue* const me);
    int (*isEmpty)(CachedQueue* const me);
    int (*getSize)(CachedQueue* const me);
    void (*insert)(CachedQueue* const me, int k);
    int (*remove)(CachedQueue* const me);
    /* new virtual functions */
    void (*flush)(CachedQueue* const me);
    void (*load)(CachedQueue* const me);
};

/* Constructors and destructors:*/
void CachedQueue_Init(CachedQueue* const me, char* fName,
        int (*isFullfunction)(CachedQueue* const me),
        int (*isEmptyfunction)(CachedQueue* const me),
```

```
       int (*getSizefunction)(CachedQueue* const me),
       void (*insertfunction)(CachedQueue* const me, int k),
       int (*removefunction)(CachedQueue* const me),
       void (*flushfunction)(CachedQueue* const me),
       void (*loadfunction)(CachedQueue* const me));

void CachedQueue_Cleanup(CachedQueue* const me);

/* Operations */
int CachedQueue_isFull(CachedQueue* const me);
int CachedQueue_isEmpty(CachedQueue* const me);
int CachedQueue_getSize(CachedQueue* const me);
void CachedQueue_insert(CachedQueue* const me, int k);
int CachedQueue_remove(CachedQueue* const me);
void CachedQueue_flush(CachedQueue* const me);
void CachedQueue_load(CachedQueue* const me);

CachedQueue * CachedQueue_Create(void);
void CachedQueue_Destroy(CachedQueue* const me);
#endif /*CACHEDQUEUE_H_*/
```

Code Listing 8: CachedQueue header file

Code Listing 9 shows the implementation (sans file I/O, just to simplify the example a bit). You can see how the `CachedQueue_Init()` function constructs the subclass instance (with a call to `Queue_Create()`) and then sets up its own attributes and virtual member functions. You can also see how the `CachedQueue_getSize()` function computes the number of elements held: it is the sum of the data held in the `queue`, the `outputQueue`, and the number of elements stored on disk. While the implementation is not quite up to shippable standards (I'd want to add error handling, for example), it does illustrate one way to create classes and instances of those classes in the C language.

```
#include <stdio.h>
#include <stdlib.h>
#include <string.h>
#include "cachedQueue.h"

void CachedQueue_Init(CachedQueue* const me, char* fName,
          int (*isFullfunction)(CachedQueue* const me),
          int (*isEmptyfunction)(CachedQueue* const me),
          int (*getSizefunction)(CachedQueue* const me),
          void (*insertfunction)(CachedQueue* const me, int k),
          int (*removefunction)(CachedQueue* const me),
          void (*flushfunction)(CachedQueue* const me),
          void (*loadfunction)(CachedQueue* const me)) {

    /* initialize base class */
       me->queue = Queue_Create(); /* queue member must use its original functions */

    /* initialize subclass attributes */
       me->numberElementsOnDisk = 0;
       strcpy(me->filename, fName);
```

```
        /* initialize aggregates */
    me->outputQueue = Queue_Create();

    /* initialize subclass virtual operations ptrs */
        me->isFull = isFullfunction;
        me->isEmpty = isEmptyfunction;
        me->getSize = getSizefunction;
        me->insert = insertfunction;
        me->remove = removefunction;
    me->flush = flushfunction;
    me->load = loadfunction;
}

/* operation Cleanup() */
void CachedQueue_Cleanup(CachedQueue* const me) {
        Queue_Cleanup(me->queue);
}

/* operation isFull() */
int CachedQueue_isFull(CachedQueue* const me){
    return    me->queue->isFull(me->queue) &&
              me->outputQueue->isFull(me->outputQueue);
}

/* operation isEmpty() */
int CachedQueue_isEmpty(CachedQueue* const me){
    return    me->queue->isEmpty(me->queue) &&
              me->outputQueue->isEmpty(me->outputQueue) &&
              (me->numberElementsOnDisk == 0);
}

/* operation getSize() */
int CachedQueue_getSize(CachedQueue* const me) {
    return    me->queue->getSize(me->queue) +
              me->outputQueue->getSize(me->outputQueue) +
              me->numberElementsOnDisk;
}

/* operation insert(int) */
// Imsert Algorithm:
// if the queue is full,
//        call flush to write out the queue to disk and reset the queue
// end if
// insert the data into the queue
void CachedQueue_insert(CachedQueue* const me, int k) {
    if (me->queue->isFull(me->queue)) me->flush(me);
    me->queue->insert(me->queue, k);
}

/* operation remove */
// remove algorithm
// if there is data in the outputQueue,
//        remove it from the outputQueue
```

```
// else if there is data on disk
//         call load to bring it into the outputQueue
//         remove it from the outputQueue
// else if there is data in the queue
//         remove it from there
//         (if there is no data to remove then return sentinel value)
int CachedQueue_remove(CachedQueue* const me) {
    if (!me->outputQueue->isEmpty(me->outputQueue))
        return me->outputQueue->remove(me->outputQueue);
    else if (me->numberElementsOnDisk>0) {
        me->load(me);
        return me->queue->remove(me->queue);
    }
    else
        return me->queue->remove(me->queue);
}
/* operation flush */
// Precondition: this is called only when queue is full
//         and filename is valid
// flush algorithm
// if file is not open, then open file
// while not queue->isEmpty()
//         queue->remove()
//         write data to disk
//         numberElementsOnDisk++
// end while
void CachedQueue_flush(CachedQueue* const me){
    // write file I/O statements here ...

}


/* operation load */
// Precondition: this is called only when outputQueue is empty
//     and filename is valid
// load algorithm
// while (!outputQueue->isFull() && (numberElementsOnDisk>0)
//         read from start of file (i.e., oldest datum)
//         numberElementsOnDisk--;
//         outputQueue->insert()
// end while
void CachedQueue_load(CachedQueue* const me) {
    // write file I/O statements here ...

}

CachedQueue * CachedQueue_Create(void) {
    CachedQueue* me = (CachedQueue *)
malloc(sizeof(CachedQueue));
    if(me!=NULL)
        {
            CachedQueue_Init(me, "C:\\queuebuffer.dat",
                    CachedQueue_isFull, CachedQueue_isEmpty,
```

```
                        CachedQueue_getSize, CachedQueue_insert, CachedQueue_remove,
                        CachedQueue_flush, CachedQueue_load);
        }

    return me;
    }
    void CachedQueue_Destroy(CachedQueue* const me) {
        if(me!=NULL)
            {
                CachedQueue_Cleanup(me);
            }
        free(me);
    }
```

Code Listing 9: CachedQueue Implementation

The capability to use inheritance (subclassing) will be useful in some of the design patterns I will discuss later in this book, besides being of value all on its own.

1.2.5 Finite State Machines

A finite state machine (FSM) is a machine specified by a finite set of conditions of existence (called "states") and a likewise finite set of transitions among the states triggered by events. An FSM differs from an activity diagram or flow chart in that the transitions are triggered by events (primarily) rather than being triggered when the work done in the previous state is complete. Statecharts are primarily used to model the behavior of *reactive* elements, such as classes and use cases, that wait in a state until an event of interest occurs. At that point, the event is processed, actions are performed, and the element transitions to a new state.

Actions, such as the execution of a primitive C language statement or the invocation of an operation, may be specified to be executed when a state is entered or exited, or when a transition is taken. The order of execution of actions is exit actions of the predecessor state, followed by the transition actions, followed by the entry actions of the subsequent state.

The UML uses statecharts as their formal FSM representation, which are significantly more expressive and scalable than "classical" Mealy-Moore FSMs. UML state machines, based on Dr. David Harel's statechart semantics and notation, have a number of extensions beyond Mealy-Moore state machines, including:

- Nested states for specifying hierarchical state membership
- AND-states for specifying logical independence and concurrency
- Pseudostates for annotating commonly-needed specific dynamic semantics

Figure 1-3 shows some of the basic elements of a statechart for the SecuritySupervisor class shown previously in Figure 1-1. It includes basic OR-states and transitions, as well as a few less-elementary concepts, including nested states and conditional and initial pseudostates.

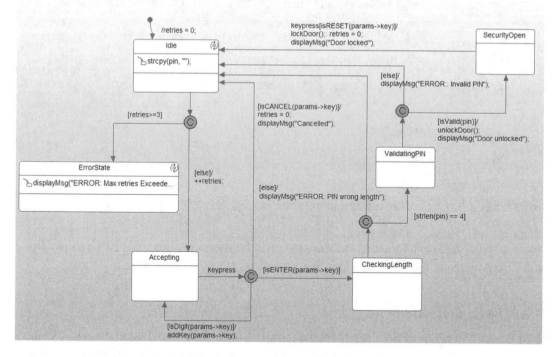

Figure 1-3: Basic State Machine

Transitions are arrowed lines coming from a predecessor state and terminating on a subsequent state. Transitions usually have the optional *event signature* and *action list*. The basic form of an event signature is

    event-name '('parameter-list')' ['guard'] '/' action-list

The event-name is simply the logical name of the event class that may be sent to an instance of the Classifier at run-time, such as keypress in Figure 1-3. The UML defines four distinct kinds of events that may be passed or handled:

- SignalEvent – an asynchronously sent event
- CallEvent – a synchronously sent event
- TimeEvent – an event due to the passage of an interval of time (most common) or arrival of an epoch
- ChangeEvent – a change in a state variable or attribute of the Classifier

Asynchronous event transfer is always implemented via queuing of the event until the element is ready to process it. That is, the sender "sends and forgets" the event and goes on about its business, ignorant of whether or not the event has been processed. Synchronous event transfer executes the state processing of the event in the thread of the sender, with the sender blocked from continuing until that state processing is complete. This is commonly implemented by invoking a class method called an *event handler* that executes the relevant part of the state machine, returning control to the sender only when the event processing is complete. Such "triggered operations" don't have a standard method body; their method body is the action list on the state machine.

Events may have parameters – typed values accepted by the state machine which may then be used in the guard and actions in the processing of the event. The statechart specifies the formal parameter list, while the object that sends the event must provide the necessary actual parameters to bind to the formal parameter list. We will use a slightly peculiar syntax for passing events. We will create a struct named *params* that contains the named parameters for every event that carries data. If an event *e* has two parameters, x and y, to use these in a guard, for example, you would dereference the params struct to access their value. So a transition triggered by event *e* with a guard that specified that x must be greater than 0 and y must be less than or equal to 10 for the transition to be taken would look like:

e[params->x>0 && params->y<=10]

Time events are almost always relative to the entry to a state. A common way to name such an event (and what we will use here) is "tm(interval)," where "interval" is the time interval parameter for the timeout event[12]. If the timeout occurs before another specified event occurs, then the transition triggered by the timeout event will be taken; if another event is sent to the object prior to the triggering of the timeout, then the time is discarded; if the state is reentered, the timeout interval is started over from the beginning.

If a transition does not provide a named event trigger, then this is activated by the "completion" or "null" event. This event occurs either as soon as the state is entered (which includes the execution of entry actions for the state) or when the state activities complete.

A guard is a Boolean expression contained within square brackets that follows the event trigger. The guard should return only TRUE or FALSE and not have side effects. If a guard is specified for a transition and the event trigger (if any) occurs, then the transition will be taken if and only if the guard evaluates to TRUE. If the guard evaluates to FALSE, then the triggering event is quietly discarded and no actions are executed.

The action list for the transition is executed if and only if the transition is taken; that is, the named event is received by the object while it is in the predecessor state and the guard, if any, evaluates to TRUE. The entire set of exit actions–transition actions–entry actions is executed in that order and is executed using run-to-completion semantics. Figure 1-3 shows actions on a number of different transitions as well as on entry into some of the states.

A run of the three "classes" from Figure 1-1 using test1 from the TestDriver (which sends '1', '2', '3', '4', 'e' (the code for the ENTER key), followed by 'r' (the code for the RESET key) is shown in Figure 1-4 in a UML sequence diagram.

---

[12] Another common way is to use the term "after(interval)."

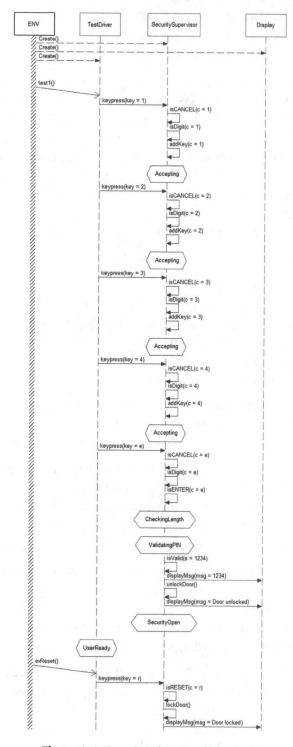

Figure 1-4: Running the state behavior

While I will later discuss design patterns for implementation of state machines, the most common implementation is to generate simple if-then or switch-case statements. For example, a common and easy implementation scheme is to

- Add a state variable (e.g., `activeState`)
- For each event to the "class," add an event receptor function and pass in any data that it needs as parameters
- Create an event dispatcher function, called by each event receptor; this processes the incoming event

The structure of the event dispatcher can be as simple as

```
switch(activeState) {
    /* for each state */
    case state1:
        /* for each event */
        Switch (eventID) {
            event1:
                /* check each guard */
                If (guard1()) {
                    action1();
                ] else if (guard2()) {
                    action2();
                };
                break;
            event2:
                if (guard3()) {
                    action3();
                } else if (guard4()) {
                    action5();
                };
                break;
        };
        break;
    case state2:
        // etc
}
```

For example, given the SecuritySupervisor class, you might implement the state machine event dispatcher as shown in Code Listing 10. Adding asynchronous events requires adding the queuing of the events and their data. Note: the Null_id represents so-called "null events" – transitions that are not triggered by explicit events, but simply "fire" when the actions in the previous state complete.

```
static eventStatus dispatchEvent(short id) {
    eventStatus res = eventNotConsumed;
    switch (activeState) {
            /* are we in the Idle state? */
```

```
case SecuritySupervisor_Idle:
{
    if(id == Null_id) /* null triggered event? */
        {
            if(retries>=3)
                {
                    activeState = SecuritySupervisor_ErrorState;
                        displayMsg("ERROR: Max retries Exceeded");
                    res = eventConsumed;
                }
            else
                {
                    ++retries;
                    activeState = SecuritySupervisor_Accepting;
                    res = eventConsumed;
                }
        }
}
break;
        /* are we in the Accepting State? */
case SecuritySupervisor_Accepting:
{
    if(id == keypress_SecuritySupervisor_Event_id)
        {
            /* params struct has the data in the attribute 'key' */
          /* transition 1 */
        if(isCANCEL(params->key))
                {
                    retries = 0;
                    displayMsg("Cancelled");
                    activeState = SecuritySupervisor_Idle;
                    {
                        /* state ROOT.Idle.(Entry) */
                        strcpy(pin, "");
                    }
                    res = eventConsumed;
                }
            else
                {
                    /* transition 3 */
                    if(isDigit(params->key))
                        {
                            /* transition 3 */
                            addKey(params->key);
                            activeState = SecuritySupervisor_Accepting;
                            res = eventConsumed;
                        }
                      else
                        {
                            /* transition 5 */
```

```c
                                if(isENTER(params->key))
                                    {
                                        activeState =
                                                SecuritySupervisor_CheckingLength;
                                      res = eventConsumed;
                                    }
                                }
                            }
                        }
                    }
                }
            break;
            case SecuritySupervisor_CheckingLength:
            {
                if(id == Null_id)
                    {
                        /* transition 10 */
                        if(strlen(pin) == 4)
                            {
                                activeState = SecuritySupervisor_ValidatingPIN;
                                res = eventConsumed;
                            }
                          else
                            {
                                {
                                    /* transition 9 */
                                    displayMsg("ERROR: PIN wrong length");
                                }
                                activeState = SecuritySupervisor_Idle;
                                {
                                    /* state ROOT.Idle.(Entry) */
                                    strcpy(pin, "");
                                }
                                res = eventConsumed;
                            }
                    }
            }
            break;
            case SecuritySupervisor_ValidatingPIN:
            {
                if(id == Null_id)
                    {
                        /* transition 13 */
                        if(isValid(pin))
                            {
                                {
                                    /* transition 13 */
                                    unlockDoor();
                                    displayMsg("Door unlocked");
                                }
                                activeState = SecuritySupervisor_SecurityOpen;
                                res = eventConsumed;
```

```
                            }
                        else
                            {
                                {
                                    /* transition 12 */
                                    displayMsg("ERROR:: Invalid PIN");
                                }
                                activeState = SecuritySupervisor_Idle;
                                {
                                    /* state ROOT.Idle.(Entry) */
                                    strcpy(pin, "");
                                }
                                res = eventConsumed;
                            }
                    }
            }
        break;
        case SecuritySupervisor_SecurityOpen:
        {
            if(id == keypress_SecuritySupervisor_Event_id)
                {
                        /* params-key has the data passed with the event */
                    /* transition 14 */
                    if(isRESET(params->key))
                        {
                            {
                                /* transition 14 */
                                lockDoor(); retries = 0;
                                displayMsg("Door locked.");
                            }
                            activeState = SecuritySupervisor_Idle;
                            {
                                /* state ROOT.Idle.(Entry) */
                                strcpy(pin, "");
                            }
                            res = eventConsumed;
                        }
                }
        }
        break;
        default:
            break;
        }
        return res;
}
```

Code Listing 10: SecuritySupervisor Event Dispatcher

There is certainly more to state machines than we've introduced so far, but those are discussed later in Chapter 5. For a more detailed discussion of UML state machines, refer to Chapter 3 of *Real-Time UML 3rd Edition*[13] and Chapters 7 and 12 in *Doing Hard Time*[14].

1.3 What Did We Learn?

This chapter discussed the special characteristics of embedded systems – working within tight design constraints while meeting functional and quality-of-service constraints. To do this, embedded developers use tools from the embedded tool chain to develop on a more powerful host computer while targeting an often smaller and less capable target computing environment. Target environments run the gamut from small eight-bit processors with a few kilobytes of memory to networked collections of 64-bit systems. At the small scale, many systems are too tightly constrained in memory, performance, or cost to permit the use of a commercial RTOS, while at the large scale, the systems may include a heterogeneous set of operating systems, middleware, and databases. Often, embedded software must be developed simultaneously with the hardware, making it difficult to test and debug since the target environments may not exist when the software is being developed.

By far, the most common language for developing embedded systems is C. C has the advantages of high availability of compilers for a wide range of target processors and a well-deserved reputation for run-time efficiency. Nevertheless, the pressures to add capabilities while simultaneously reducing costs mean that embedded developers must be continually looking for ways to improve their designs, and their ability to design efficiently. Structured methods organize the software into parallel taxonomies, one for data and the other for behavior. Object-oriented methods combine the two to improve cohesion for inherently tightly-coupled elements and encapsulation of content when loose coupling is appropriate. While C is not an object-oriented language, it can be (and has been) used to develop object-based and object-oriented embedded systems.

The next chapter will discuss the role of process and developer workflows in development. These workflows will determine when and where developers employ design and design patterns in the course of creation of embedded software. It will go on to define what design patterns are, how they will be discussed and organized within this book, and how they can be applied into your designs.

---

[13]   Douglass, B.P., 2004. *Real-Time UML: Advances in the UML for Real-Time Systems*, third ed. Addison-Wesley.

[14]   Douglass, B.P., 1999. *Doing Hard Time: Developing Real-Time Systems with UML, Objects, Frameworks, and Patterns*. Addison-Wesley.

Subsequent chapters will list a variety of design patterns that can be applied to optimize your designs. Chapter 3 provides a number of patterns that address access of hardware, such as keyboards, timers, memory, sensors, and actuators. Chapter 4 discusses concurrency patterns both for managing and executing concurrent threads as well as sharing resources among those threads. The last two chapters provide a number of patterns for the implementation and use of state machines in embedded systems: Chapter 5 provides basic state machine implementation-focused patterns, while Chapter 6 addresses the concerns of safety and high-reliability software.

CHAPTER 2

Embedded Programming with The Harmony™ for Embedded RealTime Process

---

What you will learn

- Agile process principles and workflows for embedded applications development

- The three levels of design

- The five key views of architecture

- Performing trade studies

- What are design patterns?

- How do I apply design patterns?

- Example design pattern instantiation

Design Patterns for Embedded Systems in C
DOI:10.1016/B978-1-85617-707-8.00002-9
Copyright © 2011 Elsevier Inc.

2.1 Basic Elements of the Harmony Process

Let's now turn our attention to the basic questions of how we, as embedded software developers, perform design. While I don't want to belabor process in this book, we perform design activities by following (formal or informal) workflows. To understand how to do design, we must have at least a basic agreement as to the workflow.

2.1.1 A Quick Overview of the Development Workflow

I am the author of the Harmony™ for Embedded RealTime (Harmony/ERT[1]) process from IBM Rational[2]. This process consists of a set of best practices for the development of real-time and embedded software, and is in use in many different embedded vertical markets, from medical to aerospace to consumer electronics. It is an agile, model-based approach to software development that emphasizes continual execution of software throughout its development, test-driven development, and the use of design patterns.

The Harmony™ for Embedded RealTime process is an incremental process, similar in some ways to Scrum[3]. The Harmony/ERT process focuses on three timescales, as shown in Figure 2-1. The *macrocycle* focuses on the overall project timeline and major milestones. The *microcycle* focuses on the production of a single validated increment (a.k.a. a build or prototype). The *nanocycle* timeframe is in the range of minutes to hours and concentrates on continual execution, debug, and unit test of the software.

The microcycle, which is iterated many times during the course of a typical project, is broken up into several short phases, as you can see in Figure 2-2. These phases contain the workflows relevant to the current discussion. These are briefly described in Table 2-1, where we can see a couple of key points. First, design is something that occurs only *after* we have correctly functioning software from the *Object Analysis* (or "Software Analysis" if you prefer) phase. Secondly, design focuses on *optimizing* this correctly functioning software. This latter point is important because that's what design patterns – the theme of this book – do; they optimize software. Design patterns are not about getting the software to work correctly; they are about getting the software to work optimally.

As you can see from Table 2-1, the Harmony/ERT process uses three levels of design. Architectural design looks at design decisions that optimize the system overall. Architectural design is subdivided into five key views that have the greatest impact on the structure and behavior of the software (Figure 2-3). Each of these key architectural views is described in Table 2-2.

---

[1] Formerly know as Harmony for Embedded SoftWare (Harmony/ESW. Don't ask; I didn't change the name this time).

[2] See *www-01.ibm.com/software/awdtools/rmc/* for information on tools and process content.

[3] See *www.scrumalliance.org/.*

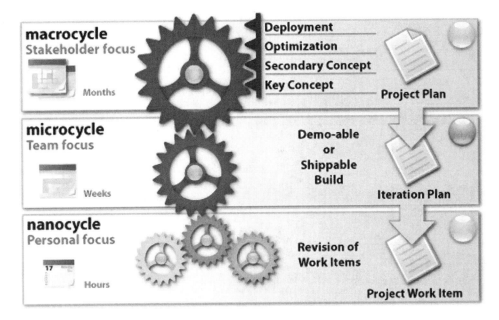

Figure 2-1: Harmony/ERT timescales

Mechanistic design optimizes at the level of the collaboration, where a *collaboration* is a collection of simpler elements (types, functions, variables, and classes) interacting to achieve a single use case or system capability. Each system has usually from as few as half a dozen to as many as several dozen such capabilities; as such, the scope of mechanistic design is an order of magnitude smaller than architectural concerns.

Detailed design limits itself to the smallest level of concern – the insides of functions, variables, types, and classes. This level of design is perhaps the most common level of focus by developers but often has the least impact on overall system performance.

At any level of focus, the basic workflow for design is provided by Figure 2-4. Because these steps will form the basis of how we use design patterns, let's discuss each of them in turn.

2.1.1.1 Construct the Initial Model

This step is crucial but often omitted in the developer's haste to optimize. The step means that before optimization is considered, a working model and code base is created. By "working" I mean not only that it compiles and runs, but also that it is robust and functionally correct. I also mean that it is debugged and unit-tested for functional correctness. It may be tested for performance to determine whether optimization is required but that is optional at this point.

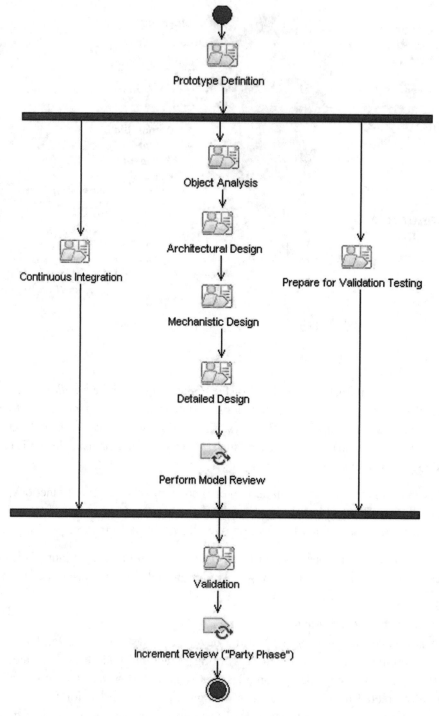

Figure 2-2: Harmony/ERT microcycle

Table 2-1: Harmony/ERT microcycle phases.

Phase	Primary Inputs	Primary Outputs	Description
Prototype definition	Stakeholder requirements	System requirements use case model	Transforms a subset of stakeholder requirements into system requirements for development in the current microcycle
Object analysis	System requirements	Functionally correct model and source code	Develops software that realizes the functional requirements; includes coding, debugging, and unit testing
Architectural design	Functionally correct code and models	Architecturally-optimized code and models	Optimizes the code based on explicit design decisions meant to optimize the system overall; includes coding, debugging, and unit testing
Mechanistic design	Architecturally-optimized code and models	Collaboration-optimized code and models	Optimizes the code based on explicit design decisions optimizing collaborations of elements meant to realize a feature or use case; includes coding, debugging, and unit testing
Detailed design	Collaboration-optimized code and models	Optimized functions and data structures	Optimizes the code based on the explicit design decisions at the function, variable, and type level; includes coding, debugging, and unit testing
Model/code review	Models and unit tested code	Change requests based on review	Reviews the analysis, design, and code for consistency, completeness, correctness
Validation	Optimized source code	Validated system and change requests based on defects identified	Tests the system against the requirements implemented so far, including regression testing of requirements from previous microcycles
Increment review	Schedule, risk management plan, reuse plan, hazard analysis, team questionnaire	Change requests to process, plans, or development environment	Evaluates progress against plan and identifies where plans must be updated; also identifies team, environment, and development issues that need to be improved
Continuous integration	Models and unit-tested source code	Evolving software baseline	Executed at least daily, this activity integrates source code from multiple developers, applies integration tests, and updates the evolving software baseline
Prepare for validation	System requirements and use case models	Test plans, test cases, test fixtures	Prepares for validation testing by developing plans, test cases, and test environments

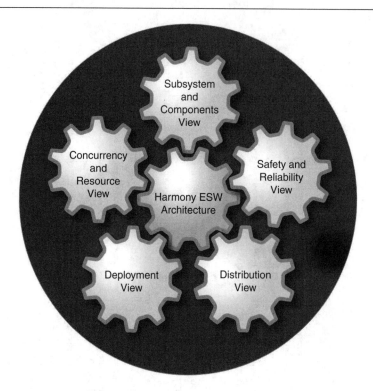

Figure 2-3: Architectural design views

Table 2-2: Key architectural views.

Key Architectural View	Description
Subsystem and components view	Specifies the set of system subsystems and components, their responsibilities, connections to other subsystems, and interfaces
Concurrency and resource view	Identifies the concurrency units, how primitive elements (functions, variables, and objects) map to the concurrency units, how those concurrency units are scheduled, and how the concurrency units synchronize and share resources
Distribution view	Describes how elements are distributed across multiple address spaces and how those elements find and collaborate with each other
Safety and reliability view	Defines how faults are identified, isolated, and corrected during system execution to maintain safety and reliability
Deployment view	Characterizes how responsibilities are split across elements from different engineering disciplines (e.g., software, electronic, mechanical, and chemical) and the interfaces between those disciplines

2.1.1.2 Identify Important Design Criteria

This is another commonly omitted step. It is important to remember that whenever some system aspects are optimized, it is almost always true that some other aspects are deoptimized. In order

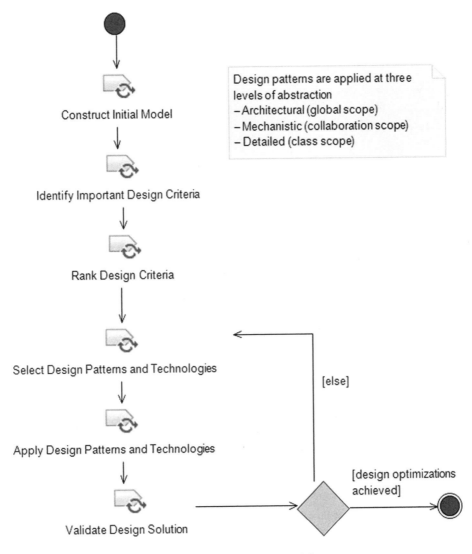

Figure 2-4: Basic design workflow

to select appropriate design patterns and technologies to achieve optimization, we need to first understand what are the important aspects to optimize, so we can trade them off appropriately. Some common design optimization criteria are:

- Performance
  - Worst case
  - Average case
- Predictability
- Schedulability

- Throughput
  - Average
  - Sustained
  - Burst
- Reliability
  - With respect to errors
  - With respect to failures
- Safety
- Reusability
- Distributability
- Portability
- Maintainability
- Scalability
- Complexity
- Resource usage, e.g., memory
- Energy consumption
- Recurring cost, i.e., hardware
- Development effort and cost

2.1.1.3 Rank Design Criteria

The ranking of the criteria is necessary for us to make appropriate tradeoffs when we select design patterns. Some patterns will optimize worst-case performance at the expense of using more memory; others will optimize robustness at the expense of average case performance. Others may optimize energy economy at the expense of responsiveness. The ranking of the design criteria makes their relative importance explicit, so that we can reason about which design solutions are best.

2.1.1.4 Select Design Patterns and Technologies

There is a wealth of effective design solutions and technologies available. A *trade study* is a comparison of technologic solutions based on their effectiveness at achieving a balanced set of optimization goals. Although we may choose to be less formal in our approach, the FAA recommends a formalized decision analysis matrix[4]. Basically, this is a spreadsheet that summarizes the "goodness" of a design solution based on the degree to which it matches our weighted design criteria. For example, Table 2-3 shows the kind of analysis that we might do formally in the spreadsheet or informally in our heads. The weight for each criteria is the relative importance of the criteria on a scale of 0 to 10. Each alternative – a combination of design patterns and technologies – is assigned a score (degree to which the criteria is optimized)

---

[4] *www.faa.gov/about/office_org/headquarters_offices/ato/service_units/operations/sysengsaf/seman/SEM3.1/Section%204.6.pdf*

Table 2-3: Design tradeoff spreadsheet.

| Design Solution | Design Criteria | | | | | Total Weighted Score |
| | Criteria 1 Weight = 7 | Criteria 2 Weight = 5 | Criteria 3 Weight = 3 | Criteria 4 Weight = 2 | Criteria 5 Weight = 1.5 | |
	Score	Score	Score	Score	Score	
Alternative 1	7	3	6	9	4	106
Alternative 2	4	8	5	3	4	95
Alternative 3	10	2	4	8	8	**120**
Alternative 4	2	4	9	7	6	84

in the range 0 to 10. Each alternative is then assigned a computed total weighted score equal to the sum of the criteria weights times their relative scores. The alternative with the largest total weighted score provides the best balanced design solution for the set of design criteria.

2.1.1.5 Apply Design Patterns and Technologies

This step refers to the application of the selected design alternative. In design pattern terms, this is known as *pattern instantiation*. As we will see, this is primarily a matter of replacing the formal pattern parameters with the relevant elements from the working analysis model.

2.1.1.6 Validate Design Solution

The validation step ensures that the solution works and checks two different facets of the solution. First, it ensures that the functionality of the system continues to be correct; that is, adding in the design solution didn't break the running model and code. Secondly, since we selected the design patterns to optimize the system, we must ensure that we've achieved our optimization goals. This means that we will rerun the previous functional and unit tests and also add quality of service tests to measure performance, memory usage, or whatever measures are appropriate to verify our design criteria are met.

2.1.2 What Is a Design Pattern?

A design pattern is "a generalized solution to a commonly occurring problem." To be a pattern, the problem must recur often enough to be usefully generalizable and the solution must be general enough to be applied in a wide set of application domains. If it only applies to a single application domain, then it is probably an *analysis pattern*[5]. Analysis patterns define ways for organizing problem-specific models and code for a particular application domain.

Remember that analysis is driven by *what* the system must do while design is driven by *how well* the system must achieve its requirements. A design pattern is a way of organizing a design that improves the optimality of a design with respect to one or a small set of design criteria, such as qualities of service (QoS).

A design pattern always has a focused purpose – which is to take one or a small set of these QoS properties and optimize them at the expense of the others. Each pattern optimizes some aspect of a system at the cost of deoptimizing some other aspects. These constitute a set of pros and cons.

A good design is one composed of a set of design patterns applied to a piece of functional software that achieves a balanced optimization of the design criteria and incurs an acceptable cost in doing so. That is, a good design optimizes the important design criteria at a cost that we are willing to pay. Note that a design always encompasses several different design patterns, so

---

[5] A coherent set of analysis patterns is also known as a *reference model*.

each alternative in our trade study must also examine how well the patterns work together, in addition to how each works in isolation.

As mentioned previously, patterns may be applied at the different levels of design abstraction. Architectural patterns have systemic scope and apply mostly to only one of the five key views of architecture. At the next level down in design abstraction, mechanistic design patterns apply to individual collaborations, optimizing the same criteria, but in a more narrow focus. Detailed design patterns – often referred to as design idioms – optimize the primitive elements of the software.

2.1.3 Basic Structure of Design Patterns

According to Gamma, et. al.[6], a pattern has four important aspects:

- Name
  The name provides a "handle" or means to reference the pattern.
- Purpose
  The purpose provides the problem context and the QoS aspects the pattern seeks to optimize. The purpose identifies the kinds of problem contexts where the pattern might be particularly appropriate.
- Solution
  The solution is the pattern itself.
- Consequences
  The consequences are the set of pros and cons of the use of the pattern.

The pattern *name* brings us two things. First, it allows us to reference the pattern in a clear, unambiguous way, with the details present, but unstated. Secondly, it gives us a more abstract vocabulary to speak about our designs. The statement "The system uses a Homogeneous Channel redundant architecture with Cyclic Executive Concurrency distributed across a set of processors with a Proxy Pattern" has a lot of information about the overall structure of the system because we can discuss the architecture in terms of these patterns.

The *purpose* of the pattern identifies the essential problem contexts required for the pattern to be applicable and what design criteria the pattern is attempting to optimize. This section specifies under which situations the pattern is appropriate and under which situations it should be avoided.

The *solution*, of course, specifies the structure and behavior of the pattern itself. It identifies the elements of the pattern and their roles in the context of the pattern. This pattern solution is integrated into your design in a process known as *pattern instantiation*.

---

[6] Gamma, E., Helm, R., Johnson, R., Vlissides, J.M., 1995. *Design Patterns: Elements of Reusable Object-Oriented Software*. Addison-Wesley.

The *consequences* are important because we always make tradeoffs when we select one pattern over another. We must understand the pros and cons of the pattern to apply it effectively. The pros and cons are usually couched in terms of improvement or degradation of some design properties, as well as a possible elaboration of problem contexts in which these consequences apply.

2.1.4 How to Read Design Patterns in This Book

All the patterns of this book are organized in the same fashion to improve the usability of the patterns. Each pattern is provided in the following format:

- Abstract
  The abstract gives a brief description of the pattern use or justification. This is meant as an overview of the problem, solution, and consequences.
- Problem
  The problem section gives a statement of the problem context and the qualities of service addressed by the pattern.
- Pattern structure
  This section provides a structural UML diagram of the pattern showing the important elements of the pattern. These elements are the places into which you will substitute your own specific application elements to instantiate the pattern. Relations among elements of the pattern are shown as well.
- Consequences
  The consequences section describes the tradeoffs made when the pattern is used.
- Implementation strategies and source code
  This section discusses issues around the implementation of the pattern on different computing platforms or in different source level languages.
- Example
  Each pattern is shown in an example, illustrating how the pattern is applied in some particular case. This usually involves the presentation of a UML structural diagram showing particular application elements fulfilling the pattern collaboration roles and relevant code snippets showing the pattern implementation.

Each pattern in this book is shown using both generic, standard UML, and C source code. The UML graphical notation provides a higher-level overview of the pattern, but if you don't find this helpful, you can skip directly to the source code. If you're interested, Appendix A gives a UML notational summary. For a more detailed understanding of the UML itself, see *Real-Time UML 3rd Edition*[7].

---

[7] Douglass, B.P., 2004. *Real-Time UML: Advances in the UML for Real-Time Systems*, third ed. Addison-Wesley.

2.1.5 Using Design Patterns in Development

By this point, you should have a reasonably good understanding of what a pattern is and how it is organized. You should also have at least a vague grasp of the Harmony/ERT process and our view of where design fits into the overall scheme of things. Earlier in this chapter, you got a grounding in what we mean by the term *architecture* and the five views of architecture. At this point, you are almost ready to read the patterns in this book and apply them to your applications development. But first, let's briefly discuss how we might use patterns in our daily work lives.

2.1.5.1 Pattern Hatching – Locating the Right Patterns

You're facing a design problem. How do you find the patterns that can be applied to solve your particular design issues? We recommend a multistep approach, as shown in Figure 2-5.

1. Familiarize yourself with the patterns

   First, before starting your design, familiarize yourself with the patterns literature[8]. There are a number of books, papers, and websites devoted to patterns in many application domains. Some of those patterns are given here and others are given in the references. Once you have increased your vocabulary to include patterns likely to be relevant to your application domain, you have more intellectual ammunition to face your design challenges.
2. Identify design criteria

   In parallel with reading about potential design alternatives, you must identify design criteria. There are many (often conflicting) optimizations you can make to your system. This step identifies the design optimizations that are important for this particular system in its operational and development contexts.
3. Rank design criteria

   Not all optimizations are equally important. For this reason, the design criteria must be ranked to enable good tradeoff decisions to be made.
4. Identify potential design alternatives

   This step finds patterns that optimize the design criteria of importance at a cost that seems reasonable. While it is possible to add a single pattern at a time in this way and repeat the workflow, it is also common to consider a set of compatible patterns together as a single design alternative.
5. Evaluate the design alternatives

   Using the trade study approach suggested earlier, this step compares the design alternatives (consisting of one or more design patterns) against each other on the basis of their effectiveness against the design criteria, weighted with the relative importance of each.

---

[8] That means read the book!

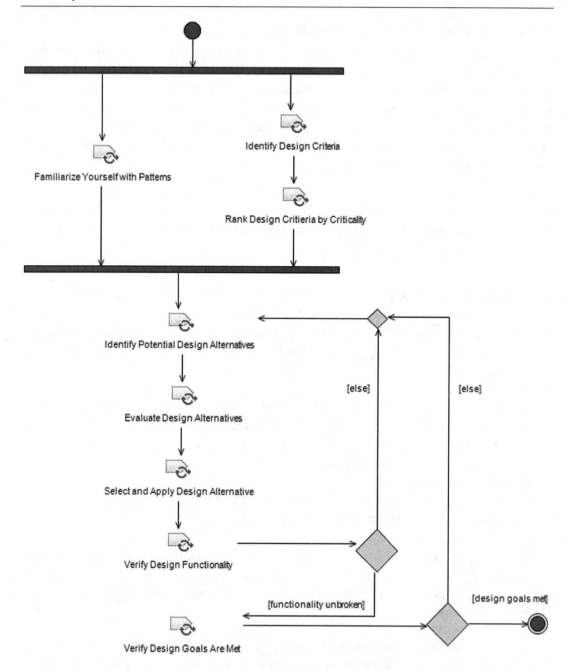

Figure 2-5: Pattern hatching

6. Select and apply design alternative
   Once a design alternative is selected, the patterns are usually added to the collaboration one at a time, including verification of the efficacy of the solution.

7. Verify design functionality

   The collaboration of software elements worked before applying the design patterns. This task ensures that it still does. Because the application of design patterns usually requires some (possibly small) amount of refactoring, some unit tests may require updates. If the pattern application has broken the functionality, the software must be repaired so that it passes its functional unit tests.

8. Verify design goals are met

   Once the design functionality is verified, we must also ensure that the optimization goals of the design decisions result in the right tradeoffs. If we've optimized the code for memory usage, the usage should be measured as the software runs. If we've optimized worst-case execution time, then that time should be measured.

2.1.5.2 Pattern Mining – Rolling Your Own Patterns

Creating your own pattern is useful, especially when you have a depth of experience to understand the optimization issues in a particular area, and sufficient breadth to understand the general properties of the solutions enough to abstract them into a generalized solution. We call this *pattern mining* (see Figure 2-6). Pattern mining isn't so much a matter of invention as it is

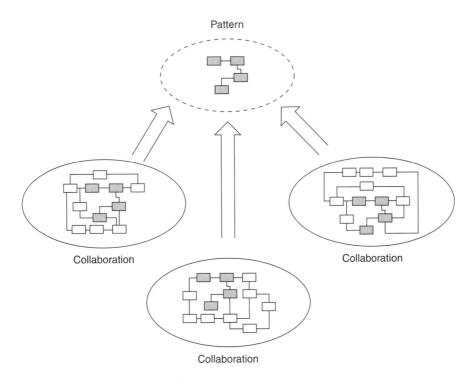

Figure 2-6: Pattern mining

discovery – seeing that this solution in some context is similar to another solution in another context and abstracting away the specifics of the solutions. Keep in mind that to be a useful pattern, it must occur in different contexts and perform a useful optimization of one or more design criteria.

2.1.5.3 Pattern Instantiation – Applying Patterns in Your Designs

Pattern instantiation is the opposite of pattern mining. It is applying the pattern to a particular collaboration to gain the benefits of the pattern (see Figure 2-7). Patterns are normally applied to a collaboration of software elements, such as classes, types, functions, and/or variables. As mentioned previously, a collaboration is a set of software elements (at some scope these may be small low-level functions or classes, while at others they may be large-grain assemblies such as subsystems and components). The purpose is to organize, and possibly elaborate, this already existing collaboration with the selected pattern.

The application or instantiation of a pattern in your design is a matter of identifying the elements in your collaboration that fulfill the pattern parameter roles. For some of the patterns you may create the role as a super class from which you subclass to instantiate that pattern. In other patterns, you may simply replace the pattern element with one from your application

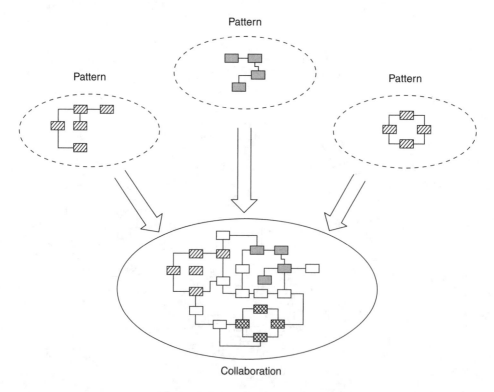

Figure 2-7: Pattern instantiation

domain, adding in the required operations and behavior. Or, you may choose to create elements just as they exist in the pattern itself.

2.1.5.4 An Example: The Observer Pattern

Let's consider a simple example. Figure 2-8 shows a system that contains an ECG module data source that pipes data into a queue for consumption of a number of different clients. The data include both a time interval marker (of type `long`) and the `dataValue`, a measurement of the potential difference between the selected ECG lead pair. The queue is known as a "leaky queue" in that data are never explicitly removed from the circular buffer, but are overwritten as the buffer inserts wraparound to the start of the queue.

The source code for a simple implementation of the elements is straightforward. The ECGPkh.h header file defines the queue size and the `boolean` type.

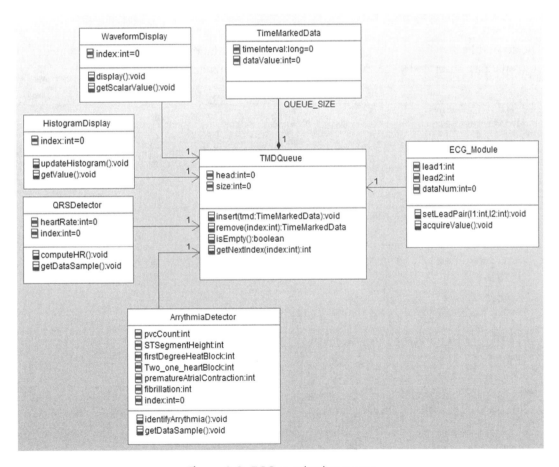

Figure 2-8: ECG monitoring system

```
#ifndef ECGPkg_H
#define ECGPkg_H

#include <stdio.h>

/* forward declarations */
struct ArrythmiaDetector;
struct ECG_Module;
struct HistogramDisplay;
struct QRSDetector;
struct TMDQueue;
struct TestBuilder;
struct TimeMarkedData;
struct WaveformDisplay;

typedef unsigned char boolean;
#define QUEUE_SIZE (20000)

#endif
```

Code Listing 2-1: ECGPkg.h Header File

The second header file (Code Listing 2-2) is the Timed Marked Data queue (TMDQueue). It holds the data inserted by the ECG_Module for its various data clients.

```
#ifndef TMDQueue_H
#define TMDQueue_H

/*## auto_generated */
#include <stdio.h>
#include "ECGPkg.h"
#include "TimeMarkedData.h"

typedef struct TMDQueue TMDQueue;
/*
This queue is meant to operate as a "leaky" queue. In this queue, data are never removed per
se, but are instead overwritten when the buffer pointer wraps around. This allows for many
clients to read the same data from the queue. */

struct TMDQueue {
    int head;
    int size;
    struct TimeMarkedData buffer[QUEUE_SIZE];
};

/* Constructors and destructors:*/
void TMDQueue_Init(TMDQueue* const me);
void TMDQueue_Cleanup(TMDQueue* const me);

/* Operations */
int TMDQueue_getNextIndex(TMDQueue* const me, int index);
```

```
    void TMDQueue_insert(TMDQueue* const me, const struct TimeMarkedData tmd);

boolean TMDQueue_isEmpty(TMDQueue* const me);

struct TimeMarkedData TMDQueue_remove(TMDQueue* const me, int index);

int TMDQueue_getBuffer(const TMDQueue* const me);

TMDQueue * TMDQueue_Create(void);

void TMDQueue_Destroy(TMDQueue* const me);

#endif
```

Code Listing 2-2: TMDQueue Header File

Code Listing 2-3 shows the header file for the ECG_Module. The code here acquires the data across the selected lead pair and calls the `TMDQueue_insert()` function to store them.

```
#ifndef ECG_Module_H
#define ECG_Module_H
#include <stdio.h>
#include "ECGPkg.h"

struct TMDQueue;

/* class ECG_Module */
typedef struct ECG_Module ECG_Module;
struct ECG_Module {
    int dataNum;
    int lead1;
    int lead2;
    struct TMDQueue* itsTMDQueue;
};

/* Constructors and destructors:*/
void ECG_Module_Init(ECG_Module* const me);
void ECG_Module_Cleanup(ECG_Module* const me);

/* Operations */
void ECG_Module_acquireValue(ECG_Module* const me);
void ECG_Module_setLeadPair(ECG_Module* const me,
        int l1, int l2);

struct TMDQueue* ECG_Module_getItsTMDQueue(const ECG_Module* const me);

void ECG_Module_setItsTMDQueue(ECG_Module* const me, struct TMDQueue* p_TMDQueue);

ECG_Module * ECG_Module_Create(void);

void ECG_Module_Destroy(ECG_Module* const me);

#endif
```

Code Listing 2-3: ECG_Module Header File

The clients, for the purpose of this discussion, look pretty much the same. They all have a get_xxx() function that pulls data out of the queue and does something interesting with them. Each client is required to maintain its own position within the queue so that it doesn't miss any data. For this reason, Code Listing 2-4 shows only the header file for the HistogramDisplay class.

```
#ifndef HistogramDisplay_H
#define HistogramDisplay_H

#include <stdio.h>
#include "ECGPkg.h"
struct TMDQueue;

/* class HistogramDisplay */
typedef struct HistogramDisplay HistogramDisplay;
struct HistogramDisplay {
    int index;
    struct TMDQueue* itsTMDQueue;
};

/* Constructors and destructors:*/
void HistogramDisplay_Init(HistogramDisplay* const me);
void HistogramDisplay_Cleanup(HistogramDisplay* const me);

/* Operations */
void HistogramDisplay_getValue(HistogramDisplay* const me);
void HistogramDisplay_updateHistogram(HistogramDisplay* const me);

struct TMDQueue* HistogramDisplay_getItsTMDQueue(const HistogramDisplay* const me);

void HistogramDisplay_setItsTMDQueue(HistogramDisplay* const me, struct TMDQueue*
p_TMDQueue);

HistogramDisplay * HistogramDisplay_Create(void);

void HistogramDisplay_Destroy(HistogramDisplay* const me);

#endif
```

Code Listing 2-4: HistogramDisplay Header File

The TMDQueue.c implementation file is likewise simple enough, as shown in Code Listing 2-5.

```
#include "TMDQueue.h"

static void initRelations(TMDQueue* const me);

static void cleanUpRelations(TMDQueue* const me);
```

```
void TMDQueue_Init(TMDQueue* const me) {
    me->head=0;
    me->size=0;
    initRelations(me);
}

void TMDQueue_Cleanup(TMDQueue* const me) {
    cleanUpRelations(me);
}

/* operation getNextIndex(int) */
int TMDQueue_getNextIndex(TMDQueue* const me, int index) {
    /* this operation computes the next index from the first using modulo arithmetic
    */
    return (index+1) % QUEUE_SIZE;
}

/* operation insert(TimeMarkedData) */
void TMDQueue_insert(TMDQueue* const me, const struct TimeMarkedData tmd) {
    /* note that because we never 'remove' data from this leaky queue, size only increases to
    the queue size and then stops increasing. Insertion always takes place at the head.
    */
    printf("Inserting at: %d Data #: %d", me->head,
    tmd.timeInterval);
    me->buffer[me->head]=tmd;
    me->head=TMDQueue_getNextIndex(me, me->head);
    if (me->size < QUEUE_SIZE) ++me->size;
    printf(" Storing data value: %d\n", tmd.dataValue);
}

/* operation isEmpty() */
boolean TMDQueue_isEmpty(TMDQueue* const me) {
    return (boolean)(me->size== 0);
}

/* operation remove(int) */
struct TimeMarkedData TMDQueue_remove(TMDQueue* const me, int index) {
    TimeMarkedData tmd;
    tmd.timeInterval=-1; /* sentinel values */
    tmd.dataValue=-9999;

    if (!TMDQueue_isEmpty(me) &&
        (index>=0) && (index < QUEUE_SIZE)
        && (index < me->size)) {
        tmd=me->buffer[index];
    }
    return tmd;
}
```

```
int TMDQueue_getBuffer(const TMDQueue* const me) {
    int iter=0;
    return iter;
}

TMDQueue * TMDQueue_Create(void) {
    TMDQueue* me = (TMDQueue *) malloc(sizeof(TMDQueue));
    if(me!=NULL)
        {
            TMDQueue_Init(me);
        }
    return me;
}

void TMDQueue_Destroy(TMDQueue* const me) {
    if(me!=NULL)
        {
            TMDQueue_Cleanup(me);
        }
    free(me);
}

static void initRelations(TMDQueue* const me) {
    {
        int iter=0;
        while (iter < QUEUE_SIZE){
            TimeMarkedData_Init(&((me->buffer)[iter]));
            TimeMarkedData__setItsTMDQueue(&((me->buffer)[iter]), me);
                iter++;
            }
        }
}

static void cleanUpRelations(TMDQueue* const me) {
    {
        int iter=0;
        while (iter < QUEUE_SIZE){
                TimeMarkedData_Cleanup(&((me->buffer)[iter]));
                iter++;
            }
        }
}
```

Code Listing 2-5: TMDQueue.c Implementation File

The ECG_Module.c implementation file shows how data are inserted into the queue (Code Listing 2-6).

```
#include "ECG_Module.h"
#include "TMDQueue.h"
```

```c
#include <stdlib.h>

static void cleanUpRelations(ECG_Module* const me);

void ECG_Module_Init(ECG_Module* const me) {
    me->dataNum=0;
    me->itsTMDQueue=NULL;
}

void ECG_Module_Cleanup(ECG_Module* const me) {
    cleanUpRelations(me);
}

/* operation acquireValue() */
void ECG_Module_acquireValue(ECG_Module* const me) {
    // in actual implementation, this would return the
    // measured voltage across the lead pair
    TimeMarkedData tmd;
    tmd.dataValue=rand();
    tmd.timeInterval=++me->dataNum;
    TMDQueue_insert(me->itsTMDQueue, tmd);
}

/* operation setLeadPair(int,int) */
void ECG_Module_setLeadPair(ECG_Module* const me, int l1, int l2)
{
    me->lead1=l1;
    me->lead2=l2;
}

struct TMDQueue* ECG_Module_getItsTMDQueue(const ECG_Module* const me) {
    return (struct TMDQueue*)me->itsTMDQueue;
}

void ECG_Module_setItsTMDQueue(ECG_Module* const me, struct TMDQueue* p_TMDQueue) {
    me->itsTMDQueue=p_TMDQueue;
}

ECG_Module * ECG_Module_Create(void) {
    ECG_Module* me=(ECG_Module *) malloc(sizeof(ECG_Module));
    if(me!=NULL)
        {
            ECG_Module_Init(me);
        }
    return me;
}

void ECG_Module_Destroy(ECG_Module* const me) {
    if(me!=NULL)
        {
            ECG_Module_Cleanup(me);
        }
    free(me);
}
```

```
static void cleanUpRelations(ECG_Module* const me) {
    if(me->itsTMDQueue != NULL)
        {
            me->itsTMDQueue = NULL;
        }
}
```

Code Listing 2-6: ECG_Module.c

The HistogramDisplay.c implementation file in Code Listing 2-7 shows the use of the TMDQueue_remove() function.

```
#include "HistogramDisplay.h"
#include "TMDQueue.h"

static void cleanUpRelations(HistogramDisplay* const me);

void HistogramDisplay_Init(HistogramDisplay* const me) {
    me->index = 0;
    me->itsTMDQueue = NULL;
}

void HistogramDisplay_Cleanup(HistogramDisplay* const me) {
    cleanUpRelations(me);
}

/* operation getValue() */
void HistogramDisplay_getValue(HistogramDisplay* const me) {
    TimeMarkedData tmd;
    tmd = TMDQueue_remove(me->itsTMDQueue, me->index);
    printf("    Histogram index: %d TimeInterval: %d DataValue:
        %d\n", me->index, tmd.timeInterval, tmd.dataValue);
    me->index = TMDQueue_getNextIndex(me->itsTMDQueue,
        me->index);
}

/* operation updateHistogram() */
void HistogramDisplay_updateHistogram(HistogramDisplay* const me)
{
    /* put some histogram stuff here... */
}

struct TMDQueue* HistogramDisplay_getItsTMDQueue(const HistogramDisplay* const me) {
    return (struct TMDQueue*)me->itsTMDQueue;
}

void HistogramDisplay_setItsTMDQueue(HistogramDisplay* const me, struct TMDQueue*
p_TMDQueue) {
    me->itsTMDQueue = p_TMDQueue;
}
```

```
HistogramDisplay * HistogramDisplay_Create(void) {
    HistogramDisplay* me = (HistogramDisplay *)
    malloc(sizeof(HistogramDisplay));
    if(me!=NULL)
        {
            HistogramDisplay_Init(me);
        }
    return me;
}

void HistogramDisplay_Destroy(HistogramDisplay* const me) {
    if(me!=NULL)
        {
            HistogramDisplay_Cleanup(me);
        }
    free(me);
}

static void cleanUpRelations(HistogramDisplay* const me) {
    if(me->itsTMDQueue != NULL)
        {
            me->itsTMDQueue = NULL;
        }
}
```

Code Listing 2-7: HistogramDisplay.c Implementation File

For testing, we can put instances of the classes into a TestBuilder class as shown in Figure 2-9.

```
#ifndef TestBuilder_H
#define TestBuilder_H

#include <stdio.h>
#include "ECGPkg.h"
#include "ArrythmiaDetector.h"
#include "ECG_Module.h"
#include "HistogramDisplay.h"
#include "QRSDetector.h"
#include "TMDQueue.h"
#include "WaveformDisplay.h"

/* class TestBuilder */
typedef struct TestBuilder TestBuilder;
struct TestBuilder {
    struct ArrythmiaDetector itsArrythmiaDetector;
    struct ECG_Module itsECG_Module;
    struct HistogramDisplay itsHistogramDisplay;
    struct QRSDetector itsQRSDetector;
    struct TMDQueue itsTMDQueue;
    struct WaveformDisplay itsWaveformDisplay;
};
```

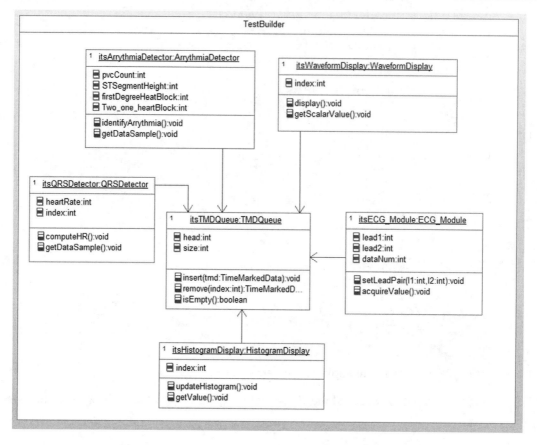

Figure 2-9: ECG system test environment builder class

```
/* Constructors and destructors:*/
void TestBuilder_Init(TestBuilder* const me);
void TestBuilder_Cleanup(TestBuilder* const me);

struct ArrythmiaDetector* TestBuilder_getItsArrythmiaDetector(const TestBuilder*
const me);

struct ECG_Module* TestBuilder_getItsECG_Module(const TestBuilder* const me);

struct HistogramDisplay* TestBuilder_getItsHistogramDisplay(const TestBuilder* const
me);

struct QRSDetector* TestBuilder_getItsQRSDetector(const TestBuilder* const me);

struct TMDQueue* TestBuilder_getItsTMDQueue(const TestBuilder* const me);

struct WaveformDisplay* TestBuilder_getItsWaveformDisplay(const TestBuilder* const
me);
```

```
TestBuilder * TestBuilder_Create(void);

void TestBuilder_Destroy(TestBuilder* const me);

#endif
```

Code Listing 2-8: TestBuilder.h Header File

```c
#include "TestBuilder.h"

static void initRelations(TestBuilder* const me);
static void cleanUpRelations(TestBuilder* const me);

void TestBuilder_Init(TestBuilder* const me) {
    initRelations(me);
}

void TestBuilder_Cleanup(TestBuilder* const me) {
    cleanUpRelations(me);
}

struct ArrythmiaDetector* TestBuilder_getItsArrythmiaDetector(const TestBuilder*
const me)
{
    return (struct ArrythmiaDetector*)&(me->itsArrythmiaDetector);
}

struct ECG_Module* TestBuilder_getItsECG_Module(const TestBuilder* const me) {
    return (struct ECG_Module*)&(me->itsECG_Module);
}

struct HistogramDisplay* TestBuilder_getItsHistogramDisplay(const TestBuilder* const
me) {
    return (struct HistogramDisplay*)&(me->itsHistogramDisplay);
}

struct QRSDetector* TestBuilder_getItsQRSDetector(const TestBuilder* const me) {
    return (struct QRSDetector*)&(me->itsQRSDetector);
}

struct TMDQueue* TestBuilder_getItsTMDQueue(const TestBuilder* const me) {
    return (struct TMDQueue*)&(me->itsTMDQueue);
}

struct WaveformDisplay* TestBuilder_getItsWaveformDisplay(const TestBuilder* const me)
{
    return (struct WaveformDisplay*)&(me->itsWaveformDisplay);
}

TestBuilder * TestBuilder_Create(void) {
    TestBuilder* me = (TestBuilder *)
malloc(sizeof(TestBuilder));
    if(me!=NULL)
```

```
        {
            TestBuilder_Init(me);
        }
    return me;
}

void TestBuilder_Destroy(TestBuilder* const me) {
    if(me!=NULL)
        {
            TestBuilder_Cleanup(me);
        }
    free(me);
}

static void initRelations(TestBuilder* const me) {
    ArrythmiaDetector_Init(&(me->itsArrythmiaDetector));
    ECG_Module_Init(&(me->itsECG_Module));
    HistogramDisplay_Init(&(me->itsHistogramDisplay));
    QRSDetector_Init(&(me->itsQRSDetector));
    TMDQueue_Init(&(me->itsTMDQueue));
    WaveformDisplay_Init(&(me->itsWaveformDisplay));
    ECG_Module_setItsTMDQueue(&(me->itsECG_Module),
        &(me->itsTMDQueue));
    HistogramDisplay_setItsTMDQueue(&(me->itsHistogramDisplay),
        &(me->itsTMDQueue));
    QRSDetector_setItsTMDQueue(&(me->itsQRSDetector),
        &(me->itsTMDQueue));
    WaveformDisplay_setItsTMDQueue(&(me->itsWaveformDisplay),
        &(me->itsTMDQueue));
    ArrythmiaDetector_setItsTMDQueue(&(me->itsArrythmiaDetector),
        &(me->itsTMDQueue));
}

static void cleanUpRelations(TestBuilder* const me) {
    WaveformDisplay_Cleanup(&(me->itsWaveformDisplay));
    TMDQueue_Cleanup(&(me->itsTMDQueue));
    QRSDetector_Cleanup(&(me->itsQRSDetector));
    HistogramDisplay_Cleanup(&(me->itsHistogramDisplay));
    ECG_Module_Cleanup(&(me->itsECG_Module));
    ArrythmiaDetector_Cleanup(&(me->itsArrythmiaDetector));
}
```

Code Listing 2-9: TestBuilder.c Implementation File

Lastly, Code Listing 2-10 shows a simple use of the various classes to insert some data into the queue and have the clients pull them out for their use.

```
int main(int argc, char* argv[]) {
    int status=0;

    TestBuilder * p_TestBuilder;
```

```
p_TestBuilder=TestBuilder_Create();
ECG_Module_acquireValue(&(p_TestBuilder->itsECG_Module));
ECG_Module_acquireValue(&(p_TestBuilder->itsECG_Module));
ECG_Module_acquireValue(&(p_TestBuilder->itsECG_Module));
ECG_Module_acquireValue(&(p_TestBuilder->itsECG_Module));
ECG_Module_acquireValue(&(p_TestBuilder->itsECG_Module));

HistogramDisplay_getValue(&(p_TestBuilder
    ->itsHistogramDisplay));
HistogramDisplay_getValue(&(p_TestBuilder
    ->itsHistogramDisplay));
HistogramDisplay_getValue(&(p_TestBuilder
    ->itsHistogramDisplay));
HistogramDisplay_getValue(&(p_TestBuilder
    ->itsHistogramDisplay));
HistogramDisplay_getValue(&(p_TestBuilder
    ->itsHistogramDisplay));

QRSDetector_getDataSample(&(p_TestBuilder->itsQRSDetector));
QRSDetector_getDataSample(&(p_TestBuilder->itsQRSDetector));
QRSDetector_getDataSample(&(p_TestBuilder->itsQRSDetector));

WaveformDisplay_getScalarValue(&(p_TestBuilder
    ->itsWaveformDisplay));
WaveformDisplay_getScalarValue(&(p_TestBuilder
    ->itsWaveformDisplay));
WaveformDisplay_getScalarValue(&(p_TestBuilder
    ->itsWaveformDisplay));
WaveformDisplay_getScalarValue(&(p_TestBuilder
    ->itsWaveformDisplay));
WaveformDisplay_getScalarValue(&(p_TestBuilder
    ->itsWaveformDisplay));

ArrythmiaDetector_getDataSample(&(p_TestBuilder
    ->itsArrythmiaDetector));
ArrythmiaDetector_getDataSample(&(p_TestBuilder
    ->itsArrythmiaDetector));
ArrythmiaDetector_getDataSample(&(p_TestBuilder
    ->itsArrythmiaDetector));
ArrythmiaDetector_getDataSample(&(p_TestBuilder
    ->itsArrythmiaDetector));
ArrythmiaDetector_getDataSample(&(p_TestBuilder
    ->itsArrythmiaDetector));

printf("Done\n");
TestBuilder_Destroy(p_TestBuilder);
return 0;
}
```

Code Listing 2-10: Sample main.c

The code presented above works in a classic client-server approach. The `ECG_Monitor` puts data into the queue when it thinks it should and the various clients all individually decide when it is time to pull the data from the queue and do their work. In this case, the TMDQueue is the server of the data for the clients, and the WaveformDisplay, HistogramDisplay, QRSDetector, and ArrythmiaDetector are the clients.

We have a functioning model (with functioning code) for the system. Let's now think about how we'd like to optimize the system. In this case, we'd like to optimize execution time because we'd like to free up processor cycles for other work. In addition, we'd like to optimize maintainability of the system, limiting the number of places we'd like to make changes to the system when it needs to be updated. We'd also like to be able to dynamically (at run-time) vary the number of clients without wasting extra memory by allocating a potentially large array of pointers (1 per client) that might never be used. So our design criteria (with scaled criticality weights) are: 1) execution efficiency (weight=7), 2) maintainability (weight=5), 3) flexibility to add new clients easily at run-time (weight=4), and 4) memory efficiency at the larger scale (weight=7). Let's consider three alternative strategies (design patterns).

First, we have the current client-server approach:

1. Execution efficiency
   This solution is not particularly optimal because the clients need to check frequently if there are new data, even if there are not. On a scale of 0 to 10, let's call that a 3. This is the primary weakness of the approach.
2. Maintainability
   In a classic client-server architecture, the clients know about the servers (so they can invoke the services) but the servers don't know about the clients. That limitation on knowledge makes the system more maintainable because knowledge is encapsulated in one place. It earns a score of 7 here.
3. Run-time flexibility
   In this context flexibility primarily translates to the ability to create new clients and easily link them into the server. Since we can adapt it to handle those concerns, it earns an 8 on this criterion.
4. Memory usage
   In this criterion, each client must maintain its position in the list, which requires additional memory in each client. Therefore, the classic client-server approach only rates a 5 on this criterion.

The second alternative approach is to create a "push" implementation in which the responsibility for notifying the clients is given to the TMDQueue. To implement this, the direction of the associations in Figure 2-8 must be reversed, and the TMDQueue must explicitly know about all of its clients to notify them when data become available. The client list would probably be maintained as a list of pointers to clients.

1. Execution efficiency

   This approach is efficient in the sense that the clients are notified when data are available and don't need to poll to find discover new data. Let's give this approach an 8 here.

2. Maintainability

   The push approach breaks the classic client-server architecture (the servers know about the clients), so it only rates a 4 on this criterion. It will be harder to maintain if we add more clients.

3. Run-time flexibility

   The push approach has reasonably good run-time flexibility because it means that the server must be modified to accept new clients. It rates a 7 here.

4. Memory usage

   The approach uses memory only when a client is added, so it gets a 9 on this criterion.

The last alternative is the observer pattern[9]. The observer pattern is shown in Figure 2-10.

Before we use the pattern, let's look at its description and properties. Note that in the description that follows, the classic observer pattern has been modified slightly to fit into the example problem easier.

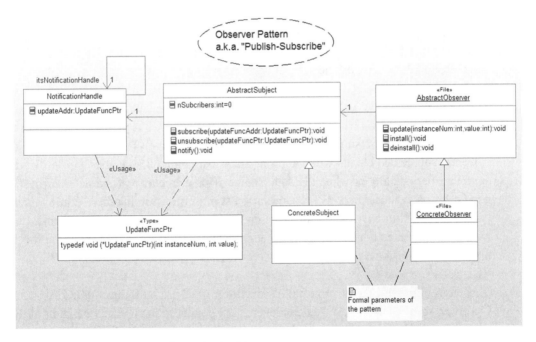

Figure 2-10: Observer pattern structure

---

[9] See, for example, Douglass, B.P., 2002. *Real-Time Design Patterns: Robust Scalable Architecture for Real-Time Systems*. Addison-Wesley; or Gamma, E., Helm, R., Johnson, R., Vlissides, J.M., 1995. *Design Patterns: Elements of Reusable Object-Oriented Software*. Addison-Wesley.

2.1.6 Observer Pattern

- Abstract

  The observer pattern addresses the concern of dynamically adding and removing clients of data or services, as well as optimizing the processing associated with updating those clients in a timely way.

- Problem

  This pattern maintains the class client-server architecture in which the clients know about a server that has either data or services of interest, but that knowledge is one way. In spite of this, the responsibility for updating the clients in an efficient and timely way is given to the server. In addition, because the server maintains a list of interested clients, it notifies all clients when it is appropriate to do so, such as periodically or when the data changes. Clients can be dynamically added to or removed from the server as the system runs, while not requiring the server to have any knowledge of the clients (other than the location of an operation of the correct signature).

- Pattern structure

  This pattern defines the following roles:

  - Abstract Observer

    This element has a one-way association (link) to the Abstract Subject so that it can request to be added to the list of interested clients (a process known as *subscribing*) or removed from the list (known as *unsubscribing*). It provides an operation with a specific signature to be invoked by the subject when appropriate.

  - Abstract Subject

    This element will serve as the base class for the server in the client-server architecture; it has information or services needed by the Abstract Observer. It provides operations to add or remove clients (subscribe and unsubscribe, respectively) and can notify all currently subscribed clients when appropriate. Most common strategies for notification are either periodically or when the data change. This element creates a new notification handle for each subscribed client, forming a list of notification handles. While this list can be managed in any number of ways, in this case it will be handled as a linked list. Notification of the clients is simply a matter of "walking the list" and invoking the update operation for each Abstract Observer currently subscribed.

  - Notification handle

    This element provides a dynamic callback link for the Abstract Subject to the Abstract Observer; specifically, it has, as a data element, a pointer to the update operation of the server. It also has a link to the next element in the list which is NULL at the end of the list.

  - Concrete Observer (formal parameter)

    This element is the subclass of the Abstract Observer; it inherits the link to the Abstract Subject and provides a (nonimplemented) update operation that will be invoked by the Concrete Subject.

- Concrete Subject (formal parameter)
  This element is the subclass of the Abstract Subject; it inherits the ability to add and remove clients and notify the clients when appropriate. It also provides the application-specific data or services of interest to the Concrete Observer.
- Consequences
  This pattern simplifies the process for dynamically adding and removing clients, especially ones that may not have even been defined when the server was created. In addition, because the server can decide when it is appropriate to update the clients, it results in timely delivery of information and services. This delivery is also efficient because the clients need not poll the server for updated data only to find the data are the same.

Now, let's look at how this solution stacks up.

1. Execution efficiency
   This approach is efficient in the sense that the client gets data functionality only when there is a need similar to the push approach. It rates an 8 here.
2. Maintainability
   This solution maintains the classic client-server architecture data encapsulation, improving its maintainability. The solution earns a score of 7 here.
3. Run-time flexibility
   The observer pattern has good run-time flexibility; the only requirement on the client is that it provides an update method with the appropriate signature. It rates a 9 here.
4. Memory usage
   Similar to the push approach, this pattern uses less memory because the clients need not track their position within the data buffer, so it gets a 9 on this criterion.

The results of the analysis are shown in Table 2-4. Based on the weighted criteria, the observer pattern looks to be the best of the design alternatives examined.

Table 2-4: Design tradeoffs for ECG monitor system.

Design Solution	Design Criteria				Total Weighted Score
	Efficiency	Maintainability	Flexibility	Memory Usage	
	Weight = 7	Weight = 5	Weight = 4	Weight = 7	
	Score	Score	Score	Score	
Client Server	3	7	8	5	123
Push	8	4	7	9	167
Observer	**8**	**7**	**9**	**9**	**190**

Once selected, the next step is to instantiate the pattern and, as a part of that, to make decisions about how to best implement the pattern. Instantiation is a matter of replacing the formal parameters of the pattern with elements from your original model. In this case, formal parameter Concrete Subject will be replaced by the TMDQueue and the formal parameter Concrete Observer will be replaced by the clients WaveformDisplay, HistogramDisplay, QRSDetector, and ArrythmiaDetector. The resulting design model is shown in Figure 2-11. In this diagram you can see the original environmental control elements, the pattern elements, and their relations.

The relations, attributes, and operations of the Abstract Subject are inherited by the TMDQueue class while the features of the Abstract Observer are inherited by the various clients of the TMDQueue. Inheritance can be implemented in C in a number of ways, such as:

- Copying from the super class implementation into the subclass implementation
- Using a "mix-in"; that is, delegate service requests from an instance of the subclass to an instance of the super class
- Implementing a virtual function table, as discussed in Chapter 1

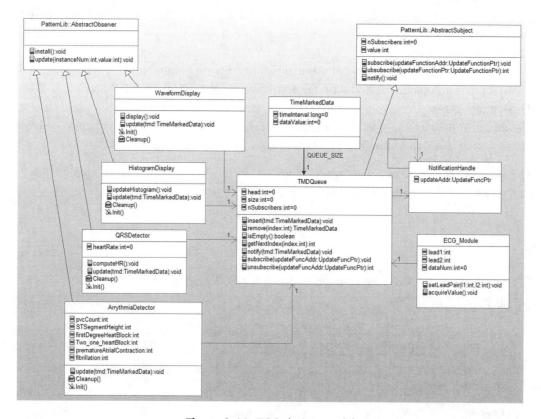

Figure 2-11: ECG design model

In our code implementation, we will take the simplest possible solution just to illustrate the pattern; we will copy and implement code from the Abstract Subject into the TMDQueue and do the same for code from the Abstract Observer into the clients. Other implementation strategies can be used, if desired.

The net result can be seen in Figure 2-11. Note that the clients have lost their index attribute and gained an update() member function, as well as an Init() function (that calls subscribe() on the TMDQueue) and a Cleanup() function that calls unsubscribe(). The TMDQueue has added subscribe(), unsubscribe(), and notify() operations and an association to the list of NotificationHandle classes. The notification handle holds the address for the relevant class's update() function. Calling notify() walks the linear linked list of notification handles.

Code Listing 2-11 and Code Listing 2-12 show the header and implementation of the TMDQueue class updated to include the server side of the observer pattern. You can see the addition of the link for the NotificationHandle list to manage the subscriptions of interested observers.

```c
#ifndef TMDQueue_H
#define TMDQueue_H

/*## auto_generated */
#include <stdio.h>
#include "ECGPkg.h"
#include "TimeMarkedData.h"

struct NotificationHandle;

typedef struct TMDQueue TMDQueue;
/*
This queue is meant to operate as a "leaky" queue. In this queue,
data are never removed per se, but are instead overwritten when the
buffer pointer wraps around. This allows for many clients to read
the same data from the queue. */
struct TMDQueue {
    int head;
    int nSubscribers;
    int size;
    struct TimeMarkedData buffer[QUEUE_SIZE];
    struct NotificationHandle* itsNotificationHandle;
};

/* Constructors and destructors:*/

void TMDQueue_Init(TMDQueue* const me);
void TMDQueue_Cleanup(TMDQueue* const me);

/* Operations */

int TMDQueue_getNextIndex(TMDQueue* const me, int index);

void TMDQueue_insert(TMDQueue* const me, const struct TimeMarkedData tmd);
```

```
boolean TMDQueue_isEmpty(TMDQueue* const me);

void TMDQueue_notify(TMDQueue* const me, const struct TimeMarkedData tmd);

struct TimeMarkedData TMDQueue_remove(TMDQueue* const me, int index);

/* The NotificationHandle is managed as a linked list, with insertions coming at the end. */
void TMDQueue_subscribe(TMDQueue* const me, const UpdateFuncPtr updateFuncAddr);

int TMDQueue_unsubscribe(TMDQueue* const me, const UpdateFuncPtr updateFuncAddr);

int TMDQueue_getBuffer(const TMDQueue* const me);

struct NotificationHandle*
TMDQueue_getItsNotificationHandle(const TMDQueue* const me);
void TMDQueue_setItsNotificationHandle(TMDQueue* const me, struct NotificationHandle*
p_NotificationHandle);

TMDQueue * TMDQueue_Create(void);

void TMDQueue_Destroy(TMDQueue* const me);

#endif
```

Code Listing 2-11: TMDQueue Header File with Observer Pattern

```
#include "TMDQueue.h"
#include "NotificationHandle.h"

static void initRelations(TMDQueue* const me);
static void cleanUpRelations(TMDQueue* const me);

void TMDQueue_Init(TMDQueue* const me) {
    me->head=0;
    me->nSubscribers=0;
    me->size=0;
    me->itsNotificationHandle=NULL;
    initRelations(me);
}

void TMDQueue_Cleanup(TMDQueue* const me) {
    cleanUpRelations(me);
}

int TMDQueue_getNextIndex(TMDQueue* const me, int index) {
    /* this operation computes the next index from the
    first using modulo arithmetic
    */
    return (index+1) % QUEUE_SIZE;
}

void TMDQueue_insert(TMDQueue* const me, const struct TimeMarkedData tmd) {
    /* note that because we never 'remove' data from this
    leaky queue, size only increases to the queue size and
    then stops increasing. Insertion always takes place at the head.
    */
```

```
        printf("Indenting at: %d Data #: %d", me->head, tmd.timeInterval);
        me->buffer[me->head]=tmd;
        me->head=TMDQueue_getNextIndex(me, me->head);
        if (me->size < QUEUE_SIZE) ++me->size;
        printf(" Storing data value: %d\n", tmd.dataValue);
        TMDQueue_notify(me, tmd);
}

boolean TMDQueue_isEmpty(TMDQueue* const me) {
        return (boolean)(me->size == 0);
}

void TMDQueue_notify(TMDQueue* const me, const struct TimeMarkedData tmd) {
        NotificationHandle *pNH;
        pNH= me->itsNotificationHandle;
        while (pNH) {
            printf("----->> calling updateAddr on pNH %d\n",pNH);
            pNH->updateAddr(NULL, tmd);
            pNH=pNH->itsNotificationHandle;
            };
}

struct TimeMarkedData TMDQueue_remove(TMDQueue* const me, int index) {
            TimeMarkedData tmd;
            tmd.timeInterval=-1; /* sentinel values */
            tmd.dataValue=-9999;

            if (!TMDQueue_isEmpty(me) && (index>=0) && (index < QUEUE_SIZE)
                && (index < me->size)) {
                tmd=me->buffer[index];
            }
            return tmd;
}

void TMDQueue_subscribe(TMDQueue* const me, const UpdateFuncPtr updateFuncAddr) {
        struct NotificationHandle *pNH;
        pNH=me->itsNotificationHandle;

        if (!pNH) { /* empty list? */
            /* create a new Notification Handle, initialize it, and point to it */
            printf("-----> Added to a new list\n");
            me->itsNotificationHandle=NotificationHandle_Create();
            printf("-----> Called NH_Create()\n");
            pNH=me->itsNotificationHandle;
            }
        else {
            /* search list to find end */
            printf("-----> Adding to an existing list\n");
            while (pNH->itsNotificationHandle != NULL) {
              printf("Getting ready to augment ptr %d to %d\n", pNH, pNH->itsNotificationHandle);
                pNH=pNH->itsNotificationHandle; /* get next element in list */
                printf("-----> augmenting ptr\n");
```

```
            };
        printf("-----> calling NH_Create\n");
            pNH->itsNotificationHandle=NotificationHandle_Create();
            pNH=pNH->itsNotificationHandle; /* pt to the new instance */
            printf("-----> called NH_Create()\n");
    }; /* end if */

    /* pNH now points to an constructed Notification Handle */
    pNH->updateAddr=updateFuncAddr; /* set callback address */
    ++me->nSubscribers;
    printf("-----> wrote updateAddr \n");
    if (pNH->itsNotificationHandle)
        printf("xxxxxxx> next Ptr not null!\n\n");
    else
        printf("-----> next ptr null\n\n");
}

int TMDQueue_unsubscribe(TMDQueue* const me, const UpdateFuncPtr updateFuncAddr) {
    struct NotificationHandle *pNH, *pBack;
    pNH=pBack=me->itsNotificationHandle;

    if (pNH== NULL) { /* empty list? */
        return 0; /* can't delete it from an empty list */
        }
    else { /* is it the first one? */
        if (pNH->updateAddr == updateFuncAddr) {
            me->itsNotificationHandle=pNH->itsNotificationHandle;
            free(pNH);
            printf(">>>>>> Removing the first element\n");
            --me->nSubscribers;
            return 1;
            }
        else { /* search list to find element */
            printf(">>>>>> Searching....\n");
            while (pNH != NULL) {
                if (pNH->updateAddr != updateFuncAddr) {
                    pBack->itsNotificationHandle=pNH->itsNotificationHandle;
                    free(pNH);
                    printf(">>>>>> Removing subscriber in list\n");
                    --me->nSubscribers;
                    return 1;
                    }; /* end if found */
                pBack=pNH; /* points to the list element before pNH */
                pNH=pNH->itsNotificationHandle; /* get next element in list */
                }; /* end while */
            }; /* end else */
        printf(">>>>>> Didn't remove any subscribers\n");
        return 0;
        }; /* non-empty list */
}
```

```
int TMDQueue_getBuffer(const TMDQueue* const me) {
    int iter=0;
    return iter;
}

struct NotificationHandle*
TMDQueue_getItsNotificationHandle(const TMDQueue* const me) {
    return (struct NotificationHandle*)me->itsNotificationHandle;
}
void TMDQueue_setItsNotificationHandle(TMDQueue* const me, struct NotificationHandle*
p_NotificationHandle) {
        me->itsNotificationHandle=p_NotificationHandle;
}

TMDQueue * TMDQueue_Create(void) {
    TMDQueue* me=(TMDQueue *) malloc(sizeof(TMDQueue));
    if(me!=NULL)
        {
            TMDQueue_Init(me);
        }
    return me;
}

void TMDQueue_Destroy(TMDQueue* const me) {
    if(me!=NULL)
        {
            TMDQueue_Cleanup(me);
        }
    free(me);
}

static void initRelations(TMDQueue* const me) {
    {
        int iter=0;
        while (iter < QUEUE_SIZE){
            TimeMarkedData_Init(&((me->buffer)[iter]));
            TimeMarkedData__setItsTMDQueue(&((me->buffer)[iter]), me);
            iter++;
        }
    }
}

static void cleanUpRelations(TMDQueue* const me) {
    {
        int iter=0;
        while (iter < QUEUE_SIZE){
            TimeMarkedData_Cleanup(&((me->buffer)[iter]));
            iter++;
        }
    }
}
```

```
        if(me->itsNotificationHandle != NULL)
          {
              me->itsNotificationHandle=NULL;
          }
   }
```

Code Listing 2-12: TMDQueue Implementation File with Observer Pattern

The code for the notification handle class is straightforward, as can be seen in Code Listing 2-13 and Code Listing 2-14.

```c
#ifndef NotificationHandle_H
#define NotificationHandle_H

#include <stdio.h>
#include "ECGPkg.h"
typedef struct NotificationHandle NotificationHandle;
struct NotificationHandle {
    UpdateFuncPtr updateAddr;
    struct NotificationHandle* itsNotificationHandle;
};

/* Constructors and destructors:*/
void NotificationHandle_Init(NotificationHandle* const me);

void NotificationHandle_Cleanup(NotificationHandle* const me);

struct NotificationHandle*
NotificationHandle_getItsNotificationHandle(const
NotificationHandle* const me);

void
NotificationHandle_setItsNotificationHandle(NotificationHandle*
const me, struct NotificationHandle* p_NotificationHandle);

NotificationHandle * NotificationHandle_Create(void);

void NotificationHandle_Destroy(NotificationHandle* const me);

#endif
```

Code Listing 2-13: NotificationHandle.h Header File

```c
#include "NotificationHandle.h"

static void cleanUpRelations(NotificationHandle* const me);

void NotificationHandle_Init(NotificationHandle* const me) {
    me->itsNotificationHandle=NULL;
}
```

```
void NotificationHandle_Cleanup(NotificationHandle* const me) {
    cleanUpRelations(me);
}

struct NotificationHandle*
NotificationHandle_getItsNotificationHandle(const
NotificationHandle* const me) {
    return (struct NotificationHandle*)me->itsNotificationHandle;
}

void
NotificationHandle_setItsNotificationHandle(NotificationHandle*
const me, struct NotificationHandle* p_NotificationHandle) {
    me->itsNotificationHandle=p_NotificationHandle;
}

NotificationHandle * NotificationHandle_Create(void) {
    NotificationHandle* me = (NotificationHandle *)
malloc(sizeof(NotificationHandle));
    if(me!=NULL)
        {
            NotificationHandle_Init(me);
        }
    return me;
}

void NotificationHandle_Destroy(NotificationHandle* const me) {
    if(me!=NULL)
        {
            NotificationHandle_Cleanup(me);
        }
    free(me);
}

static void cleanUpRelations(NotificationHandle* const me) {
    if(me->itsNotificationHandle != NULL)
        {
            me->itsNotificationHandle=NULL;
        }
}
```

Code Listing 2-14: NotificationHandle.c Implementation File

Finally, let's look at just one of the clients to see how the client side of the observer pattern is implemented. Code Listing 2-15 shows the implementation file for the HistogramDisplay class with the client side of the observer pattern (specifically, it calls the TMDQueue_subscribe() operation in its initializer).

```
#include "HistogramDisplay.h"
#include "TimeMarkedData.h"
#include "TMDQueue.h"
```

```
static void cleanUpRelations(HistogramDisplay* const me);

void HistogramDisplay_Init(HistogramDisplay* const me) {
    me->itsTMDQueue=NULL;
    {
        /* call subscribe to connect to the server */
        TMDQueue_subscribe(me->itsTMDQueue,
HistogramDisplay_update);
    }
}

void HistogramDisplay_Cleanup(HistogramDisplay* const me) {
    /* remove yourself from server subscription list */
    TMDQueue_unsubscribe(me->itsTMDQueue,
HistogramDisplay_update);
    cleanUpRelations(me);
}

void HistogramDisplay_update(HistogramDisplay* const me, const struct TimeMarkedData
tmd) {
        printf(" Histogram -> TimeInterval: %d DataValue: %d\n", tmd.timeInterval, tmd.
dataValue);
}

void HistogramDisplay_updateHistogram(HistogramDisplay* const me) {
}

struct TMDQueue* HistogramDisplay_getItsTMDQueue(const HistogramDisplay* const me) {
    return (struct TMDQueue*)me->itsTMDQueue;
}

void HistogramDisplay_setItsTMDQueue(HistogramDisplay* const me,
struct TMDQueue* p_TMDQueue) {
    me->itsTMDQueue=p_TMDQueue;
}

HistogramDisplay * HistogramDisplay_Create(void) {
    HistogramDisplay* me=(HistogramDisplay *)
malloc(sizeof(HistogramDisplay));
    if(me!=NULL)
        {
            HistogramDisplay_Init(me);
        }
    return me;
}

void HistogramDisplay_Destroy(HistogramDisplay* const me) {
    if(me!=NULL)
        {
            HistogramDisplay_Cleanup(me);
        }
        free(me);
}
```

```
static void cleanUpRelations(HistogramDisplay* const me) {
    if(me->itsTMDQueue != NULL)
    {
        me->itsTMDQueue=NULL;
    }
}
```

Code Listing 2-15: HistogramDisplay.c Implementation File with Observer Pattern

To summarize, we created a correctly functioning code base and then decided how we wanted to optimize it. We did a miniature trade study to look at some different alternatives and how differently they optimized our weighted set of design criteria. In this case, the observer pattern had the best results, so we updated our code base to include the pattern. Simple, eh?

2.2 The Approach

This book provides many different patterns for use in embedded systems, as you will see in the following chapters. In each chapter, a number of patterns are presented. Each pattern follows the same basic presentation style and is presented with nine basic elements:

- Abstract
  The abstract provides an overview of the pattern and why you might care, including an overview of the problems it addresses, the solution structure, and the key consequences.
- Problem
  This section discusses the problem context and the optimizations provided by the pattern.
- Applicability
  This section gives a list of conditions required for the pattern to be applicable.
- Pattern structure
  The pattern structure shows the structural elements within the pattern, their roles, and their relations. Each pattern will be shown using a UML class diagram.
- Collaboration role definitions
  This section details the role each structural element plays in the pattern and its responsibilities.
- Consequences
  The consequences are a key aspect of the pattern description. This section identifies both the benefits and the costs of using the pattern.
- Implementation strategies
  Patterns can be implemented in different ways. This section discusses some implementation alternatives.

- Related patterns
  This section provides references to other patterns in the book that address similar problems, possibly with different applicability criteria and consequences.
- Example
  Each pattern will be illustrated with an example. Both a UML diagram and source code will be provided with the pattern.

2.3 What's Coming Up

The remaining chapters of this book present design patterns in a number of subject areas of particular interest to embedded C developers. Chapter 3 focuses on design patterns for controlling hardware. Chapter 4 is about concurrency with embedded C, including not only basic scheduling patterns but patterns to share resources across threads and to avoid deadlock. Chapter 5 deals with state machines, an important but underutilized tool in the embedded toolbox, while Chapter 6 looks at safety and reliability.

Design Patterns for Accessing Hardware

Design Patterns for Embedded Systems in C

DOI:10.1016/B978-1-85617-707-8.00003-0

Copyright © 2011 Elsevier Inc.

Patterns in this chapter:

Hardware Proxy Pattern – Encapsulate the hardware into a class or struct

Hardware Adapter Pattern – Adapt between a required and a provided interface

Mediator Pattern – Coordinate complex interactions

Observer Pattern – Support efficient sensor data distribution

Debouncing Pattern – Reject intermittent hardware signals

Interrupt Pattern – Handle high-ugency hardware signals

Polling Pattern – Periodically check for new sensor data

3.1 Basic Hardware Access Concepts

Probably the most distinguishing property of embedded systems is that they must access hardware directly. The whole point of an embedded system is that the software is embedded in a "smart device" that provides some specific kind of service, and that requires accessing hardware. Broadly, software-accessible hardware can be categorized into four kinds – infrastructure, communications, sensors, and actuators.

Infrastructure hardware refers to the computing infrastructure and devices on which the software is executing. This includes not only the CPU and memory, but also storage devices, timers, input devices (such as keyboards, knobs, and buttons), user output devices (printers, displays, and lights), ports, and interrupts. Most of this hardware is not particularly application specific and may be located on the same motherboard, on daughterboards with high-speed access connections, or as other boards sharing a common backplane.

Communications hardware refers to hardware used to facilitate communication between different computing devices, such as standard (nonembedded) computers, other embedded systems, sensors, and actuators. Usually this means wired or wireless communications subsystems such as RS-232, RS-485, Ethernet, USB, 802.11x, or some other such facility. In embedded systems, DMA (Direct Memory Access) and multiported memory is sometimes used as well for this purpose, somewhat blurring the distinction between infrastructure and communications hardware.

Lastly, we have devices that monitor or manipulate the physical universe – these are the sensors and actuators. Sensors use electronic, mechanical, or chemical means for monitoring the status of physical phenomena, such as the rate of a heart beat, position of an aircraft, the mass of an object, or the concentration of a chemical. Actuators, on the other hand, change the physical state of some real-world element. Typical actuators are motors, heaters, generators, pumps, and switches.

All of these kinds of hardware are usually initialized, tested, and configured – tasks normally performed by the embedded software either on start-up or during execution, or both. All of these hardware elements are managed by the embedded software, provide commands or data to the software, or receive commands or data from the software.

This chapter will assume that you have enough hardware background to understand the hardware terms used in this book such as bus, CPU, RAM, ROM, EPROM, EEPROM, and Flash. If you want to know more, there any many excellent books on the topic, including books by David Simons[1], Michael Barr and Anthony Massa[2], Michael Pont[3], and Jack Ganssle[4]. Instead, we will focus on common ways to develop embedded C applications to solve problems that commonly arise with the management and manipulation of such hardware elements. Some hardware-related concepts that we will talk briefly about are bit fields, ports, and interrupts.

It is exceedingly common to have hardware that uses bit fields to specify commands to the hardware or to return data. A bit field is a continuous block of bits (one or more) within an addressable memory element (e.g., a byte or word) that together has some semantic meaning to the hardware device. For example, an 8-bit byte might be mapped to a hardware device into four different fields:

0 0 0000 00

The bits represent the following information to or from the memory-mapped hardware device:

Bit Range	Access	Name	Description
0	Write only	Enable bit	0 = disable device, 1 = enable device
1	Read only	Error status bit	0 = no error, 1 = error present
2–5	Write only	Motor speed	Range 0000 = no speed to 1111 (16d) top speed
6–7	Write only	LED color	00 (0d) = OFF 01 (1d) = GREEN 10 (2d) = YELLOW 11 (3d) = RED

---

[1] Simon, D., 1999. *An Embedded Software Primer.* Addison-Wesley.
[2] Barr, M. Massa, A., 2007. *Programming Embedded Systems with C and GNU Development Tools.* O'Reilly.
[3] Pont, M.J., 2002. *Embedded C.* Addison-Wesley.
[4] Ganssle, J., 2007. *Embedded Systems: World Class Designs.* Newnes.

Bit fields are manipulated with C's bit-wise operators & (bit-wise AND), | (bit-wise OR), ~ (bit-wise NOT), ^ (bit-wise XOR), >> (right shift), and << (left shift).

Here are some examples of the results of using these bit-wise operators with different masks on a specific value:

```
Value    01010101    Value    01010101    Value    01010101
Mask     11111111    Mask     00000000    Mask     00001111
ANDED    01010101    ANDED    00000000    ANDED    00000101

Value    01010101    Value    01010101    Value    01010101
Mask     11111111    Mask     00000000    Mask     00001111
ORED     11111111    ORED     01010101    ORED     01011111

Value    01010101    Value    01010101    Value    01010101
Mask     11111111    Mask     00000000    Mask     00001111
XORED    10101010    XORED    01010101    XORED    01011010
```

A common related C idiom is to use #define to create bit masks for ANDing and ORing and give them meaningful names. For example, we might create a set of bit masks for common manipulations of the memory mapped device above, such as:

```c
#include <stdlib.h>
#include <stdio.h>
#define TURN_OFF (0x00)
#define INITIALIZE (0x61)
#define RUN (0x69)
#define CHECK_ERROR (0x02)
#define DEVICE_ADDRESS (0x01FFAFD0)

void emergencyShutDown(void){
    printf("OMG We're all gonna die!\n");
}

int main() {
    unsigned char* pDevice;

    pDevice = (unsigned char *)DEVICE_ADDRESS; // pt to device
    // for testing you can replace the above line with
    // pDevice = malloc(1);
    *pDevice = 0xFF;     // start with all bits on
    printf ("Device bits %X\n", *pDevice);
    *pDevice = *pDevice & INITIALIZE; // and the bits into
    printf ("Device bits %X\n", *pDevice);
    if (*pDevice & CHECK_ERROR) { // system fail bit on?
            emergencyShutDown();
            abort();
            }
    else    {
            *pDevice = *pDevice & RUN;
```

```
        printf ("Device bits %X\n", *pDevice);
        };
    return 0;
  };
```

Code Listing 3-1: Manipulating bit-oriented memory mapped hardware

The left and right shift operators are useful for setting or testing particular bits in isolation as well as for processing serial bit data. For example, the expression "1<<3" sets bit 3 resulting in the value 8. This is sometimes used in the definition of bits in a mask. For example, the mask CHECK_ERROR in Code Listing 3-1 could have been written:

```
#define CHECKERROR (1<<1)
```

Bit fields in C offer another way to represent bit-mapped fields in device interfaces. This syntax represents variable-length bit fields as fields within a struct. The ':' operator separates the name of the field from its length in the declaration. For example, see Code Listing 3-2.

```
#include <stdlib.h>
#include <stdio.h>

int main() {
    typedef struct _statusBits {
            unsigned enable : 1;
            unsigned errorStatus : 1;
            unsigned motorSpeed : 4;
            unsigned LEDColor : 2;
    } statusBits;
    statusBits status;

    printf("size = %d\n",sizeof(status));
    status.enable = 1;
    status.errorStatus = 0;
    status.motorSpeed = 3;
    status.LEDColor = 2;

    if (status.enable) printf("Enabled\n");
        else printf ("Disabled\n");
    if (status.errorStatus) printf("ERROR!\n");
        else printf("No error\n");
    printf ("Motor speed %d\n",status.motorSpeed);
    printf ("Color %d\n",status.LEDColor);
    return 0;
  };
```

Code Listing 3-2: Bit fields in C

There are a couple of problems with bit fields in C. First, bit ordering is compiler- and processor-dependent. Further, the compiler may enforce byte padding rules, meaning that it may not map onto a memory mapped device as you expect. For example, the GNU C compiler returns size of `status` in Code Listing 3-2 to be 4 bytes even though the size of an unsigned char is only 1 byte.

Even worse, because most CPUs must write a byte or word at a time, bit fields may not be written in an atomic step, leading to thread safety issues if separate mutexes are used for different bit fields.

Another potential problem with using bit fields is that it is impossible to cast between a scalar and a user-defined struct. Thus, the following is disallowed[5]:

```
Unsigned char f;
f = 0xF0;
status = (statusBits) f;
```

Let's turn our attention now to a number of design patterns that have proven themselves useful for the manipulation of hardware. The Hardware Proxy Pattern, discussed next, is an archetypal pattern for the abstraction of hardware for the purpose of encapsulating details that are likely to change from the usage of the information provided to or by the hardware. The Hardware Adapter Pattern extends the Hardware Proxy Pattern to provide the ability to support different hardware interfaces. The Mediator Pattern supports coordination of multiple hardware devices to achieve a system level behavior. The Observer Pattern is a way of distributing sensed data to the software elements that need it. The Debouncing and Interrupt Patterns are simple reusable approaches to interface with hardware devices. The Timer Interrupt Pattern extends the Interrupt timer to provide accurate timing for embedded systems.

3.2 Hardware Proxy Pattern

The *Hardware Proxy Pattern* creates a software element responsible for access to a piece of hardware and encapsulation of hardware compression and coding implementation.

3.2.1 Abstract

The *Hardware Proxy Pattern* uses a class (or struct) to encapsulate all access to a hardware device, regardless of its physical interface. The hardware may be memory, port, or interrupt

---

[5] Although you *can* get around that by doing something as circuitous as

```
status = *(statusBits*)&f
```

(that is, casting the address of f (unsigned char) to be an address of the structure type statusBits) but I can hardly recommend that!

mapped, or may even be mapped via a serial connection, bus, network, or wireless link. The proxy publishes services that allow values to be read from and written to the device, as well as initialize, configure, and shut down the device as appropriate. The proxy provides an encoding and connection-independent interface for clients and so promotes easy modification should the nature of the device interface or connection change.

3.2.2 Problem

If every client accesses a hardware device directly, problems due to hardware changes are exacerbated. If the bit encoding, memory address, or connection technology changes, then every client must be tracked down and modified. By providing a proxy to sit between the clients and the actual hardware, the impact of hardware changes is greatly limited, easing such modifications. For easiest maintenance, the clients should be unaware of bit encoding, encryption, and compression used by the device; these details should be managed by the Hardware Proxy with internal private functions.

3.2.3 Pattern Structure

The pattern structure is quite simple as can be seen in Figure 3-1[6]. There may be many clients but a single Hardware Proxy per device being controlled. The proxy contains both public functions and private, encapsulated functions and data. The device is represented as a `void*` in the figure, but it should be appropriately typed for the device.

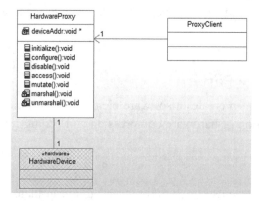

Figure 3-1 Hardware Proxy Pattern

---

[6] Each of these patterns will be shown as a UML diagram and code will be provided for the examples. The Appendix provides a short overview of the UML notation.

3.2.4 Collaboration Roles

This section describes the roles for this pattern.

3.2.4.1 Hardware Device

This element represents the actual hardware. As such, you will not write C code for this element, but it is present just to aid with understanding the diagram. The association shown between the Hardware Device and the Hardware Proxy is realized via a software-addressable hardware interface, such as a port address, memory address, or interrupt.

3.2.4.2 Hardware Proxy

This is the primary class (or function) in the system. It has both data and functions that are customized to the device at hand, although usually each one has an initialize(), configure(), and disable() function. The other public functions provide read access to values from the device or set values sent to the device. Although the pattern structure has a single access() and mutate() function, it is common to have several such functions, each bearing the name of the value being read or set.

The key features[7] of the class include:

access()

This public function returns a particular value from the device. In most cases, the class will contain a separate function for each separately-identifiable piece of information desired from the device. This operation usually calls the `unmarshal()` function prior to returning the retrieved values to the client.

configure()

This public function provides a means by which the device may be configured. Although in the general pattern no parameter list can be provided, this function almost always takes a list of parameters that specify the proper operating state of the device.

disable()

This public function provides a means by which the device may be safely turned off or disabled. It may or may not have parameters depending upon the device characteristics.

deviceAddr

This private variable provides low-level direct access to the hardware. In the pattern it is shown as a `void*` but it may be a memory mapped integer (`int*`) or other primitive type, a port-mapped device port number, or some other location identified. If more complex means are used

---

[7] By the term *feature*, I am referring primarily to data elements (attributes or variables), operations and functions, states, and event receptions.

to access the device, such as an RS232 serial port or an Ethernet connection, then this data type and its access method will be more complex. In any event, the public functions provided by the Hardware Proxy completely hide how the proxy connects to the actual device. This variable is not directly accessible by the client.

initialize()

This public function enables and initializes the device prior to first use. It is common for this function to have no parameters but in some cases it may need values provided by the client.

marshal()

This private function takes parameters from various other functions and performs any required encryption, compression, or bit-packing required to send the data to the device. This ensures that the peculiarities of the device interface are hidden from the client. Data in the format required by the actual device are known as "marshaled data" or are said to be in "native format." Data that are in a form that is easily manipulated by software are said to be "unmarshaled" or in "presentation format." This function is not accessible by the client since native format is hidden from the clients.

mutate()

This public function writes data values to the device. This function always has one or more input parameters. It usually calls the `marshal()` operation prior to writing the value to the device.

unmarshal()

This private function performs any necessary unpacking, decryption, and decompression of the data retrieved from the device prior to returning them to the client in presentation format. This ensures that the peculiarities of the device interface are hidden from the client. This function is not accessible by the client.

3.2.4.3 Proxy Client

The proxy client "knows about" the Hardware Proxy and invokes its services to access the hardware device.

3.2.5 Consequences

The pattern is extremely common and provides all the benefits of encapsulation of the hardware interface and encoding details. It provides flexibility for the actual hardware interface to change radically with absolutely no changes in the clients. This is because the hardware details are all encapsulated within the Hardware Proxy. This means that the proxy clients are usually unaware of the native format of the data and manipulate them only in presentation format.

This can, however, have a negative impact on run-time performance. It may sometimes be more time efficient for the clients to know the encoding details and manipulate the data in their native format. However, this degrades the maintainability of the system because now the clients need to be modified should the hardware interface or encoding change.

3.2.6 Implementation Strategies

As discussed in Chapter 1, classes may be implemented in different ways in C ranging from simple files to using struct to store class attributes, to using virtual function tables that support true polymorphism. All of those approaches can be used to implement this very simple pattern.

Usually, a Hardware Proxy supports all features of a specific device and a different Hardware Proxy is used for each separate device, but this isn't an absolute rule. Separation into different devices means that the devices can follow independent maintenance paths and so is very flexible for the future.

In general, it is preferable to hide the bit-encoding, encryption, and data compression of the actual hardware away from the application software[8]. However, it is possible to implement the pattern without the `marshal()` and `unmarshal()` functions by making the native format visible to the clients.

3.2.7 Related Patterns

The simple implementation of this pattern doesn't implement anything for thread safety. It can be combined with the *Critical Region, Guarded Call*, or *Queuing* patterns to provide thread safety. For deadlock avoidance, it can be combined with the *Ordered Locking* or *Simultaneous Locking* patterns.

3.2.8 Example

The example, shown in Figure 3-2, is of a system with a motor with a memory-mapped interface. The interface is 16 bits wide and is detailed in the figure comment. In addition to the motor proxy, there are two clients in the example. The first of these is a motor controller than makes high-level decisions about what to do with the motor direction, speed, and how to handle errors. The motor display client is concerned with providing user feedback of the speed, direction, and status of the motor. Most of the action, of course, is in the Hardware Proxy *per se*.

The intention of the motor proxy is to provide the services to access the hardware in a hardware interface-independent fashion so that the two clients needn't worry about – or even be aware of

---

[8] It may not be in high-security applications however. In those cases, it may be necessary to always have the data stored in encrypted formats to make it more difficult to violate data security protocols.

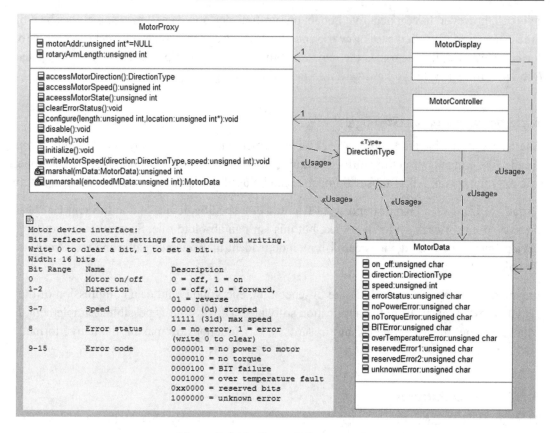

Figure 3-2 Hardware Proxy example

– the specific encoding of the bits. To this end, several types of services are offered by the motor proxy:

Motor Management Functions

- configure() – this function sets up the memory-mapped address of the motor and the length of the rotary arm; this function must be called first
- disable() – this function turns off the motor but keeps the set values intact
- enable() – this function turns the motor on with the current set values
- initialize() – this function turns on the motor to default set values (off)

Motor Status Functions

- accessMotorDirection() – this function returns the current motor direction (off, forward, or reverse)
- accessMotorSpeed() – this function returns the speed of motor

Motor Control Functions

- writeMotorSpeed() – this function sets the speed of the motor and adjusts for the length of the rotary arm (if set)

Motor Error Management Functions

- clearErrorStatus() – this function clears all error bits
- accessMotorState() – this function returns the error status

Internal Data Formatting Functions (private)

These functions aren't provided to the clients, but are used internally to exchange the data between native and presentation formats.

- marshal() – converts presentation (client) data format to native (motor) format
- unmarshal() – converts native (motor) data format to presentation (client) format

The C source code for the data types and MotorProxy class are shown in the next several code segments. Code Listing 3-3 shows the data typing for the DirectionType while Code Listing 3-4 shows the presentation client format for MotorData.

```c
#ifndef HWProxyExample_H
#define HWProxyExample_H

struct MotorController;
struct MotorData;
struct MotorDisplay;
struct MotorProxy;

typedef enum DirectionType {
        NO_DIRECTION,
        FORWARD,
        REVERSE
} DirectionType;
#endif
```

Code Listing 3-3: HardwareProxyExample.h

```c
#ifndef MotorData_H
#define MotorData_H

#include "HWProxyExample.h"
typedef struct MotorData MotorData;
struct MotorData {
        unsigned char on_off;
        DirectionType direction;
        unsigned int speed;
        unsigned char errorStatus;
        unsigned char noPowerError;
```

```
            unsigned char noTorqueError;
            unsigned char BITError;
            unsigned char overTemperatureError;
            unsigned char reservedError1;
            unsigned char reservedError2;
            unsigned char unknownError;
    };
    #endif
```

Code Listing 3-4: MotorData.h

The next two code listings show the header and implementation files for the MotorProxy class itself including both the data structure and the member functions.

```
#ifndef MotorProxy_H
#define MotorProxy_H

#include "HWProxyExample.h"
#include "MotorData.h"
/* class MotorProxy */
typedef struct MotorProxy MotorProxy;
/* This is the proxy for the motor hardware.       */
/* Note that the speed of the motor is adjusted for the length of the rotary arm */
/* to keep a constant speed at the end of the arm. */
struct MotorProxy {
        unsigned int* motorAddr;
        unsigned int rotaryArmLength;
};

void MotorProxy_Init(MotorProxy* const me);
void MotorProxy_Cleanup(MotorProxy* const me);
DirectionType* MotorProxy_accessMotorDirection(MotorProxy* const me);
unsigned int MotorProxy_accessMotorSpeed(MotorProxy* const me);
unsigned int MotorProxy_aceessMotorState(MotorProxy* const me);
/* keep all settings the same but clear error bits */ void MotorProxy_clearErrorStatus
(MotorProxy* const me);
/* Configure must be called first, since it sets up the */
/* address of the device.     */
void MotorProxy_configure(MotorProxy* const me, unsigned int length, unsigned int*
location);

/* turn motor off but keep original settings */
void MotorProxy_disable(MotorProxy* const me);
/* Start up the hardware but leave all other settings of the */ /* hardware alone */ void
MotorProxy_enable(MotorProxy* const me);
/* precondition: must be called AFTER configure() function.     */ /* turn on the hardware to
a known default state. */ void MotorProxy_initialize(MotorProxy* const me);
/* update the speed and direction of the motor together */
void MotorProxy_writeMotorSpeed(MotorProxy* const me, const DirectionType* direction,
unsigned int speed);
```

```
MotorProxy * MotorProxy_Create(void); void MotorProxy_Destroy(MotorProxy* const me);
#endif
```

Code Listing 3-5: MotorProxy.h

```c
#include "MotorProxy.h"
/* class MotorProxy */
/* This function takes a MotorData structure and creates     */
/* a device-specific unsigned int in device native format. */
static unsigned int marshal(MotorProxy* const me, const struct MotorData* mData);
static struct MotorData* unmarshal(MotorProxy* const me, unsigned int encodedMData);
void MotorProxy_Init(MotorProxy* const me) { me->motorAddr = NULL;
}

void MotorProxy_Cleanup(MotorProxy* const me) {
}

DirectionType* MotorProxy_accessMotorDirection(MotorProxy* const me) {
        MotorData mData;
        if (!me->motorData) return 0;
        mData = unmarshall(*me->motorAddr);
        return mData.direction;
}

unsigned int MotorProxy_accessMotorSpeed(MotorProxy* const me) {
        MotorData mData;
        if (!me->motorData) return 0;
        mData = unmarshall(*me->motorAddr);
        return mData.speed;
}

unsigned int MotorProxy_aceessMotorState(MotorProxy* const me) {
        MotorData mData;
        if (!me->motorData) return 0;
        mData = unmarshall(*me->motorAddr);
        return mData.errorStatus;
}

void MotorProxy_clearErrorStatus(MotorProxy* const me) {
        if (!me->motorData) return;
        *me->motorAddr &= 0xFF;
}

void MotorProxy_configure(MotorProxy* const me, unsigned int length, unsigned int* location) {
        me->rotaryArmLength = length;
        me->motorAddr = location;
}

void MotorProxy_disable(MotorProxy* const me) {
        // and with all bits set except for the enable bit
        if (!me->motorData) return;
        me->MotorAddr & = 0xFFFE;
}
```

```
void MotorProxy_enable(MotorProxy* const me) {
        if (!me->motorData) return;
        *me->motorAddr |= 1;
}

void MotorProxy_initialize(MotorProxy* const me) {
        MotorData mData;
        if (!me->motorData) return;
        mData.on_off = 1;
        mData.direction = 0;
        mData.speed = 0;
        mData.errorStatus = 0;
        mData.noPowerError) = 0;
        mData.noTorqueError) = 0;
        mData.BITError) = 0;
        mData.overTemperatureError) = 0;
        mData.reservedError1) = 0;
        mData.reservedError2) = 0;
        Data.unknownError) = 0;
        *me->motorAddr = marshall(mData);
}

void MotorProxy_writeMotorSpeed(MotorProxy* const me, const DirectionType* direction,
unsigned int speed) {
        MotorData mData
        double dPi, dArmLength, dSpeed, dAdjSpeed;
        if (!me->motorData) return; mData = unmarshall(*me->motorAddr); mData.direction =
direction;
        // ok, let's do some math to adjust for
        // the length of the rotary arm times 10
        if (me->rotaryArmLength > 0) {
                dSpeed = speed;
                dArmLength = me->rotaryArmLength;
                dAdjSpeed = dSpeed / 2.0 / 3.14159 / dArmLength * 10.0;
                mData.speed = (int)dAdjSpeed;
        }
        else
            mData.speed = speed;
        *me->motorData = marshal(mData);
        return;
}

static unsigned int marshal(MotorProxy* const me, const struct MotorData* mData) {
        unsigned int deviceCmd;
        deviceCmd = 0; // set all bits to zero
        if (mData.on_off) deviceCmd |= 1; // OR in the appropriate bit
        if (mData.direction == FORWARD)
            deviceCmd |= (2 << 1);
        else if (mData.direction == REVERSE)
            deviceCmd |= (1 << 1);
        if (mData.speed < 32 && mData.speed >= 0)
            deviceCmd |= mData.speed << 3;
```

```
            if (mData.errorStatus) deviceCmd |= 1 << 8;
            if (mData.noPowerError) deviceCmd |= 1 << 9;
            if (mData.noTorqueError) deviceCmd |= 1 << 10;
            if (mData.BITError) deviceCmd |= 1 << 11;
            if (mData.overTemperatureError) deviceCmd |= 1 << 12;
            if (mData.reservedError1) deviceCmd |= 1 << 13;
            if (mData.reservedError2) deviceCmd |= 1 << 14;
            if (mData.unknownError) deviceCmd |= 1 << 15;
            return deviceCmd;
}

static struct MotorData* unmarshal(MotorProxy* const me, unsigned int encodedMData) {
        MotorData mData
        int temp;

        mData.on_off = encodedMData & 1;
        temp = (encodedMData & (3 << 1)) >> 1;      if (temp == 1)
            mData.direction = REVERSE;
        else if (temp == 2)
            mData.direction = FORWARD;
        else
            mData.direction = NO_DIRECTION;
        mData.speed = encodedMData & (31 << 3);
        mData.errorStatus = encodedMData & (1 << 8);
mData.noPowerError) = encodedMData & (1 << 9);
mData.noTorqueError) = encodedMData & (1 << 10); mData.BITError) = encodedMData & (1 <<
11); mData.overTemperatureError) = encodedMData & (1 << 12); mData.reservedError1) =
encodedMData & (1 << 13);      mData.reservedError2) = encodedMData & (1 << 14);
        Data.unknownError) = encodedMData & (1 << 15);
        return mData;
}

MotorProxy * MotorProxy_Create(void) {
        MotorProxy* me = (MotorProxy *) malloc(sizeof(MotorProxy));
        if(me!=NULL)
                {
                        MotorProxy_Init(me);
                }
        return me;
}

void MotorProxy_Destroy(MotorProxy* const me) {
        if(me!=NULL)
                {
                        MotorProxy_Cleanup(me);
                }
        free(me);
}
```

Code Listing 3-6: MotorProxy.c

3.3 Hardware Adapter Pattern

The Hardware Adapter Pattern provides a way of adapting an existing hardware interface into the expectations of the application. This pattern is a straightforward derivative of the Adapter Pattern.

3.3.1 Abstract

The Hardware Adapter Pattern is useful when the application requires or uses one interface, but the actual hardware provides another. The pattern creates an element that converts between the two interfaces.

3.3.2 Problem

While hardware that performs similar functions tend to have similar interfaces, often the information they need and the set of services differ. Rather than rewrite the clients of the hardware device to use the provided interface, an adapter is created that provides the expected interface to the clients while converting the requests to and from the actual hardware interface.

The Hardware Adapter Pattern is useful when you have hardware that meets the semantic need of the system but that has an incompatible interface. The goal of the pattern is to minimize the reworking of code when one hardware design or implementation is replaced with another.

3.3.3 Pattern Structure

The pattern structure is shown in Figure 3-3. The pattern extends the Hardware Proxy Pattern by adding a Hardware Adapter and explicitly showing the interface the client expects the hardware to support.

3.3.4 Collaboration Roles

This section describes the roles for this pattern.

3.3.4.1 Adapter Client

The adapter client expects to be invoking the services of a software element that represents the hardware, as specified by the Hardware Interface to Client interface. It is an element in the application system that expects to control, monitor, or use the hardware device.

3.3.4.2 Hardware Adapter

The Hardware Adapter performs "impedance matching" between the client and the Hardware Proxy. That is, service requests made by the Adapter Client are converted into a sequence of services that are available from the Hardware Proxy. This potentially includes both factoring of the service invocations and reformatting and restructuring the data.

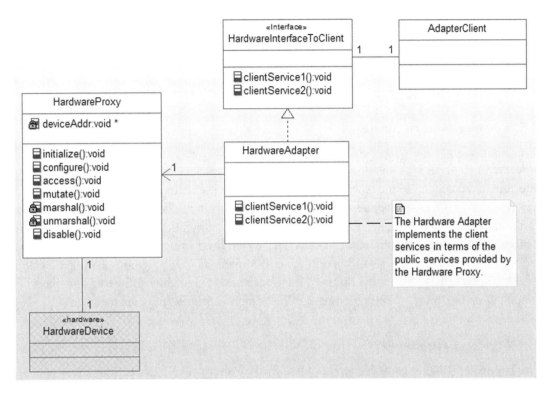

Figure 3-3 Hardware Adapter Pattern

3.3.4.3 Hardware Interface to Client
This interface represents the set of services and parameter lists that the client expects the Hardware Proxy to provide. As an interface, it is a collection of service specifications and has no implementation. The implementation in this case is provided by the Hardware Adapter Class.

3.3.4.4 Hardware Device
This element is described in the Hardware Proxy Pattern. See Section 3.2.4.1.

3.3.4.5 Hardware Proxy
This element is described in the Hardware Proxy Pattern. See Section 3.2.4.2.

3.3.5 Consequences

The use of this pattern allows various Hardware Proxies and their related hardware devices to be used as-is in different applications, while at the same time allowing existing applications to use different hardware devices without change. The key is that the adapter provides the connective glue to match the Hardware Proxy to the application. This means that it will be easier, less error-prone, and faster to change hardware devices for an application or to reuse an existing hardware device in a new application.

The cost of using the pattern is that it adds a level of indirection and therefore decreases run-time performance slightly.

3.3.6 Implementation Strategies

The classic *Design Patterns*[9] book by Gamma, et. al., identifies two forms for this pattern. The *object adapter* form is as shown in Figure 3-3. In this form, the adapter subclasses along one interface and delegates to the other element (although we use realization in this case rather than inheritance). In the *class adapter* form, the Hardware Adapter subclasses from both interfaces and reimplements the client services in terms of the proxy services that it inherits. This requires a more elaborate implementation of classes than we've been using so far in this book since it requires the creation of a virtual function table to implement true polymorphism (see Section 1.2.3). The tradeoffs for the two different implementations are subtly different. The class adapter form is a bit more reusable but requires a more complex implementation in C.

3.3.7 Related Patterns

The Hardware Adapter extends the Hardware Proxy Pattern. The latter pattern encapsulates hardware interface detail but does not translate service requests into radically different ones. The Hardware Adapter Pattern adds a level of indirection between the client and the Hardware Proxy. This allows the clients to be unchanged in the application while at the same time permitting reuse of existing Hardware Proxies that may also have been created for other systems. The implementation of the Hardware Proxy and the Hardware Adapter can be merged, but that undermines the reusability of the Hardware Proxy.

3.3.8 Example

The example is shown graphically in Figure 3-4. This example shows a system that tracks both oxygen concentration and oxygen flow rates, such as might be found in a medical ventilator. Two clients are shown. The Gas Display client displays these data for the attending medical staff. The Gas Mixer client uses these data for the closed loop control of gas delivery. Both are implemented to use the interface as shown in the iO2Sensor interface. The two services the clients expect to see are gimmeO2Conc(), which returns an integer in the range of 0–100, and gimmeO2Flow(), which returns an integer type with units of cc/min.

---

[9] Gamma, E., Helm, R., Johnson, R., Vlissides, J., 1995. *Design Patterns: Elements of Reusable Object-Oriented Software*. Addison-Wesley.

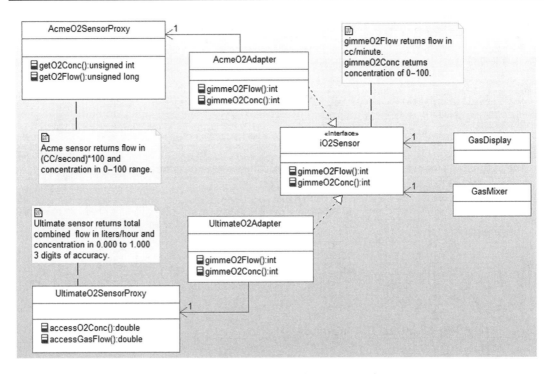

Figure 3-4 Hardware Adapter example

The system can be delivered with either of two physical sensors. These sensors do much the same job but they provide different programmatic interfaces. The Acme O2 Sensor Proxy provides two services. getO2Conc() returns the oxygen concentration in a range of 0–100 as an unsigned integer. getO2Flow() returns the flow oxygen flow as a scaled integer of 100* cc/sec. The Ultimate O2 sensor has two similar functions. accessO2Conc() returns concentration as a double in the range of 0.000 to 1.000 (3 digits of accuracy) and the accessGasFlow() returns the total gas flow (of which oxygen is just a part) as a double in liters/hour.

Each Hardware Proxy requires its own adapter to convert to the expected client interface.

The previous pattern showed code for the proxy classes, so this section will only show code for the adapters. The header files aren't very interesting, so we will concentrate solely on the implementation files of the two adapters. The implementation of the Acme O2 Adapter is shown in Code Listing 3-7 and the implementation for the Ultimate O2 Adapter is given in Code Listing 3-8.

```
int AcmeO2Adapter_gimmeO2Conc(AcmeO2Adapter* const me) {
    return me->itsAcmeO2SensorProxy->getO2Conc();
}
```

```
int AcmeO2Adapter_gimmeO2Flow(AcmeO2Adapter* const me) {
        return (me->itsAcmeO2SensorProxy->getO2Flow()*60)/100;
```

Code Listing 3-7: AcmeO2Adapter.c

```
int UltimateO2Adapter_gimmeO2Conc(UltimateO2Adapter* const me) {
        return int(me->getItsUltimateO2SensorProxy
                ->accessO2Conc()*100);
}
int UltimateO2Adapter_gimmeO2Flow(UltimateO2Adapter* const me) {
        double totalFlow;
        // convert from liters/hr to cc/min
        totalFlow = me->itsUltimateO2SensorProxy->accessGasFlow() *
            1000.0/60.0;
        // now return the portion of the flow due to oxygen
        // and return it as an integer
        return (int)(totalFlow *
                me->itsUltimateO2SensorProxy->accessO2Conc());
}
```

Code Listing 3-8: UltimateO2Adapter.c

3.4 Mediator Pattern

The Mediator Pattern provides a means of coordinating a complex interaction among a set of elements.

3.4.1 Abstract

The Mediator Pattern is particularly useful for managing different hardware elements when their behavior must be coordinated in well-defined but complex ways. It is particularly useful for C applications because it doesn't require a lot of specialization (subclassing), which can introduce its own complexities into the implementation.

3.4.2 Problem

Many embedded applications control sets of actuators that must work in concert to achieve the desired effect. For example, to achieve a coordinated movement of a multi-joint robot arm, all the motors must work together to provide the desired arm movement. Similarly, using reaction wheels or thrusters in a spacecraft in three dimensions requires many different such devices acting at precisely the right time and with the right amount of force to achieve attitude stabilization.

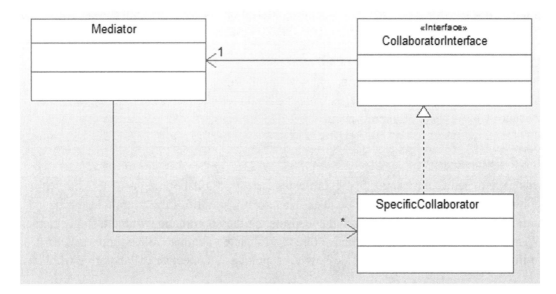

Figure 3-5 Mediator Pattern

3.4.3 Pattern Structure

The Mediator Pattern uses a mediator class to coordinate the actions of a set of collaborating devices to achieve the desired overall effect. The Mediator class coordinates the control of the set of multiple Specific Collaborators (their number is indicated by the '*' multiplicity on the association between the Mediator and the Specific Collaborator in Figure 3-5). Each Specific Collaborator must be able to contact the Mediator when an event of interest occurs.

3.4.4 Collaboration Roles

This section describes the roles for this pattern.

3.4.4.1 Collaborator Interface

The Collaborator Interface is a specification of a set of services common to all Specific Collaborators that may be invoked by the Mediator. For example, it is common to have reset(), shutdown, initialize() style operations for all the hardware devices to facilitate bringing them all to a known initial, recovery and/or shutdown state. Each of the common services must be implemented in each Specific Collaborator (although, of course, the implementation will usually be unique to each collaborator). If there are no services common to all Specific Collaborators, then this interface may be omitted.

3.4.4.2 Mediator

The Mediator class coordinates the Specific Collaborators in the pattern. It has a link to each Specific Collaborator so that it can send messages to it. In addition, each Specific Collaborator must be able to send the Mediator messages when events of interest occur. The Mediator provides the coordination logic that would otherwise require Specific Collaborators to coordinate among themselves.

3.4.4.3 Specific Collaborator

The Specific Collaborator represents one device and so may be a device driver or Hardware Proxy. It receives command messages from the Mediator and also sends messages to the Mediator when events of interest occur.

3.4.5 Consequences

This pattern creates a mediator that coordinates the set of collaborating actuators but without requiring direct coupling of those devices. This greatly simplifies the overall design by minimizing the points of coupling and encapsulating the coordination within a single element. Whenever the Collaborator would have directly contacted another Collaborator, instead it notifies the Mediator who can decide how to respond as a collective collaborative whole.

Since many embedded systems must react with high time fidelity, delays between the actions may result in unstable or undesirable effects. It is important that the mediator class can react within those time constraints. This is of a particular concern when there is two-way collaboration between the actuators and the mediator.

3.4.6 Implementation Strategies

The Mediator must be able to link to each Specific Collaborator. This can be done with an array of pointers (the obvious implementation strategy), a separate pointer for each Specific Collaborator, or a combination of the two. A common rule of thumb is that if a set of Specific Collaborators are indistinguishable in terms of their use other than their position in the list (e.g., they provide a common interface), then an array of pointers is best. If the Specific Collaborators serve different purposes, then use a separate link for each such Specific Collaborator. The example in Section 3.4.8 shows the use of both an array and of specific pointers.

It is even possible to have grouped sets of different Specific Collaborators in which each class within a group provides a common interface.

3.4.7 Related Patterns

This pattern is similar to the Strategy Pattern from the classic Design Patterns book[10] and the architectural Rendezvous Pattern from my Real-Time Design Patterns book[11]. In this case we are focusing on its use for detailed hardware coordination.

[10]   Gamma, E., Helm, R., Johnson, R., Vlissides, J., 1995. *Design Patterns: Elements of Reusable Object-Oriented Software*. Addison-Wesley.

[11]   Douglass, B.P., 2003. *Real-Time Design Patterns: Robust Scalable Architecture for Real-Time Systems*. Addison-Wesley.

The Observer Pattern may be used instead of a direct link to the Mediator when there are multiple clients of the Specific Collaborator or when you wish to hide the Mediator from the Specific Collaborator.

3.4.8 Example

In this example of a robot system (See Figure 3-6), the Robot Arm Manager receives requests to grasp objects at specific points in space (x, y, z) and time (t). It first computes an arm trajectory via the `computeTrajectory()` function and produces a set of `nSteps` actions (up to 100). Each action is composed of a set of commands to each of the seven servos. If the goal is achievable, then Robot Arm Manager calls `executeStep()` function `nSteps` times to step through the computed action sequence. At each step, any command can return an error code (non-zero) which causes the `graspAt()` function to abort with an error code. Note that the computation of the arm path via the sum of the movements of the various arm joints is a complex exercise in geometry which, while interesting, is outside the scope of concern for this book. In addition, many paths

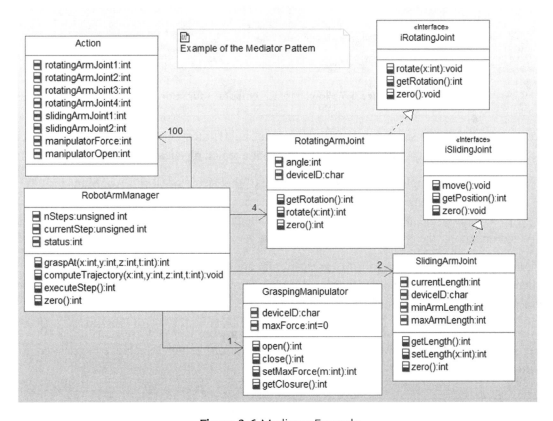

Figure 3-6 Mediator Example

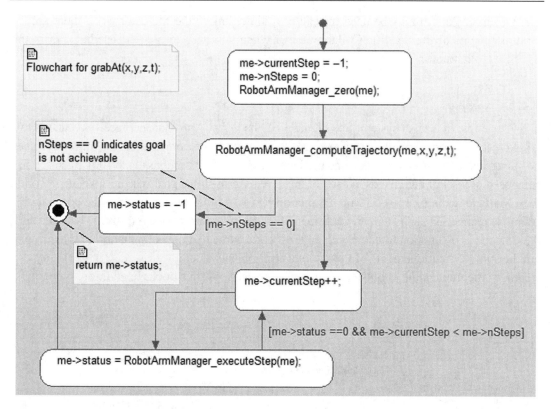

Figure 3-7 Flowchart for graspAt() function

may be disallowed for safety reasons (Rule 1: You Can't Hit the Operator)[12], because they may require a path through a physical object, or may not even be physically possible.

The point of the example is to illustrate the value of the Mediator (RobotArmManager). Without its coordinating influences, all of the actuators would have to collaborate directly with their peers. The computation of allowable versus illegal movement paths of the arm would be divided up among the collaborators and would be very difficult to manage.

Figure 3-7 shows a flowchart for the graspAt() function. This is the primary function that clients of the RobotArmManager will invoke when they want the arm to move to a position and grasp (or let go of) something.

The code for the RobotArmManager is given in Code Listing 3-9 and Code Listing 3-10. You can see that the RobotArmManager coordinates the various servos to achieve the desired arm actions.

---

[12] An entirely different Rule #1 than in *Zombieland*. See www.zombieland.com (Sony Pictures, 2009), possibly the most awesome zombie movie *ever*.

```
#include "GraspingManipulator.h"
#include "RotatingArmJoint.h"
#include "SlidingArmJoint.h"
#include "Action.h"

/*## class RobotArmManager */
typedef struct RobotArmManager RobotArmManager;
struct RobotArmManager {
        unsigned int currentStep;
        unsigned int nSteps;
        struct GraspingManipulator* itsGraspingManipulator;
        struct RotatingArmJoint *itsRotatingArmJoint[4];
        struct SlidingArmJoint *itsSlidingArmJoint[2];
        struct Action *itsAction[100];
        int status;
};
/* Constructors and destructors:*/
void RobotArmManager_Init(RobotArmManager* const me);
void RobotArmManager_Cleanup(RobotArmManager* const me);
/* Operations */
void RobotArmManager_computeTrajectory(RobotArmManager* const me, int x, int y, int z, int t);

int RobotArmManager_executeStep(RobotArmManager* const me);
int RobotArmManager_graspAt(RobotArmManager* const me, int x, int y, int z, int t);
int RobotArmManager_zero(RobotArmManager* const me);
struct GraspingManipulator*
RobotArmManager_getItsGraspingManipulator(const
RobotArmManager* const me);
void RobotArmManager_setItsGraspingManipulator(RobotArmManager* const me, struct
GraspingManipulator* p_GraspingManipulator);
int RobotArmManager_getItsRotatingArmJoint(const RobotArmManager* const me);
void RobotArmManager_addItsRotatingArmJoint(RobotArmManager* const me, struct
RotatingArmJoint * p_RotatingArmJoint);
void RobotArmManager_removeItsRotatingArmJoint(RobotArmManager* const me, struct
RotatingArmJoint * p_RotatingArmJoint);
void RobotArmManager_clearItsRotatingArmJoint(RobotArmManager* const me);
int RobotArmManager_getItsSlidingArmJoint(const RobotArmManager* const me);
void RobotArmManager_addItsSlidingArmJoint(RobotArmManager* const me, struct
SlidingArmJoint * p_SlidingArmJoint);
void RobotArmManager_removeItsSlidingArmJoint(RobotArmManager* const me, struct
SlidingArmJoint * p_SlidingArmJoint);
void RobotArmManager_clearItsSlidingArmJoint(RobotArmManager* const me);
RobotArmManager * RobotArmManager_Create(void);
void RobotArmManager_Destroy(RobotArmManager* const me);
int RobotArmManager_getItsAction(const RobotArmManager* const me);
void RobotArmManager_addItsAction(RobotArmManager* const me, struct Action * p_Action);
void RobotArmManager_removeItsAction(RobotArmManager* const me, struct Action * p_Action);
void RobotArmManager_clearItsAction(RobotArmManager* const me);
#endif
```

Code Listing 3-9: RobotArmManager.h

```c
#include "RobotArmManager.h"
static void cleanUpRelations(RobotArmManager* const me);
void RobotArmManager_Init(RobotArmManager* const me) {
        int pos;
        for(pos = 0; pos < 100; ++pos) {
            me->itsAction[pos] = NULL;
        }

        me->itsGraspingManipulator = NULL;
        for(pos = 0; pos < 4; ++pos)    {
            me->itsRotatingArmJoint[pos] = NULL;
        }

        for(pos = 0; pos < 2; ++pos) {
            me->itsSlidingArmJoint[pos] = NULL;
            }
}

void RobotArmManager_Cleanup(RobotArmManager* const me) {
        cleanUpRelations(me);
}

/* operation computeTrajectory(x,y,z,t)
    This function computes a path for the robot arm to follow to position the manipulator at
    the desired end point. It produces a set of up to 100 actions, each action is a set of
    commands to the various servos to which the RobotArmManager connects. In actual practice
    this is a complex job. The implementation here is just a placeholder.
*/
void RobotArmManager_computeTrajectory(RobotArmManager* const me, int x, int y, int z,
int t) {
        Action* ap;
        int j;

        me->nSteps = 0;
        RobotArmManager_clearItsAction(me);
        /* move the arm to the position with manipulator open*/
        ap = Action_Create();
        RobotArmManager_addItsAction(me, ap);
        ap->rotatingArmJoint1=1;
        ap->rotatingArmJoint2=2;
        ap->rotatingArmJoint3=3;
        ap->rotatingArmJoint4=4;
        ap->slidingArmJoint1=10;
        ap->slidingArmJoint2=20;
        ap->manipulatorForce=0;
        ap->manipulatorOpen=1;

        /* grab the object */
        ap = Action_Create();
        RobotArmManager_addItsAction(me, ap);
        ap->rotatingArmJoint1=1;
        ap->rotatingArmJoint2=2;
        ap->rotatingArmJoint3=3;
```

```
        ap->rotatingArmJoint4=4;
        ap->slidingArmJoint1=10;
        ap->slidingArmJoint2=20;
        ap->manipulatorForce=10;
        ap->manipulatorOpen=0;

        /* return to zero position */
        ap = Action_Create();
        RobotArmManager_addItsAction(me, ap);
        ap->rotatingArmJoint1=0;
        ap->rotatingArmJoint2=0;
        ap->rotatingArmJoint3=0;
        ap->rotatingArmJoint4=0;
        ap->slidingArmJoint1=0;
        ap->slidingArmJoint2=0;
        ap->manipulatorForce=10;
        ap->manipulatorOpen=0;

        me->nSteps = 3;
}

/*     operation executeStep()
    This operation executes a single step in the chain of actions by
    executing all of the commands within the current action
*/
int RobotArmManager_executeStep(RobotArmManager* const me) {
        int actionValue = 0;
        int step = me->currentStep;
        int status = 0;

        if (me->itsAction[step]) {
            actionValue = me->itsAction[step]->rotatingArmJoint1;
            status = RotatingArmJoint_rotate(me->itsRotatingArmJoint[0],actionValue);
            if (status) return status;
             actionValue = me->itsAction[step]->rotatingArmJoint2;
             status = RotatingArmJoint_rotate(me->itsRotatingArmJoint[1],actionValue);
            if (status) return status;
             actionValue = me->itsAction[step]->rotatingArmJoint3;
             status = RotatingArmJoint_rotate(me->itsRotatingArmJoint[2],actionValue);
            if (status) return status;
             actionValue = me->itsAction[step]->rotatingArmJoint4;
             status = RotatingArmJoint_rotate(me->itsRotatingArmJoint[3],actionValue);
            if (status) return status;
             actionValue = me->itsAction[step]->slidingArmJoint1;
             status = SlidingArmJoint_setLength(me->itsSlidingArmJoint[0],actionValue);
            if (status) return status;
             actionValue = me->itsAction[step]->rotatingArmJoint2;
             status = SlidingArmJoint_setLength(me->itsSlidingArmJoint[0],actionValue);
            if (status) return status;
                actionValue = me->itsAction[step]->manipulatorForce;
                status  =  GraspingManipulator_setMaxForce(me->itsGraspingManipulator,
actionValue);
```

```
             if (status) return status;
             if (me->itsAction[step]->manipulatorOpen)
                   status = GraspingManipulator_open(me->itsGraspingManipulator);
             else
                 status = GraspingManipulator_close(me->itsGraspingManipulator);
        };
        return status;
}
/*    operation graspAt(x,y,z,t) is the main function called by clients of the
   RobotArmManager. This operation:
      1. zeros the servos
      2. computes the trajectory with a call to computeTrajectory()
      3. executes each step in the constructed action list
*/
int RobotArmManager_graspAt(RobotArmManager* const me, int x, int y, int z, int t) {
        me->currentStep = -1;
        me->nSteps = 0;
        RobotArmManager_zero(me);
        RobotArmManager_computeTrajectory(me,x,y,z,t);

        if ( me->nSteps == 0 ) {
          me->status = -1;
        }
        else {
            do {
                me->currentStep++;
                me->status = RobotArmManager_executeStep(me);
            }
            while (me->status ==0 && me->currentStep < me->nSteps);
        }
        return me->status;
}
/*    operation zero()
   This operation returns all servos to a starting default position
*/
int RobotArmManager_zero(RobotArmManager* const me) {
        /* zero all devices */
        int j;
        for (j=0; j<4; j++){
                if (me->itsRotatingArmJoint[j] == NULL) return -1;
            if (RotatingArmJoint_zero(me->itsRotatingArmJoint[j])) return -1;
            }
        for (j=0; j<2; j++) {
                if (me->itsSlidingArmJoint[j] == NULL) return -1;
            if (SlidingArmJoint_zero(me->itsSlidingArmJoint[j])) return -1;
            }
        if (me->itsGraspingManipulator == NULL) return -1;
        if (GraspingManipulator_open(me->itsGraspingManipulator)) return -1;
        return 0;
}
```

```
struct GraspingManipulator*
RobotArmManager_getItsGraspingManipulator(const RobotArmManager* const me) {
        return (struct GraspingManipulator*)me->itsGraspingManipulator;
}

void RobotArmManager_setItsGraspingManipulator(RobotArmManager* const me, struct
GraspingManipulator* p_GraspingManipulator) {
        me->itsGraspingManipulator = p_GraspingManipulator;
}

int RobotArmManager_getItsRotatingArmJoint(const RobotArmManager* const me) {
        int iter = 0;
        return iter;
}

void RobotArmManager_addItsRotatingArmJoint(RobotArmManager* const me, struct
RotatingArmJoint * p_RotatingArmJoint) {
        int pos;
        for(pos = 0; pos < 4; ++pos) {
            if (!me->itsRotatingArmJoint[pos]) {
                me->itsRotatingArmJoint[pos] = p_RotatingArmJoint;
                break;
            }
        }
}

void RobotArmManager_removeItsRotatingArmJoint(RobotArmManager* const me, struct
RotatingArmJoint * p_RotatingArmJoint) {
        int pos;
        for(pos = 0; pos < 4; ++pos) {
            if (me->itsRotatingArmJoint[pos] == p_RotatingArmJoint) {
                me->itsRotatingArmJoint[pos] = NULL;
            }
        }
}

void RobotArmManager_clearItsRotatingArmJoint(RobotArmManager* const me) {
        {
                int pos;
                for(pos = 0; pos < 4; ++pos)
                {
                    me->itsRotatingArmJoint[pos] = NULL;
                }
        }
}

int RobotArmManager_getItsSlidingArmJoint(const RobotArmManager* const me) {
        int iter = 0;
        return iter;
}

void RobotArmManager_addItsSlidingArmJoint(RobotArmManager* const me, struct
SlidingArmJoint * p_SlidingArmJoint) {
```

```
            int pos;
            for(pos = 0; pos < 2; ++pos) {
                if (!me->itsSlidingArmJoint[pos]) {
                    me->itsSlidingArmJoint[pos] = p_SlidingArmJoint;
                    break;
                }
            }
    }
    void RobotArmManager_removeItsSlidingArmJoint(RobotArmManager* const me, struct
    SlidingArmJoint * p_SlidingArmJoint) {
            int pos;
            for(pos = 0; pos < 2; ++pos) {
                if (me->itsSlidingArmJoint[pos] == p_SlidingArmJoint) {
                    me->itsSlidingArmJoint[pos] = NULL;
                }
            }
    }

    void RobotArmManager_clearItsSlidingArmJoint(RobotArmManager* const me) {
            {
                    int pos;
                    for(pos = 0; pos < 2; ++pos)
                    {
                        me->itsSlidingArmJoint[pos] = NULL;
                    }
            }
    }

    RobotArmManager * RobotArmManager_Create(void) {
            RobotArmManager* me = (RobotArmManager *)
            malloc(sizeof(RobotArmManager));
            if(me!=NULL)
                    {
                            RobotArmManager_Init(me);
                    }
            return me;
    }
    void RobotArmManager_Destroy(RobotArmManager* const me) {
            if(me!=NULL) {
                            RobotArmManager_Cleanup(me);
            }
            free(me);
    }
    static void cleanUpRelations(RobotArmManager* const me) {
            if(me->itsGraspingManipulator != NULL) {
                            me->itsGraspingManipulator = NULL;
            }
    }

    int RobotArmManager_getItsAction(const RobotArmManager* const me)
    {
            int iter = 0;
```

```
                return iter;
        }

void RobotArmManager_addItsAction(RobotArmManager* const me, struct Action * p_Action) {
        int pos;
        for(pos = 0; pos < 100; ++pos) {
            if (!me->itsAction[pos]) {
                me->itsAction[pos] = p_Action;
                break;
            }
        }
}

void RobotArmManager_removeItsAction(RobotArmManager* const me, struct Action *
p_Action) {
        int pos;
        for(pos = 0; pos < 100; ++pos) {
            if (me->itsAction[pos] == p_Action) {
                me->itsAction[pos] = NULL;
            }
        }
}

void RobotArmManager_clearItsAction(RobotArmManager* const me) {
        int pos;
        for(pos = 0; pos < 100; ++pos) {
                me->itsAction[pos] = NULL;
        }
}
```

Code Listing 3-10: RobotArmManager.c

3.5 Observer Pattern

The Observer Pattern is one of the most common patterns around. When present, it provides a means for objects to "listen in" on others while requiring no modifications whatsoever to the data servers. In the embedded domain, this means that sensor data can be easily shared to elements that may not even exist when the sensor proxies are written.

3.5.1 Abstract

The Observer Pattern (also known as the "Publish-Subscribe Pattern") provides notification to a set of interested clients that relevant data have changed. It does this without requiring the data server to have any *a priori* knowledge about its clients. Instead, the clients simply offer a subscription function that allows clients to dynamically add (and remove) themselves to the notification list. The data server can then enforce whatever notification policy it desires. Most commonly, data are sent whenever new data arrive, but clients can also be updated periodically, or with a minimum or maximum frequency. This reduces the computational burden on the clients to ensure that they have timely data.

3.5.2 Problem

In a naïve situation, each client can request data periodically from a data server in case the data have changed, but that is wasteful of compute and communication resources as the clients generally cannot know when new data are available. If the data server pushes the data out, then it must know who all of its clients are, breaking the basic rule of client-server relations[13] requiring changes to the server to add new clients.

The Observer Pattern addresses this concern by adding subscription and unsubscription services to the data server. Thus a client can dynamically add itself to the notification list without *a priori* knowledge of the client on the part of the server. On the server side, the server can enforce the appropriate update policy to the notification of its interested clients. In addition, the pattern allows dynamic modification of subscriber lists and so adds a great deal of flexibility to the software.

3.5.3 Pattern Structure

Figure 3-8 shows the basic structure of the pattern. The AbstractSubject is the data server plus the machinery to maintain a list of interested subscribers. A client adds itself to the notification list by passing a pointer to an accept(Datum) function and removes itself by calling unsubscribe with the same pointer. When the AbstractSubject decides to notify its clients, the notify() function walks through the client list, calling the pointed-to function and passing the relevant data. The AbstractClient provides the accept(Datum) function to receive and process the incoming data.

3.5.4 Collaboration Roles

This section describes the roles for this pattern.

3.5.4.1 AbstractClient Interface

The AbstractClient associates with the AbstractSubject so that it can invoke the latter's various services. It contains an accept(Datum) function to be called when the AbstractClient subscribes as well as whenever the AbstractSubject thinks it is appropriate to send data. The AbstractClient is *abstract* in the sense that while it specifies the functions, it does not provide any implementation. It is realized by a set of ConcreteClients that provide implementation that do things with the received data.

In addition to the accept(Datum) function, the AbstractClient associates with the data. This is normally realized with a pointer but it can also be a stack, global, or static variable.

---

[13] That is, the clients know about the server but the server doesn't know who its clients are.

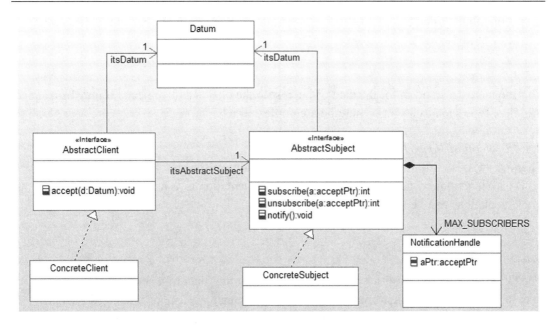

Figure 3-8 Observer Pattern

3.5.4.2 AbstractSubject Interface
The AbstractSubject is the data server in this pattern. Relevant to this pattern it provides three services. The subscribe(acceptPtr) service adds the pointer to the accept function to the notification list. Its return value is zero if the add is successful or non-zero if not. The unsubscribe(acceptPtr) function removes the accept function from the notification list, again returning zero if successful. Finally, the notify() function walks the notification list to notify the subscribed clients. This is done by invoking the function pointed to by each notification handle.

The AbstractSubject also contains a link to the Datum to be passed as well as a list of NotificationHandles. In the representation here, the NotificationHandle is implemented as an array of pointers, but it can also be a linked list or other data structure. Similarly, the Datum can be on the heap, stack, or static as appropriate.

3.5.4.3 ConcreteClient
The ConcreteClient is a concrete realization of the AbstractClient interface. That is, it provides an implementation of the acceptPtr(Datum) function as well as other functionality not relevant to the pattern *per se*.

3.5.4.4 ConcreteSubject
The ConcreteSubject is a concrete realization of the AbstractSubject interface. It provides implementation of the functions but also provides means to acquire and manage the

data it distributes. The ConcreteSubject is often a HardwareProxy in addition to an AbstractSubject.

3.5.4.5 Datum
This element is a stand-in for the data of interest to the clients and subjects. It may be a simple primitive type such as an int or may be a complex struct.

3.5.4.6 NotificationHandle
The NotificationHandle is a representation of a means to invoke a client's accept (Datum) function. By far, the most common implementation is as a function pointer, but other implementations can be used as well.

3.5.5 Consequences

The Observer Pattern simplifies the process of distributing data to a set of clients who may not be known at design-time, and dynamically managing lists of interested clients during run-time. This pattern maintains the fundamental client-server relation while providing run-time flexibility through its subscription mechanism. Compute efficiency is maintained because clients are only updated when appropriate; the most common policy is that the clients are updated when the data change, but any appropriate policy can be implemented.

3.5.6 Implementation Strategies

The only complex aspects of this pattern are the implementation of the notification handle and the management of the notification handle list. The notification handle itself is almost always a function callback, that is, a pointer to a function with the right signature. Of course, this signature varies with the data being returned, but it also depends on the implementation of the software "classes" as discussed in Chapter 1. If using the struct approach with the me pointers, this will be the first parameter of the accept function. If you don't use the approach but write a more traditional C, then the realization of the ConcreteClients is more manual but you won't have to include the me pointer in the accept function parameter list. In the example later in this section, we shall use the former approach.

The easiest approach for the notification list is to declare an array big enough to hold all potential clients. This wastes memory in highly dynamic systems with many potential clients. An alternative is to construct the system as a linked list; that is, add another variable to each notification handle which is a pointer to the next one. Obviously, the last element in the list has a NULL next pointer value.

3.5.7 Related Patterns

This pattern can be freely mixed with the previous patterns in this chapter since its concerns are orthogonal. In embedded systems it is very common to add Observer functionality to a Hardware Proxy or Hardware Adapter, for example.

3.5.8 Example

Figure 3-9 provides a straightforward example of this pattern. In this example, the `GasSensor` is the `ConcreteClient` of the pattern; it provides the subscribe() and unsubscribe functions and manages the array of `GasNotificationHandles`. Because we implemented the pattern with the usage of our pseudo-object–oriented approach, all the operations have a first parameter that identifies the instance data. These "hidden parameters" are not shown in Figure 3-9 but they do show up in the code. This impacts the `GasNotificationHandle` in that not only do we need the function pointer for the acceptor function, we also need its instance data point to pass as its first argument.

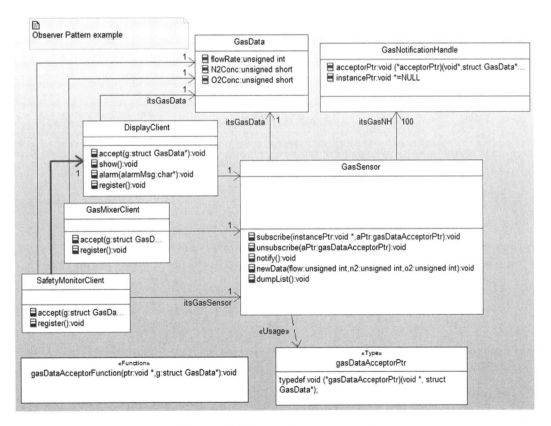

Figure 3-9 Observer Pattern example

In addition to the `subscribe()` and `unsubscribe()` operations, the `notify()` function walks the list of registered subscribers and invokes the `accept()` operation using the simple statement

```
me->itsGasNH[pos]->acceptorPtr(me->itsGasNH[pos]->instancePtr,
me->itsGasData);
```

In this way, the function being pointed to is accessed via the function pointer, where the first parameter is the instance data of the client and the second parameter is a pointer to the new data.

The `GasSensor` class also includes a `newData()` function that calls the `notify()` function and a `dumpList()` function that prints out a list of the currently subscribed clients. The code for the `GasSensor` is shown in Code Listing 3-11 and Code Listing 3-12.

```
#ifndef GasSensor_H
#define GasSensor_H

#include "GasData.h"
/* the function pointer type
    The first value of the function pointer is to the instance
    data of the class. The second is a ptr to the new gas data
*/
typedef void (*gasDataAcceptorPtr)(void *, struct GasData*);
struct GasNotificationHandle;
/* class GasSensor */
typedef struct GasSensor GasSensor;
struct GasSensor {
        struct GasData* itsGasData;
        struct GasNotificationHandle *itsGasNH[100];
};

/* Constructors and destructors:*/
void GasSensor_Init(GasSensor* const me);
void GasSensor_Cleanup(GasSensor* const me);

/* Operations */
void GasSensor_dumpList(GasSensor* const me);
void GasSensor_newData(GasSensor* const me, unsigned int flow, unsigned int n2, unsigned
int o2);
void GasSensor_notify(GasSensor* const me);
void GasSensor_subscribe(GasSensor* const me, void * instancePtr, const gasDataAccep-
torPtr* aPtr);
void GasSensor_unsubscribe(GasSensor* const me, const gasDataAcceptorPtr* aPtr);
struct GasData* GasSensor_getItsGasData(const GasSensor* const me);
void GasSensor_setItsGasData(GasSensor* const me, struct GasData* p_GasData);
int GasSensor_getItsGasNH(const GasSensor* const me);
void  GasSensor_addItsGasNH(GasSensor*  const  me,  struct  GasNotificationHandle  *
p_GasNotificationHandle);
void  GasSensor_removeItsGasNH(GasSensor*  const  me,  struct  GasNotificationHandle  *
p_GasNotificationHandle);
void GasSensor_clearItsGasNH(GasSensor* const me);
```

```
GasSensor * GasSensor_Create(void);
void GasSensor_Destroy(GasSensor* const me);
#endif
```

Code Listing 3-11: GasSensor.h

```
#include "GasSensor.h"
#include "GasData.h"
#include "GasNotificationHandle.h"

static void cleanUpRelations(GasSensor* const me);
void GasSensor_Init(GasSensor* const me) {
        me->itsGasData = NULL;
        int pos;
        for(pos = 0; pos < 100; ++pos) {
            me->itsGasNH[pos] = NULL;
            }
}

void GasSensor_Cleanup(GasSensor* const me) {
        cleanUpRelations(me);
}

void GasSensor_dumpList(GasSensor* const me) {
        int pos;
        char s[100];
        printf("Dumping registered elements/n");
        for (pos=0; pos<100; pos++)
            if (me->itsGasNH[pos])
                if (me->itsGasNH[pos]->acceptorPtr) {
                    printf("Client %d: InstancePtr=%p, acceptPtr=%p/n",
                        pos,me->itsGasNH[pos]->instancePtr, me->itsGasNH[pos]-
>acceptorPtr);
                };
}

void GasSensor_newData(GasSensor* const me, unsigned int flow, unsigned int n2, unsigned
int o2) {
        if (!me->itsGasData)
            me->itsGasData = GasData_Create();
        me->itsGasData->flowRate = flow;
        me->itsGasData->N2Conc = n2;
        me->itsGasData->O2Conc = o2;
        GasSensor_notify(me);
}

void GasSensor_notify(GasSensor* const me) {
        int pos;
        char s[100];
        for (pos=0; pos<100; pos++)
            if (me->itsGasNH[pos])
                if (me->itsGasNH[pos]->acceptorPtr)
```

```
                        me->itsGasNH[pos]->acceptorPtr(me->itsGasNH[pos]->instancePtr,
me->itsGasData);
}

void GasSensor_subscribe(GasSensor* const me, void * instancePtr, const gasDataAccep-
torPtr* aPtr) {
        struct GasNotificationHandle* gnh;
        gnh = GasNotificationHandle_Create();
        gnh->instancePtr = instancePtr;
        gnh->acceptorPtr = aPtr;
        GasSensor_addItsGasNH(me, gnh);
}

void GasSensor_unsubscribe(GasSensor* const me, const gasDataAcceptorPtr* aPtr) {
        int pos;
        for(pos = 0; pos < 100; ++pos) {
            if (me->itsGasNH[pos])
                if (me->itsGasNH[pos]->acceptorPtr == aPtr) {
                    GasNotificationHandle_Destroy(me->itsGasNH[pos]);
                    me->itsGasNH[pos] = NULL;
                }
        }
}

struct GasData* GasSensor_getItsGasData(const GasSensor* const me) {
        return (struct GasData*)me->itsGasData;
}

void GasSensor_setItsGasData(GasSensor* const me, struct GasData* p_GasData) {
        me->itsGasData = p_GasData;
}

int GasSensor_getItsGasNH(const GasSensor* const me) {
        int iter = 0;
        return iter;
}

void GasSensor_addItsGasNH(GasSensor* const me, struct GasNotificationHandle *
p_GasNotificationHandle) {
        int pos;
        for(pos = 0; pos < 100; ++pos) {
            if (!me->itsGasNH[pos]) {
                me->itsGasNH[pos] = p_GasNotificationHandle;
                break;
            }
        }
}

void GasSensor_removeItsGasNH(GasSensor* const me, struct GasNotificationHandle *
p_GasNotificationHandle) {
        int pos;
        for(pos = 0; pos < 100; ++pos) {
            if (me->itsGasNH[pos] == p_GasNotificationHandle) {
```

```
                        me->itsGasNH[pos] = NULL;
            }
        }
    }
    void GasSensor_clearItsGasNH(GasSensor* const me) {
        int pos;
        for(pos = 0; pos < 100; ++pos)
        {
            me->itsGasNH[pos] = NULL;
        }
    }
    GasSensor * GasSensor_Create(void) {
        GasSensor* me = (GasSensor *) malloc(sizeof(GasSensor));
        if(me!=NULL) {
                    GasSensor_Init(me);
        }
        return me;
    }
    void GasSensor_Destroy(GasSensor* const me) {
        if(me!=NULL) {
                    GasSensor_Cleanup(me);
        }
        free(me);
    }
    static void cleanUpRelations(GasSensor* const me) {
        if(me->itsGasData != NULL) {
                    me->itsGasData = NULL;
                }
    }
```

Code Listing 3-12: GasSensor.c

Figure 3-9 shows that our example system has three clients – the DisplayClient, the GasMix-erClient, and the SafetyMonitorClient. In addition to the accept() function, the DisplayMoni-torClient class includes a show() function that displays the data when its accept() function is called, an alarm function (invoked by the SafetyMonitorClient when appropriate), and a register() function that calls the subscribe function of the GasSensor. The code for the DisplayClient is given in Code Listing 3-13 and Code Listing 3-14.

```
#ifndef DisplayClient_H
#define DisplayClient_H
#include "GasSensor.h"

typedef struct DisplayClient DisplayClient;
struct DisplayClient {
        struct GasData* itsGasData;
        struct GasSensor* itsGasSensor;
};
```

```
/* Constructors and destructors:*/
void DisplayClient_Init(DisplayClient* const me);
void DisplayClient_Cleanup(DisplayClient* const me);
/* Operations */
void DisplayClient_accept(DisplayClient* const me, struct GasData* g);
void DisplayClient_alarm(DisplayClient* const me, char* alarmMsg);
void DisplayClient_register(DisplayClient* const me);
void DisplayClient_show(DisplayClient* const me);
struct GasData* DisplayClient_getItsGasData(const DisplayClient* const me);
void DisplayClient_setItsGasData(DisplayClient* const me, struct GasData* p_GasData);
struct GasSensor* DisplayClient_getItsGasSensor(const DisplayClient* const me);
void DisplayClient_setItsGasSensor(DisplayClient* const me, struct GasSensor* p_GasSensor);
DisplayClient * DisplayClient_Create(void);
void DisplayClient_Destroy(DisplayClient* const me);
#endif
```

Code Listing 3-13: DisplayClient.h

```
#include "DisplayClient.h"
#include "GasData.h"
#include "GasSensor.h"

static void cleanUpRelations(DisplayClient* const me);
void DisplayClient_Init(DisplayClient* const me) { me->itsGasData = NULL;
me->itsGasSensor = NULL;
}

void DisplayClient_Cleanup(DisplayClient* const me) {
        cleanUpRelations(me);
}

void DisplayClient_accept(DisplayClient* const me, struct GasData* g) {
        if (!me->itsGasData)
            me->itsGasData = GasData_Create();
        if (me->itsGasData) {
            me->itsGasData->flowRate = g->flowRate;
            me->itsGasData->N2Conc = g->N2Conc;
            me->itsGasData->O2Conc = g->O2Conc;
            DisplayClient_show(me);
        };
}

void DisplayClient_alarm(DisplayClient* const me, char* alarmMsg)
{
        printf("ALERT! ");
        printf(alarmMsg);
        printf("/n/n");
}
void DisplayClient_register(DisplayClient* const me) {
        if (me->itsGasSensor)
            GasSensor_subscribe(me->itsGasSensor, me,&DisplayClient_accept);
}
```

```c
void DisplayClient_show(DisplayClient* const me) {
        if (me->itsGasData) {
            printf("Gas Flow Rate = %5d/n", me->itsGasData->flowRate);
            printf("O2 Concentration = %2d/n", me->itsGasData->N2Conc);
            printf("N2 Concentration = %2d/n/n", me->itsGasData->N2Conc);
        }
        else
            printf("No data available/n/n");
}
struct GasData* DisplayClient_getItsGasData(const DisplayClient* const me) {
        return (struct GasData*)me->itsGasData;
}
void DisplayClient_setItsGasData(DisplayClient* const me, struct GasData* p_GasData) {
        me->itsGasData = p_GasData;
}
struct GasSensor* DisplayClient_getItsGasSensor(const DisplayClient* const me) {
        return (struct GasSensor*)me->itsGasSensor;
}
void DisplayClient_setItsGasSensor(DisplayClient* const me, struct GasSensor* p_GasSen-
sor) {
        me->itsGasSensor = p_GasSensor;
}
DisplayClient * DisplayClient_Create(void) {
        DisplayClient* me = (DisplayClient *)
malloc(sizeof(DisplayClient));
        if(me!=NULL) {
                    DisplayClient_Init(me);
        }
        return me;
}
void DisplayClient_Destroy(DisplayClient* const me) {
        if(me!=NULL) {
                    DisplayClient_Cleanup(me);
        }
        free(me);
}
static void cleanUpRelations(DisplayClient* const me) {
        if(me->itsGasData != NULL) {
                    me->itsGasData = NULL;
        }
        if(me->itsGasSensor != NULL) {
                    me->itsGasSensor = NULL;
        }
}
```

Code Listing 3-14: DisplayClient.c

Lastly, let's look at the code for the GasData class. This is a very straightforward data class with just a few helper functions, as seen in Code Listing 3-15 and Code Listing 3-16.

```
ifndef GasData_H
#define GasData_H

typedef struct GasData GasData;
struct GasData {
        unsigned short N2Conc;
        unsigned short O2Conc;
        unsigned int flowRate;
};

/* Constructors and destructors:*/
oid GasData_Init(GasData* const me);
void GasData_Cleanup(GasData* const me);
GasData * GasData_Create(void);
void GasData_Destroy(GasData* const me);
#endif
```

Code Listing 3-15: GasData.h

```
#include "GasData.h"
void GasData_Init(GasData* const me) {
}

void GasData_Cleanup(GasData* const me) {
}

GasData * GasData_Create(void) {
        GasData* me = (GasData *) malloc(sizeof(GasData));
        if(me!=NULL) {
                    GasData_Init(me);
        }
        return me;
}

void GasData_Destroy(GasData* const me) {
        if(me!=NULL) {
                    GasData_Cleanup(me);
        }
        free(me);
}
```

Code Listing 3-16: GasData.c

3.6 Debouncing Pattern

This simple pattern is used to reject multiple false events arising from intermittent contact of metal surfaces.

3.6.1 Abstract

Push buttons, toggle switches, and electromechanical relays are input devices for digital systems that share a common problem – as metal connections make contact, the metal deforms or "bounces", producing intermittent connections during switch open or closure. Since this happens very slowly (order of milliseconds) compared to the response speed of embedded systems (order of microseconds or faster), this results in multiple electronic signals to the control system. This pattern addresses this concern by reducing the multiple signals into a single one by waiting a period of time after the initial signal and then checking the state.

3.6.2 Problem

Many input devices for embedded systems use metal-on-metal contact to indicate events of interest, such as button presses, switch movement, and activating or deactivating relays. As the metal moves into contact, physical deformation occurs resulting in an intermittent bouncing contact until the vibrations dampen down. This results in an intermediate contact profile such as that shown in Figure 3-10.

3.6.3 Pattern Structure

The basic solution is to accept the initial event, wait for the vibrations to dampen out, and then sample the data source for its state. See Figure 3-11 for the pattern structure.

3.6.4 Collaboration Roles

This section describes the roles for this pattern.

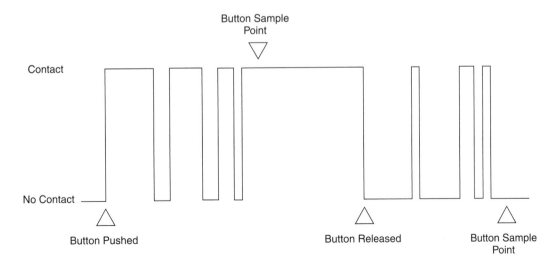

Figure 3-10 Bouncing electrical contacts

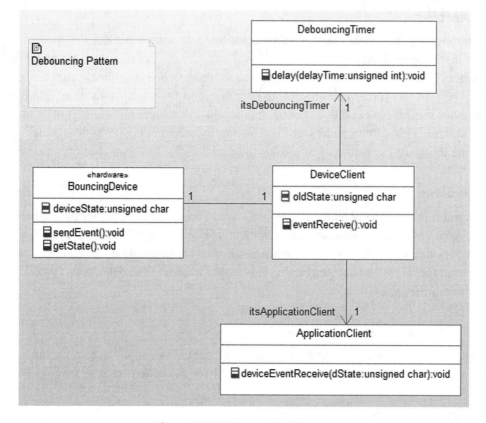

Figure 3-11 Debouncing Pattern

3.6.4.1 ApplicationClient

This element is the ultimate recipient of the debounced event. Its operation deviceEventReceive () is only activated when the event is real (i.e., results in a changed device state).

3.6.4.2 BouncingDevice

The BouncingDevice represents the hardware for the device itself. The most common implementation of this device is completely in hardware; in the case, the sendEvent() operation is simply an activation of an interrupt in the interrupt vector table of the embedded processor and the getState operation is implemented via a read of a memory location or IO port. DeviceState is normally a bivalued attribute, ON or OFF.

3.6.4.3 Debouncer

The Debouncer is the software element that processes the incoming event, debounces it, and makes sure that it represents an actual device state change. Its eventReceive() function is activated by the sendEvent() service of the BouncingDevice. In turn, it sets the delay timer (disabling interrupts from the device if necessary) and then checks the device state. If, after the debouncing time, the state is different then the event must have been real, so it sends the appropriate message to the ApplicationClient. The old state of the button is stored in its variable oldState; this variable is updated whenever there is a detected change of state of the device.

3.6.4.4 DebouncingTimer

This timer provides a nonbusy wait via its `delay()` service. This is often done with an OS call but might be done with specialty timer hardware as well.

3.6.5 Consequences

This pattern is not necessary for hardware that debounces before announcing events. In my experience, however, it often falls on the software to perform this task. This is a simple pattern that performs the debouncing so that the application need only concern itself with true events resulting from a change in device state.

3.6.6 Implementation Strategies

It is common for the device to use an interrupt vector to contact its client. Indeed, it is common for the BouncingDevice to be entirely implemented in hardware so that when the event occurs, a hardware interrupt is created, and the device state must be verified by reading a memory location or port address. When the event interface is via an interrupt, the address of the DeviceClient:: eventReceive() operation must be installed in the proper location in the interrupt vector table when the DeviceClient is initiated, and the address must be removed if the DeviceClient is deactivated. Care must be taken to ensure that the eventReceive operation has no parameters (not even hidden ones!) so that the Return from Interrupt instruction will work properly. If the hidden me parameter is used as in the other patterns here, then making the operation static will ensure there is no hidden parameter because static operations are class-wide and not instance-wide functions (that is, they use no instance data and so don't require a me pointer).

If the interrupt vector table is not used, this pattern can be mixed with the Observer Pattern to distribute the event signal to multiple clients, if desired.

The DebouncingTimer may use special timer hardware such as a programmable 555 timer. If the RTOS (Real-Time Operating System) timer is used, care must be taken with the time unit resolution. Windows, for example, has a standard timer resolution of between 10 and 25 milliseconds. You can only get a delay time as a multiple of the basic timer resolution so if you want a 45 ms delay, you will have to use the closest timer resolution that is greater than or equal to your desired time. For debouncing applications, this is usually fine but your needs may vary with your specific hardware and application.

If you don't mind fully occupying your embedded processor while you wait for the debounce timeout, it is a simple matter, as shown in Code Listing 3-17.

```
/* LOOPS_PER_MS is the # of loops in the delay() function
   required to hit 1 ms, so it is processor and
   compiler-dependent
*/
#define LOOPS_PER_MS 1000
void delay(unsigned int ms) {
    long nLoops = ms * LOOPS_PER_MS;
    do {
    while (nLoops-);
    }
```

Code Listing 3-17: Busy wait timing

3.6.7 Related Patterns

As mentioned, the pattern is often used in conjunction with the Interrupt Pattern discussed elsewhere in this chapter. Timeouts are often shown within a tm() event on the state machine.

3.6.8 Example

In the example shown in Figure 3-12, we are modeling a toggle button for a microwave oven; press it once to turn on the microwave emitter and press it again to turn it off. This means that we need to call different operations in the MicrowaveEmitter depending if this is an even-numbered press of the button or an odd-numbered press. In addition, we have to debounce both the press and release of the button, although we only act on a button release.

```
The code for the button driver is given in Code Listing 18 and Code Listing 19. #ifndef
ButtonDriver_H
#define ButtonDriver_H

#define LOOPS_PER_MS (1000)
#define DEBOUNCE_TIME (40)

struct Button;
struct MicrowaveEmitter;
struct Timer;

typedef struct ButtonDriver ButtonDriver;
struct ButtonDriver {
      unsigned char oldState;
      unsigned char toggleOn;
      struct Button* itsButton;
      struct MicrowaveEmitter* itsMicrowaveEmitter;
      struct Timer* itsTimer;
};
```

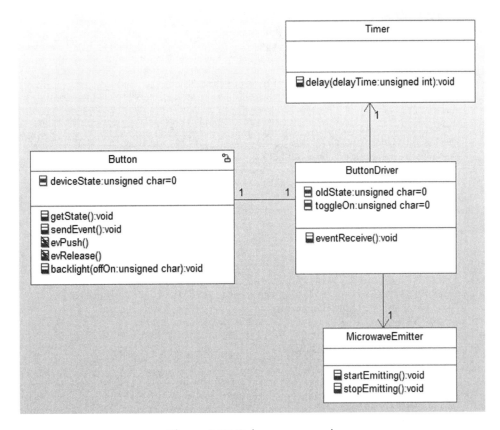

Figure 3-12 Debounce example

```
void ButtonDriver_Init(ButtonDriver* const me);
void ButtonDriver_Cleanup(ButtonDriver* const me);
/* Operations */ void ButtonDriver_eventReceive(ButtonDriver* const me);
struct Button* ButtonDriver_getItsButton(const ButtonDriver* const me);
void ButtonDriver_setItsButton(ButtonDriver* const me, struct Button* p_Button);
struct MicrowaveEmitter*
ButtonDriver_getItsMicrowaveEmitter(const ButtonDriver* const me);
void ButtonDriver_setItsMicrowaveEmitter(ButtonDriver* const me, struct MicrowaveEmitter*
p_MicrowaveEmitter);
struct Timer* ButtonDriver_getItsTimer(const ButtonDriver* const me);
void ButtonDriver_setItsTimer(ButtonDriver* const me, struct Timer* p_Timer);
ButtonDriver * ButtonDriver_Create(void);
void ButtonDriver_Destroy(ButtonDriver* const me);
void ButtonDriver___setItsButton(ButtonDriver* const me, struct Button* p_Button);
void ButtonDriver__setItsButton(ButtonDriver* const me, struct Button* p_Button);
void ButtonDriver__clearItsButton(ButtonDriver* const me);
#endif
```

Code Listing 3-18: ButtonDriver.h

```c
#include "ButtonDriver.h"
#include "Button.h"
#include "MicrowaveEmitter.h"
#include "Timer.h"

static void cleanUpRelations(ButtonDriver* const me);
void ButtonDriver_Init(ButtonDriver* const me) {
        me->oldState = 0;
        me->toggleOn = 0;
        me->itsButton = NULL;
        me->itsMicrowaveEmitter = NULL;
        me->itsTimer = NULL;
}

void ButtonDriver_Cleanup(ButtonDriver* const me) {
        cleanUpRelations(me);
}

void ButtonDriver_eventReceive(ButtonDriver* const me) {
        Timer_delay(me->itsTimer, DEBOUNCE_TIME);
        if (Button_getState(me->itsButton) != me->oldState) {
            /* must be a valid button event */
            me->oldState = me->itsButton->deviceState;
            if (!me->oldState) {
                /* must be a button release, so update toggle value */
                if (me->toggleOn) {
                    me->toggleOn = 0; /* toggle it off */
                    Button_backlight(me->itsButton, 0);
                    MicrowaveEmitter_stopEmitting(me->itsMicrowaveEmitter);
                    }
            else {
                    me->toggleOn = 1; /* toggle it on */
                    Button_backlight(me->itsButton, 1);
                    MicrowaveEmitter_startEmitting(me->itsMicrowaveEmitter);
                    }
                }

            /* if it's not a button release, then it must
               be a button push, which we ignore.
            */
        }
}

struct Button* ButtonDriver_getItsButton(const ButtonDriver* const me) {
        return (struct Button*)me->itsButton;
}

void ButtonDriver_setItsButton(ButtonDriver* const me, struct Button* p_Button) {
        if(p_Button != NULL)
                {
                        Button__setItsButtonDriver(p_Button, me);
                }
```

```
                ButtonDriver__setItsButton(me, p_Button);
}
struct MicrowaveEmitter*
ButtonDriver_getItsMicrowaveEmitter(const ButtonDriver* const me)
{
        return (struct MicrowaveEmitter*)me->itsMicrowaveEmitter;
}
void ButtonDriver_setItsMicrowaveEmitter(ButtonDriver* const me, struct MicrowaveEmit-
ter* p_MicrowaveEmitter) {
        me->itsMicrowaveEmitter = p_MicrowaveEmitter;
}
struct Timer* ButtonDriver_getItsTimer(const ButtonDriver* const me) {
        return (struct Timer*)me->itsTimer;
}
void ButtonDriver_setItsTimer(ButtonDriver* const me, struct Timer* p_Timer) {
        me->itsTimer = p_Timer;
}
ButtonDriver * ButtonDriver_Create(void) {
        ButtonDriver* me = (ButtonDriver *)
        malloc(sizeof(ButtonDriver));
        if(me!=NULL)
                {
                        ButtonDriver_Init(me);
                }
        return me;
}
void ButtonDriver_Destroy(ButtonDriver* const me) {
        if(me!=NULL)
                {
                        ButtonDriver_Cleanup(me);
                }
        free(me);
}
static void cleanUpRelations(ButtonDriver* const me) {
        if(me->itsButton != NULL)
                {
                        struct ButtonDriver* p_ButtonDriver = Button_getItsButtonDriver(me-
>itsButton);
                        if(p_ButtonDriver != NULL)
                                {
                                        Button___setItsButtonDriver(me->itsButton, NULL);
                                }
                        me->itsButton = NULL;
                }
        if(me->itsMicrowaveEmitter != NULL)
                {
                        me->itsMicrowaveEmitter = NULL;
                }
```

```
        if(me->itsTimer != NULL)
            {
                    me->itsTimer = NULL;
            }
    }
    void ButtonDriver___setItsButton(ButtonDriver* const me, struct Button* p_Button) {
        me->itsButton = p_Button;
    }
    void ButtonDriver__setItsButton(ButtonDriver* const me, struct Button* p_Button) {
        if(me->itsButton != NULL)
            {
                    Button___setItsButtonDriver(me->itsButton, NULL);
            }
        ButtonDriver___setItsButton(me, p_Button);
    }
    void ButtonDriver__clearItsButton(ButtonDriver* const me) {
        me->itsButton = NULL;
    }
```

Code Listing 3-19: ButtonDriver.c

3.7 Interrupt Pattern

The physical world is fundamentally both concurrent and asynchronous; it's nonlinear too, but that's a different story. Things happen when they happen and if your embedded system isn't paying attention, those occurrences may be lost. Interrupt handlers (a.k.a. Interrupt Service Routines, or ISRs) are a useful way to be notified when an event of interest occurs even if your embedded system is off doing other processing.

3.7.1 Abstract

The Interrupt Pattern is a way of structuring the system to respond appropriately to incoming events. It does require some processor- and compiler-specific services but most embedded compilers, and even some operating systems provide services for just this purpose. Once initialized, an interrupt will pause its normal processing, handle the incoming event, and then return the system to its original computations.

3.7.2 Problem

In many systems, events have different levels of urgency[14]. Most embedded systems have at least some events with a high urgency that must be handled even when the system is busy doing

---

[14] Urgency is a measure of the rapidity with which an event must be handled, usually expressed as the nearness of a deadline. This topic will be discussed in more detail in Chapter 4.

other processing. The Polling Pattern, discussed elsewhere in this chapter, looks for events of interest when it is convenient for the system. While this has the advantage that primary processing can proceed uninterrupted, it has the disadvantage that high urgency and high frequency events may not be handled in a timely fashion, or may be missed altogether. The Interrupt Pattern addresses this problem by immediately pausing the current processing, handling the incoming event, and then returning to the original computation.

3.7.3 Pattern Structure

Figure 3-13 shows the basic structure of the pattern. Because it is important to ensure that the interrupt handlers are parameterless, I've used the «File» stereotype to indicate that this pattern isn't using the class-structured approach with the me pointer.

3.7.4 Collaboration Roles

This section describes the roles for this pattern.

3.7.4.1 InterruptHandler

The interruptHandler is the only element that has behavior in the pattern. It has functions to install or deinstall an interrupt vector and the interrupt service routines themselves.

The install() function takes the interrupt number as an incoming parameter. When run, it copies the existing vector into the vector table and then replaces it with the address of the appropriate interrupt service routine. The deinstall() function does the reverse, restoring the original vector.

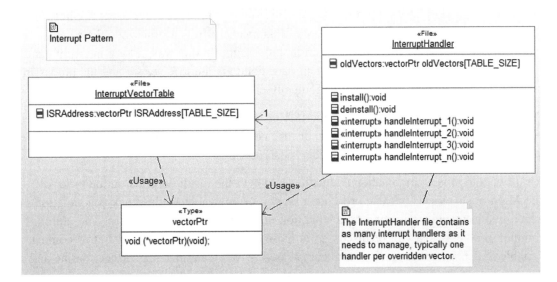

Figure 3-13 Interrupt Pattern

Each handleIinterrupt_x() function handles a specific interrupt, completing with a Return From Interrupt (RTI) statement. This statement is compiler and processor dependent. As mentioned, it is crucial that the interrupt service routines have no parameters because otherwise when they attempt to return, the wrong value will be popped off the CPU stack.

3.7.4.2 InterruptVectorTable

The InterruptVectorTable is nothing more than an array of addresses to interrupt service routines. It is located in a specific memory location that is processor dependent. When interrupt number x occurs, the CPU suspends the current processing and indirectly invokes the address corresponding to the x-th index in this table. Upon an RTI, the CPU restores execution of the suspending computation.

3.7.4.3 vectorPtr

The vectorPtr is a data type; specifically it is a pointer to a function that takes no parameters and returns no values.

3.7.5 Consequences

This pattern allows for highly responsive processing of events of interest. It interrupts normal processing (provided that interrupts have not been disabled) and so should be used carefully when time-critical processing is going on. Normally, interrupts are disabled while an interrupt service routine is executing; this means that interrupt service routines must execute very quickly to ensure that other interrupts are not missed.

Because the interrupt service routines must be short, one must take care when using them to invoke other system services. To share data signaled by an interrupt, for example, the ISR might need to queue the data and quickly RTI; at some point in the future, the application software will discover the data in the queue. This mechanism is useful when the actual acquisition of the data is more urgent than its processing. In this way, a long interrupt handle can be broken up into two parts; the urgent part, done via the ISR *per se*, and the processing part done via a second function that periodically checks for waiting data or signals.

Problems arise with this pattern when the ISR processing takes too long, when an implementation mistake leaves interrupts disabled, or race conditions or deadlocks occur on shared resources. The last of these concerns is the most insidious.

A race condition is a computational situation in which the result depends on the order of execution of statements but that order isn't known or knowable. A deadlock is a situation in which two elements are waiting on a condition that cannot in principle occur. These are common problems that will be discussed in more detail in the next chapter. In this context, a variable or data structure (such as a queue) shared between an interrupt service routine and an application service is a resource with potential race and deadlock conditions because you can never know exactly when the ISR will be executed.

Figure 3-14 ISR potential race condition

Figure 3-14 shows a typical structure when the data acquired as a part of an ISR must be shared with normal system processing. The race condition arises when the interrupt occurs when ApplicationElement is accessing the SharedResource. Imagining that the ApplicationElement is halfway through reading the data when the interrupt occurs – unless the read takes place as an atomic uninterruptible access – the ISR will pause the ApplicationElement mid-read, modify the data, and then return. The ApplicationElement will see corrupted data – partially new data and partially old data.

There are a number of solutions to this, but they all involve serializing the access to the shared resource. One way is to disable interrupts in the ApplicationElement just prior to reading the data and to reenable interrupt just after access is complete. The primary downside to that approach is the discipline necessary to ensure that this is always done. Another approach is to use a mutex semaphore to guard the data, as shown in Figure 3-15.

In this figure, the SharedResource is protected with a mutex semaphore; this results in a lock when either the getData() or setData() functions are called and in the removal of the lock when the functions are complete. Deadlock can occur if the ISR waits on the semaphore lock when it tries to access the data. Since the ISR has interrupted the ApplicationElement that owns the

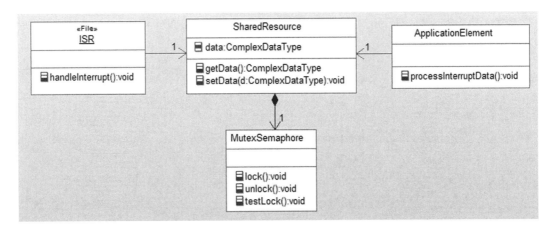

Figure 3-15 ISR potential deadlock condition

lock, if the ISR waits, the ApplicationElement will never have the opportunity to remove the lock. The solution of course, is that the ISR cannot wait on the lock. The new data can be discarded or two shared resources can be made with the proviso that the ISR or ApplicationElement can only lock one of the resources at a time. This latter solution is sometimes referred to as a "ping pong buffer."

3.7.6 Implementation Strategies

An ISR is different than a normal C function in that as long as interrupts are enabled, it can pretty much interrupt between any pair of CPU instructions. That means that the CPU registers must be saved and restored before the ISR statements can execute.

In fact, each interrupt service routine must:

*   save the CPU registers, including the CPU instruction pointer and any processor flags, such as carry, parity and zero
*   clear the interrupt bit
*   perform the appropriate processing
*   restore the CPU registers
*   return

In "standard C," it is necessary to use assembly language keyword asm to save and restore registers, something like[15]:

```
void isr (void) {
    asm {
            DI                  ; disable interrupts
            PUSH     AF    ; save registers
            PUSH     BC
            PUSH     DE
            PUSH     HL
            PUSH     IX
            PUSH     IY
            }

    /* normal C code here */
    asm {
            POP      IY
            POP      IX
            POP      HL
            POP      DE
            POP      BC
            POP      AF
            EI                   ; enable interrupts
```

---

[15] This is Z80 assembly language (my personal favorite). Your processor mileage may vary.

```
        RETI        ; return from interrupt
        }
    }
```

Some compilers support this by using a special `interrupt` keyword. When available, the keyword is used in the function declaration such as

```
interrupt void isr(void) {
    /* normal C code here */
};
```

and the compiler will insert the necessary instructions for you.

The GNU C Compiler (GCC) uses __`attribute`__ specification to indicate an interrupt handler using the syntax

```
void isr(void) __attribute__ ((interrupt ("IRQ"));
void isr(void) {
    /* normal C code here */
}
```

but again this is processor dependent.

The interrupt keyword tells the compiler to generate register-saving instructions before the body of the function is executed and to restore the registers just before the ISRs return.

Many RTOSs provide functions for installing the vectors and some CPUs only have a single vector, requiring the ISR to figure out which interrupt occurred by checking the hardware.

3.7.7 Related Patterns

An alternative means to get hardware signals and data is to periodically check for them via the Polling Pattern.

3.7.8 Example

In the example shown in Figure 3-16, a button is the hardware device. One interrupt (index 0) is generated for the push and another (index 1) is generated for the release. The ButtonHandler is the interrupt handler for the interrupts and its install() functions set up the vectors for both pointers. The main() routine does the following:

- initializes the Button and LED classes
- sets all the elements of the RobotInterruptVectorTable and ButtonHandler::oldVectors to NULL
- sets the pointer from the ButtonHandler to the LED
- calls the install() function to set up the vectors to the two ISRs

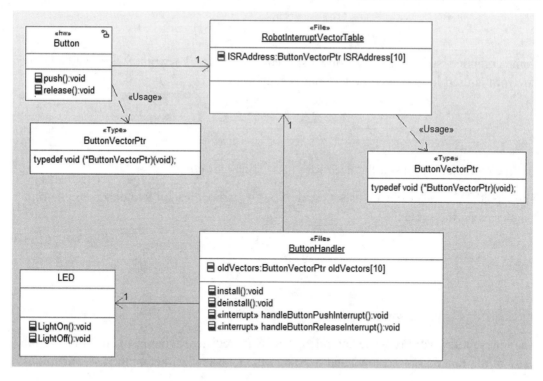

Figure 3-16 Interrupt Example

as you can see in Code Listing 20

```
int j;
Button itsButton;
LED itsLED;
itsButton = Button_Create();
itsLED = LED_Create();
for (j=0;j<9;j++) {
    ISRAddress[j] = NULL;
    oldVectors[j] = NULL;
    };
ButtonHandler_setItsLED(&itsLED);
install(); /* install interrupt vectors */
/* normal system execution stuff */
```

Code Listing 20: Interupt example main() code

The interesting part is the ButtonHandler itself. It installs the address of the two ISRs into positions 0 and 1 of the interrupt vector table so that when the interrupts occur, the appropriate

ISR is invoked. In the example, when the button is pushed, the ISR lights the LED and turns off the light when the button is released.

Note that this code uses the interrupt keyword to indicate to the compiler the interrupt service routines; you will have to use the appropriate syntax provided by your compiler. To execute this on standard desktop hardware, simply remove the interrupt keyword from the ISR's function headers and then the interrupt vector table will operate as a table of normal function pointers.

```
#ifndef ButtonHandler_H
#define ButtonHandler_H

typedef void (*ButtonVectorPtr)(void);
struct LED;

extern ButtonVectorPtr oldVectors[10];
/* Operations */
void install(void);
void deinstall(void);

interrupt void handleButtonPushInterrupt(void);
interrupt void handleButtonReleaseInterrupt(void);
struct LED* ButtonHandler_getItsLED(void); void ButtonHandler_setItsLED(struct
LED* p_LED);
#endif
```

Code Listing 21: ButtonHandler.h

```
#include "ButtonHandler.h"
#include "LED.h"
#include "RobotInterruptVectorTable.h"

ButtonVectorPtr oldVectors[10];
static struct LED* itsLED;
void deinstall(void) {
      ISRAddress[0] = oldVectors[0];
      ISRAddress[1] = oldVectors[1];
}

interrupt void handleButtonPushInterrupt(void) {
      LED_LightOn(itsLED);
}

interrupt void handleButtonReleaseInterrupt(void) {
      LED_LightOff(itsLED);
}

void install(void) {
      oldVectors[0] = ISRAddress[0];
      oldVectors[1] = ISRAddress[1];
      ISRAddress[0] = handleButtonPushInterrupt;
      ISRAddress[1] = handleButtonReleaseInterrupt;
}
```

```
struct LED* ButtonHandler_getItsLED(void) {
      return (struct LED*)itsLED;
}

void ButtonHandler_setItsLED(struct LED* p_LED) {
      itsLED = p_LED;
}
```

Code Listing 22: ButtonHandler.c

3.8 Polling Pattern

Another common pattern for getting sensor data or signals from hardware is to check periodi-cally, a process known as *polling*. Polling is useful when the data or signals are not so urgent that they cannot wait until the next polling period to be received or when the hardware isn't capable of generating interrupts when data or signals become available.

3.8.1 Abstract

The Polling Pattern is the simplest way to check for new data or signals from the hardware. Polling can be periodic or opportunistic; periodic polling uses a timer to indicate when the hardware should be sampled whereas opportunistic polling is done when it is convenient for the system, such as between major system functions or at some point in a repeated execution cycle. Opportunistic polling is, by definition, less regular but has less impact on the timeliness of other activities in which the system may be engaged.

3.8.2 Problem

The Polling Pattern addresses the concern of getting new sensor data or hardware signals into the system as it runs when the data or events are not highly urgent and the time between data sampling can be guaranteed to be fast enough.

3.8.3 Pattern Structure

This pattern comes in two flavors. Figure 3-17 shows the pattern structure for opportunistic polling while Figure 3-18 shows the pattern for periodic polling. The difference is that in the former pattern the applicationFunction will embed calls to poll() when convenient, and in the latter, a timer is created to start polling.

3.8.4 Collaboration Roles

This section describes the roles for both variants of this pattern.

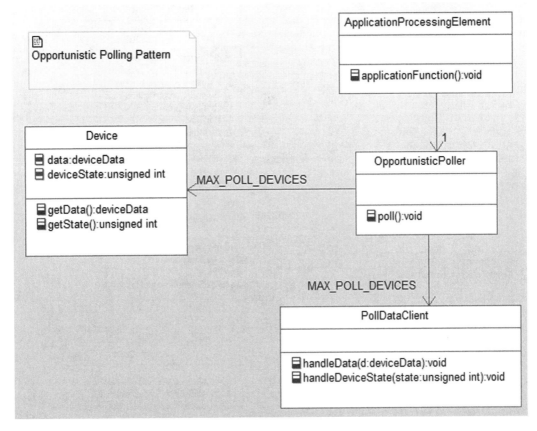

Figure 3-17 Opportunistic Polling Pattern

3.8.4.1 *ApplicationProcessingElement*

This elements has the applicationFunction that has a loop in which it invokes the poll() operation, such as

```
while (processing) {
      action1();
      action2();
      action3();
      OpportunityPoller_poll(me->itsOpportunisticPoller);
}
```

3.8.4.2 *Device*

The *device* provides data and/or device state information via accessible functions. This element can be a device driver or can simply read from memory- or port-mapped devices. This class provides two functions, one to retrieve data (typed as deviceData in the example) and one to retrieve device state (typed as an unsigned int in the example).

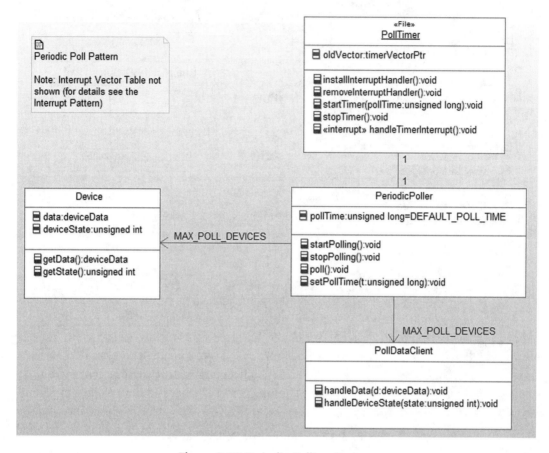

Figure 3-18 Periodic Polling Pattern

In the example, there are MAX_POLL_DEVICES connected to the polling element so that the poll() function scans them all and notifies their respective clients.

3.8.4.3 OpportunisticPoller

This element has the `poll()` function that scans the attached devices for data and device state and passes this information on to the appropriate client for each. The difference between the OpportunisticPoller and the PeriodicPoller is that the latter has timer initialization and shutdown functionality in addition to polling for the data.

3.8.4.4 PeriodicPoller

Like the `OpportunisticPoller`, the `PeriodicPoller` has the `poll()` function that scans the attached devices for data and device state and passes this information on to the appropriate client for each. In addition, it has a variable for the poll time (settable in its `setPollTime(t)` function) and services to start and stop polling. The `installInterruptTimer()` function inserts the address of the ISR into the interrupt vector table while the `removeInterruptHandler()` restores the original vector. The `startPolling()` function initializes the timer. The `stopPolling()` function stops the timer but leaves the vector in the interrupt vector table.

3.8.4.5 PollDataClient

This element is the client for data and state information from one or more of the devices. In some cases, there may be a single client for the data from all devices, but in general each device will have its own client.

3.8.4.6 PollTimer

The `PollTimer` element represents a timer and the services associated with using it. The startTimer() method installs the `handleTimerInterrupt()` interrupt service routine in the interrupt vector table and creates and initializes a timer with the `pollTime` attribute. Once the timer is started, it will set the hardware timer so that its ISR `handleTimerInterrupt()` will be invoked, which first resets the timer count and then calls the poll() function.

3.8.5 Consequences

Polling is simpler than the setup and use of the Interrupt Service Routines, although periodic polling is usually implemented with an ISR tied to a poll timer.

Polling can check many different devices at the same time for status changes but is usually less timely than interrupts. For this reason, care must be taken that if there are deadlines associated with the data or signals that the poll time plus the response time is always less than the deadlines involved. If data arrive faster than the poll time, then data will be lost. This is not a problem in many applications but is fatal (sometimes literally so) in others.

3.8.6 Implementation Strategies

The simplest implementation is to insert the hardware check in the middle of a main processing loop that executes as long as the system operates. This is called "symmetric opportunistic polling" because it operates the same way all the time, even though the length of time the processing loops take may vary widely. *Asymmetric opportunistic polling* refers to the practice of placing checks for new data at convenient but unrelated points throughout the processing. This approach allows for better tuning (if you need to be more responsive, add some more checks) but has a larger impact on the primary flow and is harder to maintain.

Periodic polling is polling that takes place with a regular time interval (known as the *period*) with a bounded variation (known as the *jitter*). To achieve this regularity, the checks for new data are initiated by a timer tied to an interrupt. Thus the periodic variant of the Polling Pattern is simply a special case of the Interrupt Pattern.

3.8.7 Related Patterns

Periodic polling is a special case of the Interrupt Pattern. In addition, the hardware checks may be done by invoking data acquisition services of Hardware Proxies or Hardware Adapters. In addition, the Observer Pattern can be merged in as well, with the polling element (OpportunisticPoller or PeriodicPoller) serving as the data server and the PollDataClients serving as the data clients.

3.8.8 Example

For the example shown in Figure 3-19, I decided to use the Periodic Polling Pattern to check for three breathing circuit sensors. The first of these sensors monitors the oxygen concentration in the breathing circuit, the second monitors the gas pressure (and holds the gas flow state), and the third monitors the pressure in the circuit. When the `BEPeriodicPoller` is created, its initializer (`BCPeriodicPoller_Init()`) calls the `BCTimer::installInterruptHandler()` service;

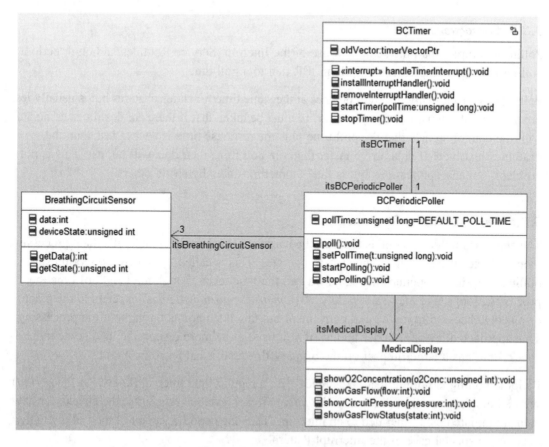

Figure 3-19 Polling example

similarly, the clean-up operation done when the BCPeriodicPoller is destroyed calls BCTimer_stopTiming() and then BCTimer_removeInterruptHandler().

At some point, the BCPeriodicPoller is commanded to start polling via a call to BCPeriodicPoller_startPolling(). This function calls BC_Timer_startTimer (timeout) service, passing the desired poll period. From then on, until commanded otherwise, the timer will fire periodically and invoke the timer interrupt handler BCTimer_handleTimeInterrupt(), which resets the timer and calls BCPeriodicPoller_poll(). The poll() function then polls the data and state information from the three devices and sends them to the MedicalDisplay.

```
#ifndef BCPeriodicPoller_H
#define BCPeriodicPoller_H

typedef int deviceData;
typedef void (*timerVectorPtr)(void);
#define MAX_POLL_DEVICES (10)
#define DEFAULT_POLL_TIME (1000)

struct BCTimer;
struct BreathingCircuitSensor;
struct MedicalDisplay;

typedef struct BCPeriodicPoller BCPeriodicPoller;
struct BCPeriodicPoller {
        unsigned long pollTime;
        struct BCTimer* itsBCTimer;
        struct BreathingCircuitSensor *itsBreathingCircuitSensor[3];
        struct MedicalDisplay* itsMedicalDisplay;
};

void BCPeriodicPoller_Init(BCPeriodicPoller* const me);
void BCPeriodicPoller_Cleanup(BCPeriodicPoller* const me);
/* Operations */ void BCPeriodicPoller_poll(BCPeriodicPoller* const me);
void BCPeriodicPoller_setPollTime(BCPeriodicPoller* const me, unsigned long t);
void BCPeriodicPoller_startPolling(BCPeriodicPoller* const me);
void BCPeriodicPoller_stopPolling(BCPeriodicPoller* const me);
struct BCTimer* BCPeriodicPoller_getItsBCTimer(const BCPeriodicPoller* const me);
void  BCPeriodicPoller_setItsBCTimer(BCPeriodicPoller*  const  me,  struct  BCTimer*
p_BCTimer);
int BCPeriodicPoller_getItsBreathingCircuitSensor(const BCPeriodicPoller* const me);
void BCPeriodicPoller_addItsBreathingCircuitSensor(BCPeriodicPoller* const me, struct
BreathingCircuitSensor * p_BreathingCircuitSensor);
void  BCPeriodicPoller_removeItsBreathingCircuitSensor(BCPeriodicPoller*  const  me,
struct BreathingCircuitSensor * p_BreathingCircuitSensor);
void BCPeriodicPoller_clearItsBreathingCircuitSensor(BCPeriodicPoller* const me);
struct MedicalDisplay* BCPeriodicPoller_getItsMedicalDisplay(const BCPeriodicPoller*
const me);
void BCPeriodicPoller_setItsMedicalDisplay(BCPeriodicPoller* const me, struct Medical-
Display* p_MedicalDisplay);
BCPeriodicPoller * BCPeriodicPoller_Create(void);
```

```
void BCPeriodicPoller_Destroy(BCPeriodicPoller* const me);
void BCPeriodicPoller___setItsBCTimer(BCPeriodicPoller* const me, struct BCTimer*
p_BCTimer);
void BCPeriodicPoller__setItsBCTimer(BCPeriodicPoller* const me, struct BCTimer*
p_BCTimer);
void BCPeriodicPoller__clearItsBCTimer(BCPeriodicPoller* const me);
#endif
```

Code Listing 3-23: BCPeriodicPoller.h

```
#include "BCPeriodicPoller.h"
#include "BCTimer.h"
#include "BreathingCircuitSensor.h"
#include "MedicalDisplay.h"

static void cleanUpRelations(BCPeriodicPoller* const me);
void BCPeriodicPoller_Init(BCPeriodicPoller* const me) { me->pollTime = DEFAULT_
POLL_TIME; me->itsBCTimer = NULL;
                int pos;
                for(pos = 0; pos < 3; ++pos) {
                    me->itsBreathingCircuitSensor[pos] = NULL;
                        me->itsMedicalDisplay = NULL;
        BCTimer_installInterruptHandler(me->itsBCTimer);
        me->pollTime = DEFAULT_POLL_TIME;
}

void BCPeriodicPoller_Cleanup(BCPeriodicPoller* const me) {
        BCTimer_stopTimer(me->itsBCTimer);
        BCTimer_removeInterruptHandler(me->itsBCTimer);
        cleanUpRelations(me);
}

void BCPeriodicPoller_poll(BCPeriodicPoller* const me) {
        int state, data;
        data = BreathingCircuitSensor_getData(me->itsBreathingCircuitSensor[0]);
        MedicalDisplay_showO2Concentration(me->itsMedicalDisplay, data);
        data = BreathingCircuitSensor_getData(me->itsBreathingCircuitSensor[1]);
        state = BreathingCircuitSensor_getState(me->itsBreathingCircuitSensor[1]);
        MedicalDisplay_showGasFlow(me->itsMedicalDisplay, data);
        MedicalDisplay_showGasFlowStatus(me->itsMedicalDisplay, state);
        data = BreathingCircuitSensor_getData(me->itsBreathingCircuitSensor[2]);
        MedicalDisplay_showCircuitPressure(me->itsMedicalDisplay, data);
}

void BCPeriodicPoller_setPollTime(BCPeriodicPoller* const me, unsigned long t) {
        me->pollTime = t;
}

void BCPeriodicPoller_startPolling(BCPeriodicPoller* const me) {
        BCTimer_startTimer(me->itsBCTimer, me->pollTime);
}
```

```
void BCPeriodicPoller_stopPolling(BCPeriodicPoller* const me) {
      BCTimer_stopTimer(me->itsBCTimer);
}
struct BCTimer* BCPeriodicPoller_getItsBCTimer(const BCPeriodicPoller* const me) {
      return (struct BCTimer*)me->itsBCTimer;
}

void BCPeriodicPoller_setItsBCTimer(BCPeriodicPoller* const me, struct BCTimer* p_
BCTimer) {
      if(p_BCTimer != NULL)
              {
                      BCTimer__setItsBCPeriodicPoller(p_BCTimer, me);
              }
      BCPeriodicPoller__setItsBCTimer(me, p_BCTimer);
}

int BCPeriodicPoller_getItsBreathingCircuitSensor(const BCPeriodicPoller* const me) {
      int iter = 0;
      return iter;
}

void BCPeriodicPoller_addItsBreathingCircuitSensor(BCPeriodicPoller* const me, struct
BreathingCircuitSensor * p_BreathingCircuitSensor) {
      int pos;
      for(pos = 0; pos < 3; ++pos) {
          if (!me->itsBreathingCircuitSensor[pos]) {
              me->itsBreathingCircuitSensor[pos] = p_BreathingCircuitSensor;
              break;
          }
      }
}

void  BCPeriodicPoller_removeItsBreathingCircuitSensor(BCPeriodicPoller*  const  me,
struct BreathingCircuitSensor * p_BreathingCircuitSensor) {
      int pos;
      for(pos = 0; pos < 3; ++pos) {
          if (me->itsBreathingCircuitSensor[pos] == p_BreathingCircuitSensor) {
              me->itsBreathingCircuitSensor[pos] = NULL;
          }
      }
}

void BCPeriodicPoller_clearItsBreathingCircuitSensor(BCPeriodicPoller* const me) {
      {
              int pos;
              for(pos = 0; pos < 3; ++pos)
              {
                  me->itsBreathingCircuitSensor[pos] = NULL;
              }
      }
}
```

```
struct MedicalDisplay* BCPeriodicPoller_getItsMedicalDisplay(const BCPeriodicPoller*
const me) {
        return (struct MedicalDisplay*)me->itsMedicalDisplay;
}

void BCPeriodicPoller_setItsMedicalDisplay(BCPeriodicPoller* const me, struct Medical-
Display* p_MedicalDisplay) {
        me->itsMedicalDisplay = p_MedicalDisplay;
}

BCPeriodicPoller * BCPeriodicPoller_Create(void) {
        BCPeriodicPoller* me = (BCPeriodicPoller *) malloc(sizeof(BCPeriodicPoller));
        if(me!=NULL)
                {
                        BCPeriodicPoller_Init(me);
                }
        return me;
}

void BCPeriodicPoller_Destroy(BCPeriodicPoller* const me) {
        if(me!=NULL)
                {
                        BCPeriodicPoller_Cleanup(me);
                }
        free(me);
}

static void cleanUpRelations(BCPeriodicPoller* const me) {
        if(me->itsBCTimer != NULL)
                {
                        struct BCPeriodicPoller* p_BCPeriodicPoller = BCTimer_
getItsBCPeriodicPoller(me->itsBCTimer);
                        if(p_BCPeriodicPoller != NULL)
                                {
                                 BCTimer___setItsBCPeriodicPoller(me->itsBCTimer, NULL);
                                }
                        me->itsBCTimer = NULL;
                }
        if(me->itsMedicalDisplay != NULL)
                {
                        me->itsMedicalDisplay = NULL;
                }
}

void  BCPeriodicPoller___setItsBCTimer(BCPeriodicPoller*  const  me,  struct  BCTimer*
p_BCTimer) {
        me->itsBCTimer = p_BCTimer;
}

void  BCPeriodicPoller__setItsBCTimer(BCPeriodicPoller*  const  me,  struct  BCTimer*
p_BCTimer) {
        if(me->itsBCTimer != NULL)
                {
```

```
                    BCTimer___setItsBCPeriodicPoller(me->itsBCTimer, NULL);
        }
      BCPeriodicPoller___setItsBCTimer(me, p_BCTimer);
  }

  void BCPeriodicPoller__clearItsBCTimer(BCPeriodicPoller* const me) {
      me->itsBCTimer = NULL;
  }
```

Code Listing 3-24: BCPeriodicPoller.c

3.9 So, What Did We Learn?

In this chapter, we've looked at a number of patterns useful for manipulating hardware. The Hardware Proxy Pattern focuses on encapsulation of hardware-specific details while the Hardware Adapter Pattern provides a straightforward means to adapt different but similar devices for the needs of the system. Many times, complex interactions of hardware are necessary for proper system functions, a benefit of the Mediator Pattern. The Observer Pattern supports dynamic addition and removal of clients for hardware data, such as from sensors.

The last three patterns provide different means for interfacing to hardware. The Debouncing Pattern handles the common problem of hardware that produces intermittent contact during open and closure of metal contact switches. The Interrupt Pattern supports the highly timely response to events of interest. The Polling Pattern repeatedly polls the hardware for new sensor data and device state information.

These are basic patterns in the sense that they all address low-level concerns about interfacing with the hardware. The more difficult design issues have to do with transforming the commands and data within the system to achieve the system goals. The next chapter provides patterns that deal with issues of concurrent execution, from basic concepts to scheduling execution, avoiding race conditions and deadlock, and synchronizing tasks when necessary.

Chapter 5 deals with finite state machines – their implementation and usage; while in Chapter 6 we look at safety and reliability.

CHAPTER 4

Design Patterns for Embedding Concurrency and Resource Management

Design Patterns for Embedded Systems in C
DOI:10.1016/B978-1-85617-707-8.00004-2
Copyright © 2011 Elsevier Inc.

Patterns in this chapter

Cyclic Executive Pattern – Schedule threads in an infinite loop

Static Priority Pattern – Schedule threads by priority

Critical Region Pattern – Protect resources by disabling task switching

Guarded Call Pattern – Protect resources through mutex semaphores

Queuing Pattern – Serialize access through queuing messages

Rendezvous Pattern – Coordinate complex task synchronization

Simultaneous Locking Pattern – Avoid deadlock by locking resources together

Ordered Locking Pattern – Avoid deadlock by locking resources only in specific orders

---

4.1 Basic Concurrency Concepts

Most embedded systems need to perform several activities simultaneously. The key to doing that is through the definition and management of the concurrency model of the system. In this section, I will provide an overview of the basic terms and concepts of concurrency to support the understanding and usage of the design patterns that come later in this chapter.

The most important terms to understand are given in the sidebar.

Concurrency Term Definitions

- *Action Sequence*: a series of actions, possibly with decision branch points, in which the order of execution is fully determined

- *Arrival Pattern*: how the event that begins a concurrency unit arrives; it may be periodic or aperiodic

- *Blocking*: see *priority inversion*

- *Blocking time*: the length of time a high-priority task is prevented from running because a lower-priority task owns a required resource

- *Concurrency*: the simultaneous execution of action sequences

- *Concurrency unit*: a set of actions executed sequentially in the same thread of execution. Also known as a *task* or *thread*

- *Criticality*: the importance of the completion of an action sequence

- *Deadline*: a point in time at which the completion of an action or action sequence becomes incorrect or irrelevant

- *Deadlock*: a situation in which multiple clients are simultaneously waiting on the same conditions but the conditions cannot, in principle, be satisfied

- *Execution time*: the length of time required for an action or action sequence

- *Interarrival time*: the time between task initiation for an aperiodic task

- *Jitter*: variation around the period of a periodic task

- *Multitasking Preemptive Scheduling*: a scheduling schema in which the highest priority task ready to run will do so, preempting any lower priority tasks that are currently running

- *Period*: the interval between periodic task invocations

- *Priority*: a numeric value used to determine which task, of a set of ready-to-run tasks, will execute preferentially

- *Priority inversion*: when a lower priority task is running even though a higher priority task is ready to run. Also known as *blocking*

- *Pseudoconcurrency*: execution of multiple tasks by executing a single task at a time and switching focus between tasks according to some scheduling policy

- *Race condition*: a computational situation in which the result depends upon the order of execution of actions, but that order is not inherently knowable

- *Real-Time Action*: an action for which timeliness is an aspect of correctness

- *Schedulability*: a task set is said to be schedulable if all of its tasks are timely

- *Synchronization Pattern*: how tasks will manage synchronization at specific points in their respective action sequences

- *Synchronization Point*: a specific action in a sequence at which the task will wait until other tasks reach corresponding synchronization points

- *Task*: see *concurrency unit*

- *Thread*: see *concurrency unit*

- *Timeliness*: the ability of a task to predictably complete prior to its deadline, or to meet its time-related performance requirements

- *Urgency*: the nearness of an action sequence's deadline

- *Worst Case Execution Time*: the longest time an action sequence execution can take

Figure 4-1 shows the relationship between actions, action sequences, and tasks[1]. Each task (also known as a concurrency unit) consists of a set of actions performed in a specific sequence. These action sequences are executed independently from the others. This means that within a task, the order of action execution is fully known; in the case of Task 1 in the figure, first Action A, then Action B, then Action C or Action D, then Action E and so on. What is not known is the order of action execution *between* tasks. Which action executes first? Action A? Action Z? Action Beta? The answer is *you don't know and you don't care* – if you care, then concurrency was a poor choice.

---

[1] Many people draw a distinction between *task, thread*, and *process*. However, these are simply the same concept at different scopes of concern. In this book, we will use these terms interchangeably.

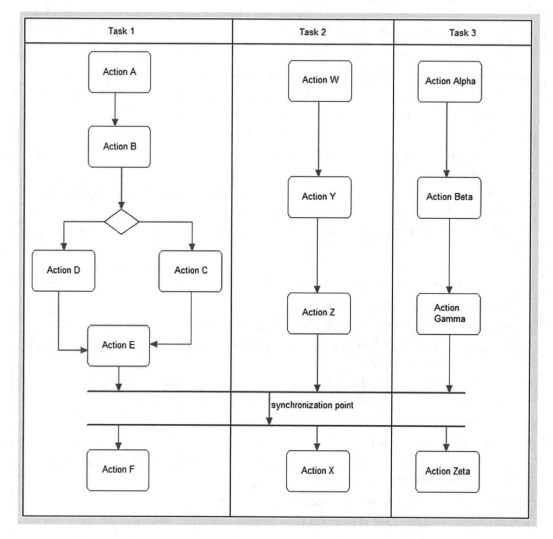

Figure 4-1: Concurrency as a set of action sequences

The only points at which you care about the order of execution between tasks are called *synchronization points*. This is indicated in the activity diagram of Figure 4-1 with a pair of horizontal bars. The first of these is called a *join*, since it joins together a set of actions from different tasks into, logically, a single thread. The second bar is called a fork, as it forks from the single thread into multiple. The semantics of using joins and forks in this way are that the flow cannot proceed past the synchronization point until all predecessor actions (in this case, Action E, Action Z, and Action Gamma) have completed. Once all three actions have completed, control goes on to subsequent actions (Action F, Action X, and Action Zeta, not necessarily in that order). Synchronization points are common places in which tasks will share data or ensure preconditions are met before proceeding.

Tasks can communicate in two fundamental ways. Synchronous communications are like a phone call – both parties are engaged at the same time and invoke services and exchange data immediately. The sender "sends and waits" for a response. Synchronous communication is usually implemented via function calls. Asynchronous communication is like a postcard. The sender of the message "sends and forgets." At some later time, the receiver gets and processes the message or data. The message processing is less timely but the sender need not wait until the receiver is ready.

For tasks to execute concurrently, they must run on different CPUs or cores within a multicore CPU. Within a given CPU core, multiple tasks are run using pseudoconcurrency in which only one task executes at any instant in time, but the processor gives the appearance of concurrency by switching among the ready tasks. There are many different policies that can be used to select which task to run, some of which will be provided as patterns later in this chapter.

The basic structure of tasks depends a bit on the scheduling pattern that will invoke them. A task for a cyclic executive is simply sequential code that returns when it is done. A task that runs in a cooperative multitasking environment usually runs in a loop but contains points at which it voluntarily releases control back to the scheduler to allow other tasks to run. Code Listing 4-1 shows a typical cyclic executive. First, the initial global data declarations are followed by system initiation, and then (if initiation succeeds), the infinite cyclic executive task loop.

```
void main(void) {
    /* global static and stack data */
    static int nTasks = 3;
    int currentTask;

    /* initialization code */
    currentTask = 0;

    if (POST()) { /* Power On Self Test succeeds */
        /* scheduling executive */
        while (TRUE) {
            task1();
            task2();
            task3();
        }; /* end cyclic processing loop */
    }
}
```

Code Listing 4-1: Cyclic executive processing loop

The different ways to schedule the tasks have both benefits and costs. *Cyclic executives* are very simple; the scheduler consists of an infinite loop that executes the tasks one after the other. While simple, it isn't flexible and can have poor response to incoming events since an event processed by a task must wait until that task is run within the cycle. In addition, task executions must be short if other tasks are going to run in a timely fashion. A variant known as *time-triggered cyclic executive* starts the cycle at a time boundary; if it completes the cycle before the next time boundary, it sleeps until the start of the next epoch. A cooperative round-robin scheduler is basically a cyclic executive in which the tasks don't run to completion but instead run to specific points at which they release control back to the scheduling loop. This approach allows long tasks to run while still permitting other tasks to make progress in their processing.

While simple, there are a number of problems with both cyclic executives and round-robin schedulers. First, they are not responsive to urgent events; the task that processes the urgent event simply runs in its sequence. Secondly, they are not flexible since the schedule order is determined at design time. Thirdly, they don't scale well to large numbers of tasks. Lastly, a single misbehaving task stops the entire system from execution. These kinds of schedulers are best-suited to very simple systems or systems that are highly safety critical (since simple systems are far easier to prove correct and certify than complex ones).

The other primary kind of scheduling is *preemptive scheduling*. In preemptive scheduling, a scheduler stops the currently executing task when it decides to do so and runs the next task. The criterion for preempting a task and the selection of the next task in sequence depends on the particular scheduling policy.

At the high level, tasks in a preemptive scheduler have four significant states. A task is said to be *waiting* when the conditions for it to be run are not currently in force. When the conditions required for a task to run are true, then the task is said to be *ready to run*. When the scheduler initiates the task, the task is said to be *running*. When a task is ready to run but is being prevented because another task owns a necessary resource, the task is said to be *blocked*.

A time-slicing scheduler (also known as *time-driven multitasking executive* or TDME) operates on a fairness doctrine; that is, every task gets an equal opportunity to execute. The tasks are preempted after they run for a specific period of time. TDME schedulers are tuned by adjusting the duration of the time slices; too long and the other tasks don't run often enough to do their jobs, too short and too high a percentage of time is spent switching the task context.

The most common scheduling pattern for larger embedded systems is *priority-based scheduling*. In this approach, each task is assigned a priority, a scalar value that will be used to determine which task, from the set of ready to run tasks, will be selected for execution.

A task running within either a time-slicing or priority-based preemptive scheduler runs in an infinite loop and will be interrupted by the scheduler when it is appropriate for other tasks to run. The basic structure for this latter task template looks like Code Listing 4-2. You can see in the code listing where the task static and stack data go. Following these data declarations, there is usually some initialization or set-up code followed by an infinite, or at least indefinite, loop where the task cycles endlessly waiting for a signal of interest – this might come from a queue insertion or event occurrence. When such a signal appears, a `switch case` statement then performs the processing. When complete, the task code loops back to wait for the next signal.

```
/*
        here is where private static data goes
*/
static int taskAInvocationCounter = 0;

void taskA(void) {
        /*  more private task data  */
        int stufftoDo;

        /*  initialization code  */
        stuffToDo  =  1;

        while (stuffToDo) {
                signal  =  waitOnSignal();
                switch (signal) {
                        case signal1:
                                /*  signal 1 processing here  */
                                break;
                        case signal2:
                                /*  signal 2 processing here  */
                        case signal3:
                                /*  signal 3 processing here  */
                };
        }; /*  end infinite while loop  */
}; /*  end task  */
```

Code Listing 4-2: Basic task structure with a preemptive scheduler

Priority can be set according to any criterion imaginable, but the two prevalent types of priority-based scheduling are urgency-based and criticality-based[2]. Many people use these terms imprecisely but the terms have precise meanings. *Urgency* refers to the nearness of a task's deadlines while *criticality* refers to how important the execution of the task is to the system. Figure 4-2 shows the difference. The curve itself is known as the utility function; this is the value to the system of the completion of an action or action sequence as a function of time.

---

[2] And, of course, these can be mixed in different ways to assign priorities, if desired.

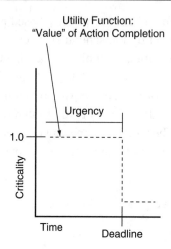

Figure 4-2: Criticality vs. urgency

The criticality is the height of the curve while urgency is the time distance to the point at which the criticality falls to zero. It should be noted that a task that is real-time does not necessarily have a deadline, but deadlines are the most common way to specify timeliness constraints for a real-time task.

When a task executes, there are several important time-related terms, summarized in Figure 4-3. For a periodic task (one that executes at a more-or-less fixed rate), two important values are the *period* (the duration of the time interval between task initiations) and the *jitter* (the variation in the period). Nonperiodic tasks (known commonly as *aperiodic tasks*) are characterized by the interarrival time (the time between task initiations) and the *minimum interarrival time* (the smallest such time). All tasks have an execution time (the time required to complete the activities of the task) although for computation of schedulability, the *worst-case execution time*

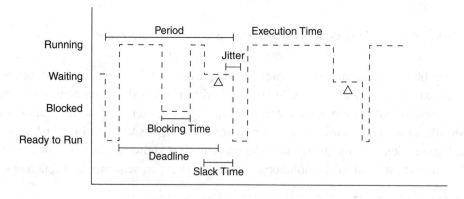

Figure 4-3: Time-related terms

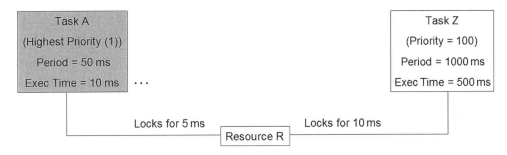

Figure 4-4: Blocking

is usually used. The *deadline* is the period of time between when the task becomes ready to run and when it must complete.

When you use semaphores to lock a resource, it can happen that a lower priority task locks a resource and then is preempted by a higher priority task that needs that resource. The high-priority task must block to allow the latter task to run and release the resource. This is known as *priority inversion* because, contrary to normal priority scheduling rules, the lower priority task is running even though there is a higher priority task ready to run.

Note that priority inversion cannot be avoided in multitasking preemptive schedule that includes the use of semaphores to guard resources. That isn't necessarily bad. What *is* almost always bad is unbounded priority inversion. Consider Figure 4-4. In this figure, there are two tasks. Task A is the higher priority[3], executing with a period of 50 ms and an execution time of 10 ms, 5 ms of which it locks resource R. Task Z is the lower priority task, with a period of 1000 ms and needing 500 ms of time to execute, 10 ms of which it locks resource R. Note that in this example, the deadline for each task occurs at the start of the next period.

Priority inversion occurs when Task Z is running with the resource R locked and then Task A becomes ready to run. It will immediately preempt Task Z but as soon as it gets to the point at which it needs to access the resource R it is "stuck"; it must block and allow Task Z to run until it releases the resource. Once the resource is release by Task Z, Task A again preempts Task Z and all is as it should be.

If schedulability of tasks is a concern (and it almost always should be), you should examine the effect of blocking on schedulability. In the case in Figure 4-4, what is the worst case? It happens when Task Z has just barely locked R and then Task A runs. In this case the total execution time for Task A is 10 ms plus the blocking time of 10 ms. Since Task A completes in 20 ms (less than the deadline) and the execution time for Z is 500 ms + 10 runs of Task A (at 10 ms each) = 600 ms, every one is, as they say "fat, dumb, and happy." Now consider Figure 4-5.

---

[3] Many Real-Time Operating Systems (RTOSs) use smaller numbers to indicate higher priority. Go figure...

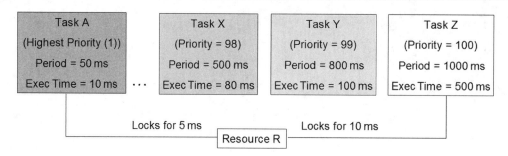

Figure 4-5: Unbounded blocking

In this example, only tasks A and Z use the resource. Task Y is the second-lowest priority task running every 800 ms for a total of 100 ms, while Task X is the third-lowest priority task with a period of 500 ms with an execution time of 80 ms. What happens to schedulability now?

The worst case is the following: Task Z runs and locks the resource and then Task A runs, preempts, runs until it needs to access R, and then blocks. At this point Task Z runs, but immediately Task Y runs. Since it is a higher priority, it preempts Z. As it runs, task X becomes ready to run and preempts Y. At this point, Task A is being blocked by three tasks. The resource will not become available for 190 ms (10+100+80), and Task A now misses its deadline. This is called *unbounded priority inversion* and is a critical problem with many embedded system designs.

There are a number of design solutions, such as using critical regions and variants of the priority inheritance pattern, that solve this problem. These patterns will be discussed later in this chapter.

4.1.1 Identifying Tasks

The entire previous discussion of what is a task and what are task properties avoids the crucial question of how to identify a good set of tasks for your particular embedded system. Generally speaking, good tasks have independent action sequences that interact with others only at specific points of interactions. However, selecting a good task set involves more than this because we create task sets to optimize performance and responsiveness. We could, in principle, run every object (or function, if you prefer) in a separate task thread or we can run them all in a single task thread. If ease of maintenance is the primary concern, then independence should be the primary task identification strategy. If schedulability is the primary concern, then the recurrence properties (e.g., the task period) should guide your choice. Table 4-1 shows common strategies for task identification[4].

In practice, the task set you select will use a combination of several of the strategies from Table 4-1.

---

[4] Borrowed (with permission) from the author's book *Real-Time Agility: The Harmony/ESW Method for Real-Time and Embedded Systems Development* (Addison-Wesley, 2009).

Table 4-1: Task Identification Strategies

Strategy	Description	Pros	Cons
Single Event Groups	Use a single event type per task	Very simple threading model	Doesn't scale to many events well; suboptimal performance
Interrupt Handler	Use a single event type to raise an interrupt	Simple to implement for handling a small set of incoming event types; highly efficient in terms of handling urgent events quickly	Doesn't scale well to many events; interrupt handles typically must be very short and atomic; possible to drop events; difficult to share data with other tasks
Event Source	Group all events from a single source so as to be handled by one task	Simple threading model	Doesn't scale to many event sources well; suboptimal performance
Related Information	Group all processing of information (usually around a resource) together in a task	Useful for "background" tasks that perform low-urgency processing	Same as Event Source
Independent Processing	Identify different sequences of actions as threads when there is no dependency between the difference sequences	Leads to simple tasking model	May result in too many or too few threads for optimality; doesn't address sharing of resources or task synchronization
Interface Device	A kind of event source	Useful for handling device (e.g., bus) interfaces	Same as Event Source
Recurrence Properties	Group together events of similar recurrence properties such as period, minimum interarrival time, or deadline	Best for schedulability; can lead to optimal scheduling in terms of timely response to events	More complex
Safety Assurance	Group special safety and reliability assurance behaviors together (e.g., watchdogs)	Good way to add on safety and reliability processing for safety-critical and high-reliability systems	Suboptimal performance

4.1.2 Concurrency in the UML

In the UML, there are three ways to represent concurrency: «active» objects, forks and joins in activity diagrams, and orthogonal regions in state machines[5]. Of these, the primary means is with «active» objects. An «active» object is one that owns a thread of execution. It is usually a structured object, meaning that it contains part objects inside itself that execute in the context of

---

[5] For more information about UML and concurrency, see the author's *Real-Time UML Third Edition: Advances in the UML for Real-Time Systems* (Addison-Wesley, 2004).

«active» object's thread. The «active» object has the explicit responsibility of managing the event queue for the set of software elements that execute in its thread's context. The icon for an «active» object is an object with a heavy border on its left and right sides although often the stereotype «active» is shown instead. A *task diagram* is a UML class (or object[6]) diagram whose primary focus is on the concurrency units and their interaction. Figure 4-6 shows an example of an object task diagram containing five tasks («active» objects), one shared resource (with guarding semaphore), and comments showing concurrency-related metadata (priority, period, and deadline in this case[7]). Note that with some tasks we use ports (named connection points) to connect directly with objects inside the tasks while in other cases we just use links

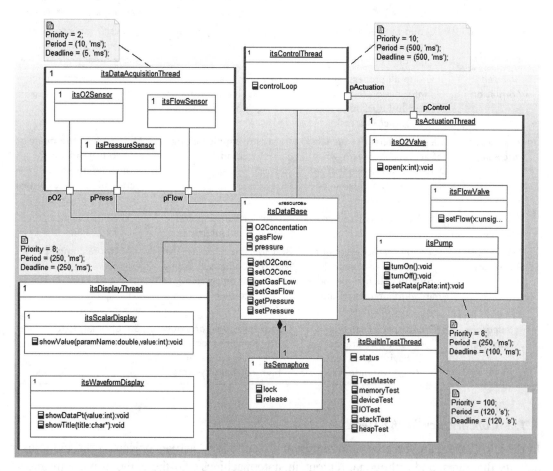

Figure 4-6: UML task diagram

---

[6] Remember from Chapter 1 that a class is a specification and an object is an instance. An «active» class is similar to a task type in Ada while an object is similar to an Ada task.

[7] To perform schedulability analysis we need additional metadata such as worst-case execution time and, since we have shared resources with blocking, worst-case blocking times.

(most likely implemented as pointers). In the case of the `ControlThread` and the `BuildInTestThread`, these «active» objects do not contain internal objects but are objects that run in their own threads.

4.1.3 Real-Time Operating Systems

In many embedded systems, a real-time operating system (RTOS[8]) provides a set of services to the application system including:

- task services such as the creation, scheduling, and destruction of tasks
- event services such as pending on an event, posting, and event and management of event queues
- creation, destruction, and usage of mutex semaphores
- time services, such as delay for a specified period of time, and the creation and destruction of timers
- file system services
- network input/output services

This is no surprise – these services are offered by most desktop operating systems as well. RTOSs are different in a number of ways. First and foremost, they are embeddable; they are designed to be burned into ROM, Flash, or some nonvolatile memory. They may execute from that memory or copy themselves into volatile memory via a boot loader. Secondly, they are tuned to be both fast and predictable[9]. Thirdly, they are usually designed to be scalable; they are often designed using the Microkernel Architecture Pattern[10] and that enables the memory image of the RTOS to include only those services that the system needs and for which it is willing to allocate space.

Figure 4-7 shows a typical structure for an RTOS. An RTOS must be specialized for the target hardware. This is done via a hardware abstraction layer that most RTOS vendors refer to as the Board Support Package (BSP). Above this are kernel services such as memory, task, and timer services. Optionally, file services and network services may be added or removed as needed by the application. The various applications and their tasks run atop the RTOS, invoking the RTOS services via an application programming interface (API) as necessary.

It is now fairly common to embedded such RTOSs, but it wasn't so long ago that the application software had to provide these services. Indeed, many embedded systems running in

---

[8]   Wind River's VxWorks™ and Green Hills' Integrity™ operating systems are popular RTOSs but there are more than 400 RTOSs in use.

[9]   Predictability is a scale with chaos on one end and determinism on the other. RTOSs tend toward the deterministic end of the spectrum.

[10] See the author's *Real-Time Design Patterns: Robust Scalable Architecture for Real-Time Systems* (Addision-Wesley, 2003) for a detailed description of the pattern.

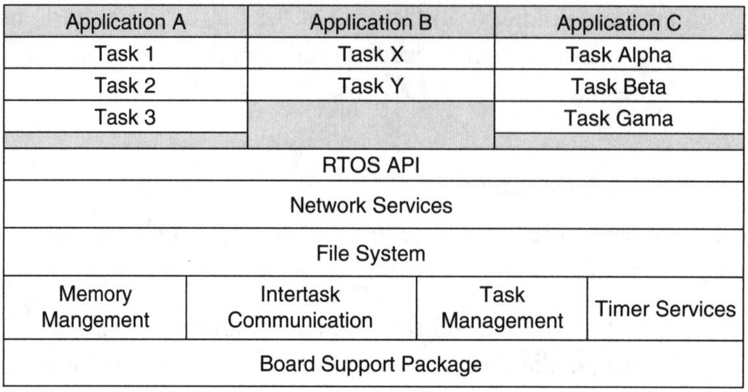

Application A	Application B	Application C
Task 1	Task X	Task Alpha
Task 2	Task Y	Task Beta
Task 3		Task Gama

RTOS API

Network Services

File System

Memory Mangement	Intertask Communication	Task Management	Timer Services

Board Support Package

Figure 4-7: Basic RTOS structure

smaller memory space and processor target environments cannot afford the time, memory, or recurring cost overhead of commercial RTOSs.

Let's now look at some patterns useful for managing the concurrency aspects of embedded systems. It should be noted that the Interrupt Pattern of the previous chapter is a concurrency pattern in addition to a hardware access pattern.

Scheduling Patterns

This section contains patterns related to the invocation of tasks. This is a service that an RTOS will provide, if available. If not, then your application software must provide these services. Even if your system uses an RTOS to provide scheduling, understanding the benefits and costs of the different scheduling approaches gives you more information to select an optimal scheduling strategy.

4.2 Cyclic Executive Pattern

As mentioned earlier in this chapter, the Cyclic Executive Pattern has the advantage of being a very simple approach to scheduling multiple tasks but isn't very flexible or responsive to urgent events. The Cyclic Executive Pattern implements a fairness doctrine for scheduling that allows all tasks an equal opportunity to run. Although suboptimal in terms of responsiveness, the Cyclic Executive Pattern is easy to analyze for schedulability (that is, the ability of the system to predictably meet all task deadlines): the deadline for each task must be greater than or equal to the sum of all the worst-case execution times for all the tasks plus the loop overhead. Assuming that the deadline for each task is the duration of one complete task execution cycle, Equation 4-1 gives the relation that must be true for the task set to be schedulable, where:

- D_j is the deadline for task j
- C_j is the worst-case execution time for task j
- K is the cyclic executive loop overhead, including the overhead of the task invocation and return

for all tasks i, $$D_i \geq \sum_{j=1}^{n} C_j + K$$

Equation 4-1: Schedulability for Cyclic Executive with *n* tasks

4.2.1 Abstract

The Cyclic Executive Pattern is used primarily in two situations. First, for very small embedded applications, it allows multiple tasks to be run pseudoconcurrently without the overhead of a complex scheduler or RTOS. Secondly, in high-safety relevant systems, the Cyclic Executive Pattern is easy to certify, hence its prevalence in avionics and flight management systems. The scheduler is a simple loop that calls each of the tasks in sequence. Each task is simply a function invoked by the scheduler that runs to completion and then returns.

4.2.2 Problem

Many embedded systems are tiny applications that have extremely tight memory and time constraints and cannot possibly host even a scaled-down microkernel RTOS. The Cyclic Executive Pattern provides a low-resource means to accomplish this goal.

4.2.3 Pattern Structure

Figure 4-1 shows the structure of this simple pattern. The CyclicExecutive has a `controlLoop()` function that iteratively invokes the `run()` operation of each of the task threads. Each `run()` operation is relatively short and runs to completion.

4.2.4 Collaboration Roles

This section describes the roles for this pattern.

4.2.4.1 AbstractCEThread

The AbstractCEThread provides a standard interface for the threads by declaring a run() function. It is this function that the Cyclic Executive will invoke to execute the task. When this function completes, the CyclicExecutive runs the next task in the list.

4.2.4.2 CyclicExecutive

As mentioned in Section 4-1, this class contains the infinite loop that runs the tasks each in turn. In addition, the CyclicExecutive contains global stack and static data that are needed either by the tasks or by the scheduler itself. The code in Figure 4-4.1 is typical for a cyclic executive.

A variant of the pattern is the time-triggered Cyclic Executive. In this variant, the CyclicExecutive will set up and use the CycleTimer to initiate each epoch (execution of the task list). Note that while the Interrupt Pattern from Chapter 3 can be and the Cyclic Executive can simply wait for the interrupt to do the next cycle, unless there is background processing to perform, it can sit in a so-called "busy loop" and poll for the timer to elapse.

4.2.4.3 CycleTimer

The CycleTimer isn't used in the most common variant of the CyclicExecutive; the fact that it is optional is indicated with the "0,1" multiplicity on the directed association in Figure 4-8. This timer can invoke an interrupt when it expires or, even more simply, can return TRUE (non-zero) to the hasElapsed() function. The CyclicExecutive would then call start() to begin the timing interval for the next epoch.

4.2.5 ConcreteCEThread

The ConcreteCEThread stands for the functions and related data that implement a logical task in the system. Each has its own implementation of the run() function.

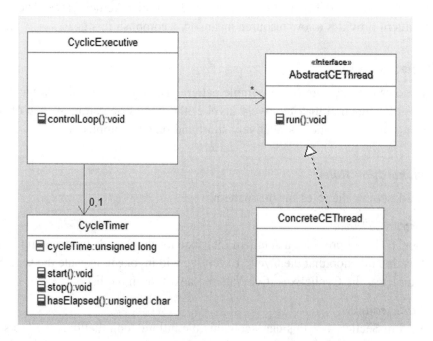

Figure 4-8: Cyclic Executive Pattern

4.2.6 Consequences

As mentioned, the primary benefit of this pattern is its simplicity. It's pretty difficult to get the scheduler wrong. On the other hand, the pattern is demonstrably less responsive to high-urgency events and this makes the pattern inappropriate for applications that have high responsiveness requirements. Another benefit of the pattern is that it is very lightweight in terms of required resources and so is appropriate for very small memory devices.

A downside of using the pattern is that the interaction of threads is made more challenging than other patterns. If an output of one task is needed by another, the data must be retained in global memory or a shared resource until that task runs (See the Task Coordination Patterns section later in this chapter). On the other hand, because of the lack of preemption, unbounded blocking is not a concern. Deadlock can only occur because of a mistake – a single misbehaved task (i.e., one that never returns control to the cyclic executive) shuts the entire system down. In a preemptive scheduling system, unless the misbehaving task has turned off task switching, the other tasks will continue to function even if one task fails.

4.2.7 Implementation Strategies

The implementation of this pattern is almost trivially simple. In most cases, the Cyclic Executive may simply be the application's `main()` function and related global data. In other cases, it will be invoked from `main()`.

4.2.8 Related Patterns

Because the pattern's responsiveness is demonstrably suboptimal, it is often augmented with some interrupt service routines (see the Interrupt Pattern in Chapter 3) for high urgency events. Another use for the Interrupt Pattern with the Cyclic Executive Pattern is to implement the epoch timer response with the time-triggered cyclic executive, although the Polling Pattern (also from Chapter 3) may be used to poll the timer as well.

The Cyclic Executive Pattern does not address data sharing between tasks. The Cyclic Executive does define global data structures that can be used for this purpose but it is possible to mix other patterns for this purpose (see the Task Coordination patterns section in this chapter). Because there is no preemption, global data do not require access serialization with this pattern.

4.2.9 Example

The example shown in Figure 4-9 shows the cyclic executive scheduler `GasControlExecutive`. Its `controlLoop()` function calls the run services in each of the threads. Because we used

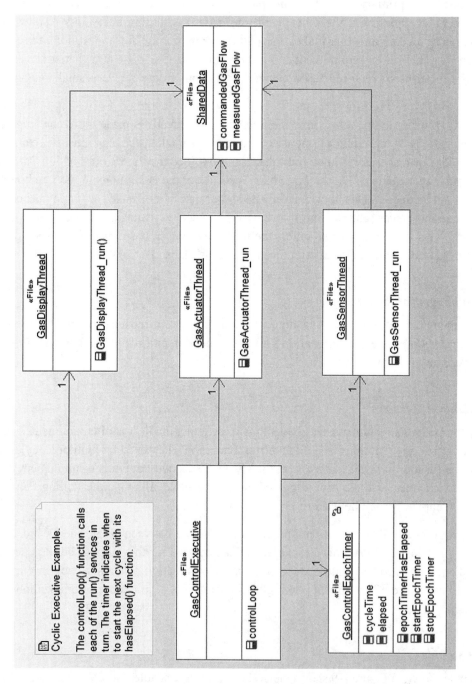

Figure 4-9: Cyclic Executive Example

the file implementation, it was necessary to encode the name of the file in the name of the run function of each to avoid global naming clash. In the example, I implemented the time-triggered variant of the pattern. As you can see in Code Listing 4-3, the `controlLoop()` function implements a busy wait for the elapse of the epoch time. Note that the timer `startEpochTimer()` function resets the elapsed variable to FALSE (0) before it starts the timer. When the timer elapses, this variable is set to TRUE (1).

```c
#include "GasControlExecutive.h"
#include "GasActuatorThread.h"
#include "GasControlEpochTimer.h"
#include "GasDisplayThread.h"
#include "GasSensorThread.h"
void controlLoop(void) {
    while (TRUE) {
        startEpochTimer();
        GasSensorThread_run();
        GasActuatorThread_run();
        GasDisplayThread_run();
        while(!epochTimerHasElapsed()) ;
    };
}
```

Code Listing 4-3: GasControlExecutive.c

Each of the code files for the other elements simply declare their data and functions and the related C file provides the implementation. For example, the implementation for the SharedData file is given in Code Listing 4-4 and Code Listing 4-5.

```c
#ifndef SharedData_H
#define SharedData_H

extern int commandedGasFlow;
extern int measuredGasFlow;

#endif
```

Code Listing 4-4: SharedData.h

```c
#include "SharedData.h"
int commandedGasFlow;
int measuredGasFlow;
```

Code Listing 4-5: SharedData.c

```
void GasDisplayThread_run(void) {
    printf("Measured Gas Flow %d\n", measuredGasFlow);
    printf("commandedGasFlow %d\n\n", commandedGasFlow);
}
```

Code Listing 4-6: GasDisplayThread.c

4.3 Static Priority Pattern

The Static Priority Pattern is supported by most real-time operating systems. It has good response to high-priority events and scales well to large numbers of tasks. Different schemas can be used to assign priorities, but, as mentioned earlier, priorities are usually assigned based on urgency, criticality, or some combination of the two. When assigned using rate monotonic scheduling (see the Implementation Strategies section below), this scheduling pattern is both optimal (you can't do better with other scheduling strategies) and stable (in an overload situation, which tasks will fail is predictable – the lowest priority tasks). Assuming that each task deadline occurs at the end of the period, the system is schedulable if the inequality in Equation 4-2 holds, where:

- C_j is the worst case execution time for task j
- T_j is the period (deadline occurs at the end of the period) for task j
- B_j is the worst-case blocking time for task j
- n is the number of tasks

$$\sum_j \frac{C_j}{T_j} + max\left(\frac{B_1}{T_1}, \cdots, \frac{B_{n-1}}{T_{n-1}}\right) \leq utilization(n) = n(2^{\frac{1}{n}} - 1)$$

Equation 4-2: Schedulability for Static Priority Pattern with RMS priorities for n tasks[11]

In the case where the task is not periodic, the minimum interarrival time for task j (the smallest time between successive invocation of the same task) is used. Note that the blocking terms only extend to the n-1 task; this is because by definition, the lowest priority task cannot be blocked.

4.3.1 Abstract

The Static Priority Pattern is a very common pattern in embedded systems because of its strong RTOS support, ease of use, and good responsiveness to urgent events. It may be overkill for very small, simple, or highly predictable systems that are not driven by urgent asynchronous events. The pattern is easy to analyze for schedulability but naïve implementation with blocking resource sharing can lead to unbounded priority inversion. Fortunately, there are resource sharing patterns that can avoid this problem or at least bound priority inversion to at

---

[11] For a proof of this equation, see Lui and Leyand, Scheduling algorithms for multiprogramming in a hard real-time environment. *Journal of ACM* 20, January 1973, 40–61.

most one level. The most common way to assign priorities is on the basis of deadline duration (that is, task urgency); the shorter the deadline, the higher the priority. If the deadline is at the end of the period, this schema is known as *rate monotonic scheduling*. It assumes that all tasks are periodic, but the minimum interarrival interval can be used for aperiodic tasks, although this often results in overdesigning the system with an impact on system recurring cost.

4.3.2 Problem

The Static Priority Pattern provides a simple approach to scheduling tasks based on priority. This pattern addresses systems with more tasks than the Cyclic Executive Pattern and also emphasizes responsiveness to urgent events over fairness.

4.3.3 Pattern Structure

Figure 4-10 shows the structure for the Static Priority Pattern. The colored classes indicate the element normally implemented for you by an RTOS. Note that the mapping constraint on the diagram indicates that there is a `StaticTaskControlBlock` for each `AbstractStaticThread`. This holds data about the task needed by the scheduler to support scheduling and preemption.

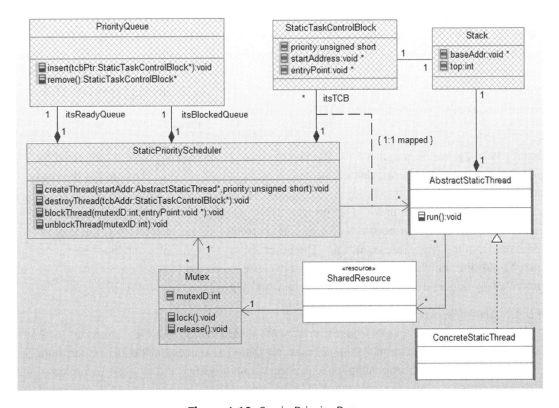

Figure 4-10: Static Priority Pattern

4.3.4 Collaboration Roles

This section describes the roles for this pattern.

4.3.4.1 AbstractStaticThread

The AbstractStaticThread class is an abstract (i.e., noninstantiable) super class for *Concrete Thread. Abstract Thread* associates with the Scheduler; specifically the Scheduler executes the run() function of the thread to run the thread. The AbstractStaticThread specifies the run interface to be realized by the ConcreteThread.

4.3.4.2 ConcreteThread

The ConcreteThread is an «active» class that serves as a container and thread context for a set of passive "semantic" objects that do the real work of the system. The ConcreteThread class provides a straightforward means of attaching these semantic objects into the concurrency architecture. ConcreteThread represents the threads in the application software.

4.3.4.3 Mutex

The Mutex is a mutual exclusion semaphore class that serializes access to a SharedResource. The guarded functions of the SharedResource invoke the lock() function of the Mutex whenever they are called and release() once the service is complete. Other client threads that attempt to invoke a service when its associated Mutex is locked become blocked until the Mutex is unlocked. The blocking is performed when the Mutex semaphore signals the StaticPriorityScheduler that a call attempt was made by the currently active thread. The mutexID identifies which Mutex is blocking the task. This element is usually provided by the RTOS.

4.3.4.4 PriorityQueue

The PriorityQueue is a queue of StaticTaskControlBlock pointers organized by priority. The pattern uses two distinct instances of PriorityQueue. The first is the ready queue. When a task becomes ready to run, a pointer to its StaticTaskControlBlock is inserted into the queue in order of its priority. If the priority of the task is higher than the currently running task (or becomes so when the current task terminates), then the task is removed from the ready queue and run. The other PriorityQueue is the blocked queue; this queue holds the set of tasks currently blocked because they tried to access SharedResources protected by a locked Mutex. This element is usually provided by the RTOS.

4.3.4.5 SharedResource

A SharedResource is a class whose instances are shared by one or more ConcreteThreads. For the system to operate properly in all cases, all shared resources must either be reentrant (meaning that corruption from simultaneous access cannot occur) or they must be protected via access serialization. Protection via mutual exclusion semaphores is discussed in the Guarded Call Pattern later in this chapter.

4.3.4.6 Stack

Each `AbstractThread` has a `Stack` for return addresses and passed parameters. This is normally only explicit at the assembly language level within the application thread, but is an important part of the scheduling infrastructure. This element is usually provided by the RTOS.

4.3.4.7 StaticPriorityScheduler

This class orchestrates the execution of multiple threads based on their priority according to a simple rule: always run the highest priority ready thread. When the «active» thread object is created, it (or its creator) calls the `createThread()` operation to create a thread for the «active» object. Whenever this thread is executed by the `StaticPriorityScheduler`, it calls the `StartAddr` address (except when the thread has been blocked or preempted, in which case it calls the `EntryPoint` address). This element is usually provided by the RTOS.

4.3.4.8 StaticTaskControlBlock

The `StaticTaskControlBlock` contains the scheduling information for its corresponding *AbstractThread* object. This includes the priority of the thread, the default start address and the current entry address, if it was preempted or blocked prior to completion. The `StaticPriorityScheduler` maintains a `StaticTaskControlBlock` object for each existing `AbstractThread`. This element is usually provided by the RTOS.

4.3.5 Consequences

As previously mentioned, the Static Priority Pattern runs well with potentially large numbers of tasks and provides responsiveness to incoming events. Typically the tasks exist in waiting for initiation most of the time and become active only when an initiating event (such as the elapse of their period) occurs. Priority is only an issue when there is more than one task ready to run; in this case the highest priority task will be run preferentially.

It is common to use blocking to serialize access to resources. While this is highly effective, naïve implementation can lead to unbounded priority inversion in the manner described in the first section of this chapter. See the section on resource sharing patterns later in this chapter for patterns to address this concern.

4.3.6 Implementation Strategies

By far, the most common implementation of this pattern is to use a commercial RTOS to provide the scheduling infrastructure. While this wasn't true 20 years ago, today there are dozens of popular RTOSs that will fit the constraints of most embedded systems.

Regardless of whether you buy a commercial RTOS or roll your own, the problem of assigning priorities must still be addressed. RMS (rate monotonic scheduling) is the most common means to assign priorities. It is optimal and stable. By *optimal*, I mean that if the task set can be scheduled by any means whatsoever, it can also be scheduled by RMS. By *stable*, I mean that in the case of an overloaded situation in which some tasks will miss their deadlines, with static priority scheduling, it is clear which tasks will fail – those with the lowest priority. RMS simply makes the priority a function of the period – the shorter the period, the higher the priority[12].

RMS makes a number of assumptions about the tasks but is reasonably robust in the presence of violations to those assumptions:

- tasks are periodic
- task deadlines occur at the end of the period
- tasks are infinitely preemptable

As to the first assumption, if tasks are not periodic, then they can occur at a maximum rate (that is, with a minimum time between task invocations). This value can be used instead of the period to determine the priority.

For the second assumption, it sometimes happens that the deadline is less than the period. In this case, simply use the deadline (D_j) instead of the period (T_j) to assign the priority.

The last assumption is that a currently running task can be preempted as soon as a higher priority task becomes ready to run. This isn't true if the currently running task is in a critical region (a period of time in which task switching is disabled – see the Critical Region Pattern later in this chapter). However, if the critical regions are short, this is usually not a problem. Such situations can be analyzed by treating the duration of the longest critical region of all other tasks as the worst case blocking time for the task under analysis.

4.3.7 Related Patterns

The Static Priority Pattern offers a more responsive alternative to fairness doctrine approaches such as the Cyclic Executive Pattern. When resources must be shared with other tasks, this pattern must be mixed with resource sharing patterns such as those in this chapter. When tasks need to coordinate, this pattern can be mixed with task coordination patterns. The task coordination patterns overlap a bit with the resource sharing patterns since one of the ways that tasks coordinate is by sharing common resources.

A very similar pattern to the Static Priority Pattern is the Dynamic Priority Pattern. In this pattern, the design does not specify the priority of the task; this is computed at run-time. The most common priority assignment strategy is known as Earliest Deadline First (EDF) scheduling. In EDF, the task must specify the duration of its deadline from which the scheduler computes the absolute time of the elapse of this task's next deadline. The task is then put into

---

[12] Note that many RTOSs use small values to indicate higher priorities so that a priority of 5 is a higher priority than a priority of 10.

the ready queue based on the nearness of the next absolute deadline for all ready tasks. That is, if a task A is being queued that has a nearer next deadline than another task B, then it will be placed in front of task B in the queue; otherwise, it will be placed later in the queue. EDF is optimal (in the same sense that RMS is) but it is not stable; in an overload situation which tasks will fail to meet their deadlines cannot be computed at design time. If you use EDF scheduling, then schedulability can be computed in a manner similar to RMS scheduling, except that the upper utilization bound is 1.00 rather than a function of the number of tasks. This is shown in Equation 4-3.

$$\sum_j \frac{C_j}{T_j} + max\left(\frac{B_1}{T_1}, \cdots, \frac{B_{n-1}}{T_{n-1}}\right) \leq 1.00$$

Equation 4-3: Earliest Deadline First schedulability

4.3.8 Example

In this example, we are going to assume the use of a commercial RTOS so we won't explicitly model or code the infrastructure elements of the pattern. Instead, the example (See Figure 4-11) focuses on concrete threads and shared resources.

In the code listings below, the calls to initialize timers, pend on the OS events, and create tasks that are all RTOS-dependent.

```
void main(void) {
        /* initialize the links and start the tasks */
        MotorPositionSensor_initRelations();
        MotorDisplay_initRelations();
        MotorController_initRelations();

        /* now let the tasks do their thing */
        while (TRUE) {
        };
};
```

Code Listing 4-7: Static Priority Example main.c

The code listings for the `MotorPositionSensor`, `MotorDisplay`, and `MotorController` are very simple, as can be seen in Code Listing 4-8 through Code Listing 4-13. The interesting code can be found in the `_Init()` functions, where the OS event queues are created and the tasks created with the `corresponding _run()` functions. A queue size of 10 events is more than enough space if the events are handled in a timely fashion. Note also that each task is created with a 1024 int sized stack. Following the task

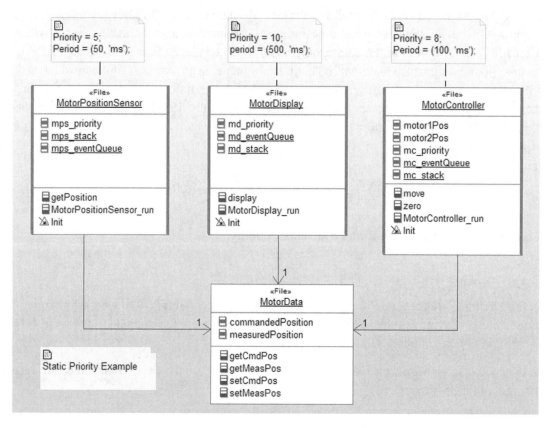

Figure 4-11: Pattern Example

creation, the corresponding _run() function creates an OS timer (with the RETRIGGER parameter so that the timer repeats automatically) and then runs an infinite loop. Within the loop, the task then pends on the event appearance. When the corresponding event with the proper id occurs, the OSQPend() function returns and the task is initiated. Since this is an infinite loop, once the loop completes, each task iterates and goes to sleep waiting for its next event.

The code listings for MotorData (Code Listing 4-14 and Code Listing 4-15) are even simpler since it is just a passive resource with services invoked by the active threads. Again note that the calls to create, initialize, and use threads are RTOS-specific.

```
#ifndef MotorPositionSensor_H
#define MotorPositionSensor_H
```

```
void MotorPositionSensor_Init(void);
void MotorPositionSensor_Cleanup(void);

void MotorPositionSensor_run(void);
void getPosition(void);

void MotorPositionSensor_initRelations(void);

#endif
```

Code Listing 4-8: MontorPositionSensor.h

```
#include "MotorData.h"
#include "MotorPositionSensor.h"
#include <stdlib.h>

static int mps_priority = 5;
static int mps_eventQueue[10];
static int mps_stack[1024];

void MotorPositionSensor_Cleanup(void) {
}
void MotorPositionSensor_Init(void) {
    /* create the OS event queue and the task */
    /* use a queue of 10 events, and a 1024 element stack */
    OSQCreate (&mps_eventQueue[0], 10);
    OSTaskCreate (MotorPositionSensor_run, NULL,
        (void *)&mps_stack[1024], 5);

    /* now run the task */
    MotorPositionSensor_run();
}
void MotorPositionSensor_run(void) {
    int os_pend_timer_id;
    os_pend_timer_id = OSCreateTimer(50, RETRIGGER);

    while (TRUE) {
      OSQPend(os_pend_timer_id, WAIT_FOREVER);
      getPosition();
    };
}
void getPosition(void) {
    int x;
    /* get position by reading the sensor */
    x = rand();
    setMeasPos(x);
}
```

```
void MotorPositionSensor_initRelations(void) {
    MotorPositionSensor_Init();
}
```

Code Listing 4-2: MotorPositionSensor.c

```
#ifndef MotorDisplay_H
#define MotorDisplay_H



void MotorDisplay_Init(void);
void MotorDisplay_Cleanup(void);

void MotorDisplay_run(void);
void display(void);

void MotorDisplay_initRelations(void);
#endif
```

Code Listing 4-3: MotorDisplay.h

```
#include "MotorData.h"
#include <stdio.h>
#include "MotorDisplay.h"

static int md_priority  =  10;
static int md_eventQueue[10];
static int md_stack[1024];

void MotorDisplay_Init(void) {
    /*  create the OS event queue and the task  */
    /*  use a queue of 10 events, and a 1024 element stack  */
    OSQCreate (&md_eventQueue[0], 10);
    OSTaskCreate (MotorDisplay_run, NULL,
        (void *)&md_stack[1024], 5);

    /*  now run the task  */
    MotorDisplay_run();
}

void MotorDisplay_Cleanup(void) {
}

void MotorDisplay_run(void) {
    int os_pend_timer_id;
    os_pend_timer_id  =  OSCreateTimer(500, RETRIGGER);
```

```
      while (TRUE) {
        OSQPend(os_pend_timer_id, WAIT_FOREVER);
        display();
        };
}

void display(void) {
    printf("Commanded position  =  %d\n", getCmdPos());
    printf("Measured position  =  %d\n\n", getMeasPos());
}

void MotorDisplay_initRelations(void) {
    MotorDisplay_Init();
}
```

Code Listing 4-4: MotorDisplay.c

```
#ifndef MotorController_H
#define MotorController_H

void MotorController_Init(void);
void MotorController_Cleanup(void);

void MotorController_run(void);
void move(void);
void zero(void);

void MotorController_initRelations(void);

#endif
```

Code Listing 4-5: MotorController.h

```
#include "MotorData.h"
#include <stdlib.h>
#include "MotorController.h"

static int mc_priority  =  8;
static int mc_eventQueue[10];
static int mc_stack[1024];

int motor1Pos;
int motor2Pos;
```

```
void MotorController_Init(void) {
    /*  use a queue of 10 events, and a 1024 element stack  */
    /*  create the OS event queue and the task  */
    OSQCreate (&mc_eventQueue[0], 10);
    OSTaskCreate (MotorController_run, NULL,
    (void  *)&mc_stack[1024], 5);

    /*  now run the task  */
    MotorController_run();
}

void MotorController_Cleanup(void) {
}

void MotorController_run(void) {
    int os_pend_timer_id, x, y;
    os_pend_timer_id  =  OSCreateTimer(100, RETRIGGER);

    while (TRUE) {
      OSQPend(os_pend_timer_id, WAIT_FOREVER);;
      move();
      };
}

void move(void) {
    int m1Pos, m2Pos;
    /*  this function would read the instrument
        panel to set the motor movement settings
        to set the position of the motors.
        Note that if you want to set only one
        of the motors, then set a negative value
        for the commanded position of the other
    */
    m1Pos  =  rand();
    m2Pos  =  rand();

    if (m1Pos  >=  0)
      motor1Pos  =  m1Pos;
    if (m2Pos  >=0)
      motor2Pos  =  m2Pos;
    setCmdPos(100*m1Pos  +  m2Pos);
}

void zero(void) {
    motor1Pos  =  0;
    motor2Pos  =  0;
    setCmdPos(0);
}

void MotorController_initRelations(void) {
    MotorController_Init();
}
```

Code Listing 4-13: MotorController.c

```c
#ifndef MotorData_H
#define MotorData_H

#define RETRIGGER (2)
#define WAIT_FOREVER (0)

extern int commandedPosition;
extern int measuredPosition;

int getCmdPos(void);
int getMeasPos(void);
void setCmdPos(int x);
void setMeasPos(int x);

#endif
```

Code Listing 4-14: MotorData.h

```c
#include "MotorData.h"

int commandedPosition = 0;
int measuredPosition = 0;

int getCmdPos(void) {
    return commandedPosition;
}

int getMeasPos(void) {
    return measuredPosition;
}

void setCmdPos(int x) {
    commandedPosition = x;
}

void setMeasPos(int x) {
    measuredPosition = x;
}
```

Code Listing 4-15: MotorData.c

Task Coordination Patterns

To be honest, when tasks are truly independent, the concurrency design is pretty simple. The problem is that independence is not the normal condition for concurrent threads. In virtually all embedded systems, tasks are "mostly independent"[13]. The four patterns of this section include the Critical Region, Guard Call, Queuing, and Rendezvous Patterns. The first three of these are often used to serialize access to shared resources as well and so could have been reasonably placed in that section of this chapter.

4.4 Critical Region Pattern

The critical region pattern is the simplest pattern for task coordination around. It consists of disabling task switching during a region where it is important that the current task executes without being preempted. This can be because it is accessing a resource that cannot be simultaneously accessed safely, or because it is crucial that the task executes for a portion of time without interruption.

4.4.1 Abstract

In a preemptive multitasking environment (see the Static Priority Pattern, above), there may be periods of time during which a task must not be interrupted or preempted. The Critical Region Pattern address these intervals by disabling task switching or even by disabling interrupts (see the Interrupt Pattern, above). This provides uninterrupted execution by the current task. Once beyond the critical region, the task must reenable task switching (or interrupts) or the system will fail.

4.4.2 Problem

There are two primary circumstances under which a task should not be interrupted or pre-empted. First, it may be accessing a resource that may not be safely accessed by multiple clients simultaneously. Secondly, the task may be performing actions that must be completed within a short period of time or that must be performed in a specific order. This pattern allows the active task to execute without any potential interference from other tasks.

4.4.3 Pattern Structure

Figure 4-12 shows just how structurally simple this pattern is. In a common variant, the protected element is a resource and not a task; this is shown in Figure 4-13. In either case,

---

[13]   In the same sense that Miracle Max reports that Wesley is "mostly dead" (Goldman, W., 1973. *Princess Bride: S, Morgenstern's Classic Tale of True Love and High Adventure.* Harcourt).

Figure 4-12: Critical Region Pattern

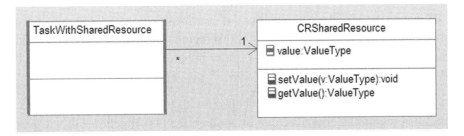

Figure 4-13: Critical Region Pattern with Resource

within the relevant services, task switching must be disabled before the critical region begins and reenabled before the service ends. Note that unlike the Guarded Call Pattern, the scheduler is not involved in the critical region begin/end process other than to provide services to disable/ enable task switching. If the scheduler does not provide such services, then the critical region can be controlled with the user of the C asm directive to turn on or off interrupt processing at the hardware level.

4.4.4 Collaboration Roles

This section describes the roles for this pattern.

4.4.4.1 CRSharedResource

This element is used in the variant of the Critical Resource Pattern that disables task switching to prevent simultaneous access to the resource. In the case of the pattern, the protected resource is the value attribute. Note that all the relevant services must implement the critical region to protect the resource. In this case, it means that both the setValue() and getValue() functions must separately implement the critical region.

4.4.4.2 TaskWithCriticalRegion

This element is used when the task wants to disable task switching for reasons unrelated to sharing resources (usually related to timing). The criticalService() disables and enables task switching or interrupts during its critical region(s).

4.4.4.3 TaskWithSharedResource

This element stands for the set of all tasks that may want to access the shared resource (hence the '*' multiplicity on the directed association). These tasks have no knowledge about the method for protecting the resource, since that is encapsulated within the shared resource.

4.4.5 Consequences

This pattern enforces the critical region policies well and can be used to turn off the scheduled task switches or, more draconianly, disable interrupts all together. Care must be taken to reenable task switching once the element is out of its critical region. Additionally, the use of this pattern can affect the timing of other tasks. Critical regions are usually short in duration for this reason. There is no problem of unbounded priority inversion because all task switching is disabled during the critical region.

Care must be taken when nesting calls to functions with their own critical regions. If such a function is called within the critical region of another, the nest function call will reenable task switching and end the critical region of the caller function.

4.4.6 Implementation Strategies

This pattern is usually implemented either by using scheduler-provided services such as `OS_disable_task_switching()` and `OS_enable_task_switching()` (or something similar) or by introducing assembly language instructions to disable or enable interrupt processing at the hardware level.

4.4.7 Related Patterns

This pattern is only appropriate when multiple tasks can access the same resource or services or when other tasks could interrupt time-critical processing of the current task. For this reason, it is not appropriate with the Cyclic Executive Pattern, but is often used with the Static Priority Pattern.

4.4.8 Example

The example in Figure 4-14 shows a single task `CRRobotArmManager` with a function `moveRobotArm` that has a critical region during the actual movement of the arm.

In either case, the critical region is encapsulated entirely with the `moveRobotArm()` and `motorZero()` services, as can be seen in Code Listing 4-16 and Code Listing 4-17.

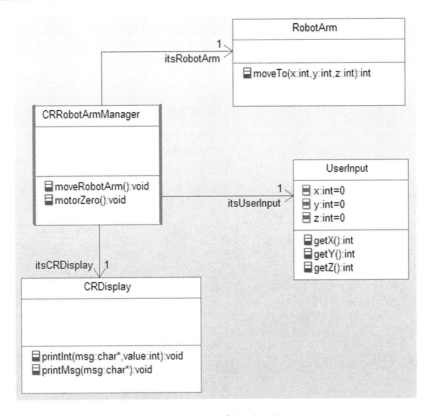

Figure 4-14: Critical Region Pattern

```
#ifndef CRRobotArmManager_H
#define CRRobotArmManager_H

struct CRDisplay;
struct RobotArm;
struct UserInput;

typedef struct CRRobotArmManager CRRobotArmManager;
struct CRRobotArmManager {
    struct CRDisplay*  itsCRDisplay;
    struct RobotArm*  itsRobotArm;
    struct UserInput*  itsUserInput;
};

/*  Constructors and destructors:*/
void CRRobotArmManager_Init(CRRobotArmManager*  const me);
void CRRobotArmManager_Cleanup(CRRobotArmManager*  const me);
```

```
/* Operations */
void CRRobotArmManager_motorZero(CRRobotArmManager* const me);
void CRRobotArmManager_moveRobotArm(CRRobotArmManager* const me);
struct CRDisplay* CRRobotArmManager_getItsCRDisplay(const
CRRobotArmManager* const me);

void CRRobotArmManager_setItsCRDisplay(CRRobotArmManager* const
me, struct CRDisplay* p_CRDisplay);

struct RobotArm* CRRobotArmManager_getItsRobotArm(const
CRRobotArmManager* const me);

void CRRobotArmManager_setItsRobotArm(CRRobotArmManager* const me,
struct RobotArm* p_RobotArm);

struct UserInput* CRRobotArmManager_getItsUserInput(const
CRRobotArmManager* const me);

void CRRobotArmManager_setItsUserInput(CRRobotArmManager* const me,
struct UserInput* p_UserInput);

CRRobotArmManager * CRRobotArmManager_Create(void);

void CRRobotArmManager_Destroy(CRRobotArmManager* const me);

#endif
```

Code Listing 4-6: CRRobotArmManager.h

```
#include "CRRobotArmManager.h"
#include "CRDisplay.h"
#include "RobotArm.h"
#include "UserInput.h"

static void cleanUpRelations(CRRobotArmManager* const me);

void CRRobotArmManager_Init(CRRobotArmManager* const me) {
    me->itsCRDisplay = NULL;
    me->itsRobotArm = NULL;
    me->itsUserInput = NULL;
}

void CRRobotArmManager_Cleanup(CRRobotArmManager* const me) {
    cleanUpRelations(me);
}

void CRRobotArmManager_motorZero(CRRobotArmManager* const me) {
    int success = 1;
```

```c
    OS_disable_task_switching();

    /* critical region code */
    success = RobotArm_moveTo(me->itsRobotArm,0,0,0);

    /* critical region ends */
    OS_enable_task_switching();

    if (!success)
        CRDisplay_printMsg(me->itsCRDisplay,"Cannot zero!");
}

void CRRobotArmManager_moveRobotArm(CRRobotArmManager* const me) {
    /* local stack variable declarations */
    int x, y, z, success = 1;
    /*noncritical region code */

    /* note that the function below has its
       own critical region and so cannot be
       called inside of the critical region
       of this function
    */
    CRRobotArmManager_motorZero(me);

    x = UserInput_getX(me->itsUserInput);
    y = UserInput_getY(me->itsUserInput);
    z = UserInput_getX(me->itsUserInput);

    /* critical region begins */
    OS_disable_task_switching();

    /* critical region code */
    success = RobotArm_moveTo(me->itsRobotArm,x,y,z);

    /* critical region ends */
    OS_enable_task_switching();

    /* more noncritical region code */
    CRDisplay_printInt(me->itsCRDisplay, "Result is ", success);
}

struct CRDisplay* CRRobotArmManager_getItsCRDisplay(const
CRRobotArmManager* const me) {
    return (struct CRDisplay*)me->itsCRDisplay;
}

void CRRobotArmManager_setItsCRDisplay(CRRobotArmManager* const
me, struct CRDisplay* p_CRDisplay) {
    me->itsCRDisplay = p_CRDisplay;
```

```
}

struct RobotArm*  CRRobotArmManager_getItsRobotArm(const
CRRobotArmManager*  const me) {
    return (struct RobotArm*)me->itsRobotArm;
}

void CRRobotArmManager_setItsRobotArm(CRRobotArmManager*  const me,
struct RobotArm*  p_RobotArm) {
    me->itsRobotArm  =  p_RobotArm;
}

struct UserInput*  CRRobotArmManager_getItsUserInput(const
CRRobotArmManager*  const me) {
    return (struct UserInput*)me->itsUserInput;
}

void CRRobotArmManager_setItsUserInput(CRRobotArmManager*  const
me, struct UserInput*  p_UserInput) {
    me->itsUserInput  =  p_UserInput;
}

CRRobotArmManager  *  CRRobotArmManager_Create(void) {
    CRRobotArmManager*  me  =  (CRRobotArmManager  *)
malloc(sizeof(CRRobotArmManager));
    if(me!=NULL)
        {
            CRRobotArmManager_Init(me);
        }
    return me;
}

void CRRobotArmManager_Destroy(CRRobotArmManager*  const me) {
    if(me!=NULL)
        {
            CRRobotArmManager_Cleanup(me);
        }
    free(me);
}

static void cleanUpRelations(CRRobotArmManager*  const me) {
    if(me->itsCRDisplay !=  NULL)
        {
            me->itsCRDisplay  =  NULL;
        }
    if(me->itsRobotArm !=  NULL)
        {
            me->itsRobotArm  =  NULL;
        }
```

```
    if(me->itsUserInput != NULL)
        {
            me->itsUserInput = NULL;
        }
    }
```

Code Listing 4-17: CRRobotArmManager.c

A little bit about semaphores

Edsger Dijkstra invented the concept of a software semaphore for the coordination of tasks wanting to share resources. The most common implementation is a *counting semaphore* which allows, at most n tasks, to use a resource and blocks additional tasks should they attempt to access the resource until one of the tasks that has successfully locked (said to *own* the semaphore) unlocks the resource. n is said to be the *capacity* of the semaphore. A *binary semaphore* is a counting semaphore for which n = 1. A *mutual exclusion (mutex) semaphore* is a binary semaphore with some extra properties, which are OS dependent. Binary semaphores are used for both mutual exclusion and event notification, while mutexes are used only to enforce mutual exclusion.

Semaphore are initialized with n when whey are created, something like

```
    semaPtr = OS_create_semaphore(int n);
```

which initializes an internal thread counter to n.

The semaphore offers two other services, lock() and release() (which, being Dutch, Dijkstra called P (for *Prolaag* or "test")[1] and V(for *verhoog* or "increase")). lock() works by decrementing the internal counter by 1. If the count has decremented to 0 or less, then the thread calling the lock() function is blocked by the scheduler. The release() function increments the semaphore counter, so that if there is are threads waiting on that lock, one of them will unblock and run (that is, its call to lock() now returns). Note that you usually can't test to see if a call to lock() will block or not before the thread calls it; it will either succeed immediately or block the thread and (hopefully) be released later. Put another way, there is usually no way to know if there are threads blocking on that semaphore or not before you call lock(). However, some RTOS give a trylock() function that locks if the resource is available or returns an error code if not.

```
void lock(Semaphore* s) {
    do {
            wait();
    } while (s->count <= 0);
    --s->count; /* this operation must be implemented as a
                    critical region */
```

---

[1] Or it could be P for "Piraat" (pirate). No one seems to know for sure.

```
};

void release(Semaphore* s) {
    ++s->count; /*  this operation must be implemented as a
                    critical region  */
}
```

Other terms for lock include *decrement* and *wait*; other terms for release include *free, increment,* and *signal,* used in symmetric pairs, (lock/free, lock/release, wait/signal, decrement/increment) depending on the usage of the semaphore.

Mutexes may be *recursive,* a useful property in which once a thread has successfully locked the semaphore, further calls it makes to `lock()` do not block the thread (that is, the thread cannot block itself through multiple calls to the same mutex). For a recursive mutex, the semaphore is freed by a thread when the number of calls it makes to free equals the number of calls it made to `lock()`.

4.5 Guarded Call Pattern

The Guarded Call Pattern serializes access to a set of services that could potentially interfere with each other in some way if called simultaneously by multiple callers. In this pattern, access is serialized by providing a locking mechanism that prevents other threads from invoking the services while locked.

4.5.1 Abstract

The Guarded Call Pattern uses semaphores to protect a related set of services (functions) from simultaneous access by multiple clients in a preemptive multitasking environment, in a process known as *mutual exclusion.* The Guarded Call Pattern provides timely access to services, provided that they are not in use by other tasks. However, if not mixed with other patterns, the use of this pattern can lead to unbounded priority inversion, such as that shown in Figure 4-5.

4.5.2 Problem

The problem this pattern addresses is the need for a timely synchronization or data exchange between threads. In such cases, it may not be possible to wait for an asynchronous (queuing) rendezvous. A synchronous (function call) rendezvous can be made more timely, but this must be done carefully to avoid data corruption and erroneous computation.

4.5.3 Pattern Structure

The pattern structure for this pattern is shown in Figure 4-15. In this case, multiple `PreemptiveTasks` access the `GuardedResource` through its functions. Inside those

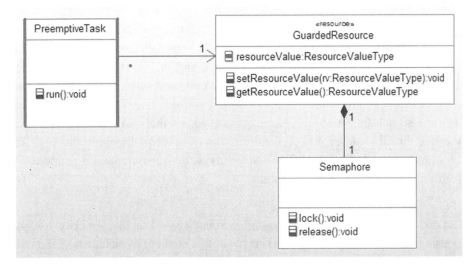

Figure 4-15: Guarded Call Pattern

functions, calls are made to lock and release the resource. The scheduler supports blocking by placing a task on the blocked queue when it invokes a locked semaphore and unblocks it when that semaphore is released. The scheduler must implement the semaphore lock() function as a critical region to eliminate the possibility of race conditions.

4.5.4 Collaboration Roles

This section describes the roles for this pattern.

4.5.4.1 GuardedResource

The GuardedResource is a shared resource that employs mutex semaphores to enforce mutual exclusion of access to its methods. Inside the relevant functions[14], and prior to the access of the resource *per se*, a call is made to the lock() function of the associated Semaphore instance. if the Semaphore is in the unlocked state, then it becomes locked; if it is in the locked state, then the Semaphore signals the StaticPriorityScheduler to block the currently running task. It is important that the relevant functions of a specific resource instance must all share the same instance of the Semaphore class. This ensures that they are protected together as a unit from simultaneous access by multiple PreemptiveTasks.

4.5.4.2 PreemptiveTask

The PreemptiveTask represents an «active» class that is run via a preemptive multitasking scheduler. It accesses the GuardedResource by invoking the latter's functions, which in turn are protected by the Semaphore.

---

[14] I say "relevant functions" because the class may offer some functions that don't require protection.

4.5.4.3 Semaphore

The Semaphore is a mutual exclusion semaphore class that serializes access to a
GuardedResource. The guarded functions of the GuardedResource invoke the lock()
function of the Semaphore whenever they are called and release()once the service is
complete. Other client threads that attempt to invoke a service when its associated Semaphore
is locked become blocked until the Semaphore is unlocked. The blocking is performed when
the Semaphore signals the StaticPriorotyScheduler that a call attempt was made by the
currently active thread. The mutexID identifies which Mutex is blocking the task. This
element is usually provided by the RTOS. Note this is the same element that appears in the
Static Priority Pattern earlier in this chapter.

4.5.4.4 StaticPriorityScheduler

This object orchestrates the execution of multiple threads based on their priority. In this pattern,
we are focusing on the responsibility (not shown) of this element to block a task if it attempts to
lock an already-locked Semaphore and to unblock all tasks blocked on the mutexID when
this lock is released. This element is usually provided by the RTOS. Note this is the same
element that appears in the Static Priority Pattern earlier in this chapter.

4.5.5 Consequences

This pattern provides a timely access to a resource and at the same time prevents multiple
simultaneous accesses that can result in data corruption and system misbehavior. If the resource
is not locked, then access proceeds without delay. If the resource is currently locked, then the
caller must block until the lock on the resource is released. Naïve use can result in unbounded
priority inversion in the manner described in Figure 4-5. See Section 4.5.7 Related Patterns for
a discussion of an RTOS-supported solution.

4.5.6 Implementation Strategies

The tricky part of implementing the pattern lies in the implementation of the Mutex. Usually,
the RTOS will provide a Semaphore instance, accessible through operations such as:

- OSSemaphore OS_create_semaphore()
  creates a semaphore and returns a pointer to it.

- void OS_destroy_semaphore(OSSemaphore*)
  destroys the semaphore pointed at.

- void OS_lock_semaphore(OSSemaphore*)
  locks the relevant semaphore. If called with an unlocked OSSemaphore, it becomes
  locked. If already locked, then it blocks the current task with its internally-stored mutexID.
  The scheduler must store the task data including the continuation entry point and stack.

- `void OS_release_semaphore(OSSemaphore*)`
  releases the lock on the relevant semaphore and unblocks all tasks blocked on its internally-stored `MutexID`.

Of course, the names and parameters of the services are OS specific.

4.5.7 Related Patterns

The Guarded Call Pattern is used with Static Priority or other preemptive multitasking environments. It is a less draconian policy than is enforced by the Critical Region Pattern (above) because it does not interfere with the execution of higher priority tasks that don't need access to the resource. It is more responsive than the Queuing Pattern (below) because it doesn't need to wait for the resource's thread to become active to check for the request.

There is one very important class of variants of this pattern that solves the problem of unbounded priority inversion that uses a concept of *priority inheritance.* The basic idea is that each resource has an addition attribute (variable) knows as its *priority ceiling,* which is equal to the priority of the highest-priority task that can ever access the resource. The underlying problem with naïve implementation of the Guarded Call Pattern is that intermediate priority tasks that don't need the resource can preempt the low-priority task that currently owns the resource needed by the blocked high-priority task. There can be any number of such intermediate-priority tasks, leading to chained blocking of the high priority task. The priority inheritance solution is to elevate the priority of the low-priority task as soon as the high-priority task attempts to lock the resource owned by the low-priority task. This can be done by creating a semaphore that implements the priority inversion. This is the approach taken, for example, by VxWorks™ from Wind River, which implements a set of mutex semaphore functions such as:

```
SEM_ID semM;
semM = semMCreate(. . .);
semTake(semM, FLAG);
semGive(semM);
```

Values for FLAG include WAIT_FOREVER, and – important to this discussion – SEM_INVERSION_SAFE. When the latter flag value is used, VxWorks' mutex semaphores implement priority inheritance.

In Figure 4-5, if the locking mutex is implemented in this way, the following sequence of events holds:

1. Task Z runs and locks Resource R at its nominal priority of 100.
2. Task A runs at priority 1, and attempts to lock the mutex guarding R and blocks. At this point, the priority of Task Z is elevated to that of Task A (i.e., priority 1).

3. Task Y becomes ready to run. With a normal naïve mutex, it would run and preempt Task Z. However, its priority (99) is less than the priority at which Task Z is running so it remains on the Task Ready queue.

4. Task Z completes its use of the Resource R and unlocks it. At this point, the priority of Task Z is demoted to its nominal priority (100).

5. Task A unblocks and now uses the resource and completes within its deadline.

6. Task Y now, being the highest ready task, runs.

7. Task Y completes, and finally, Task Z is allowed to complete its execution.

You can see in this simple example how priority inheritance prevented the unbounded priority inversion that a normal counting semaphore would allow.

4.5.8 Example

The example shown in Figure 4-16 is a portion of a flight control system. In it, the ship KinematicData (position and attitude) are managed as a shared resource for multiple clients.

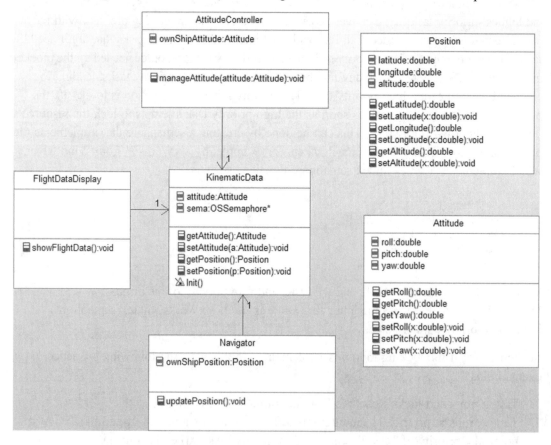

Figure 4-16: Guarded Call example

Since the setting and getting of these values is not an atomic process, there is a possibility of displaying invalid data if, for example, the `FlightDataDisplay showFlightData()` function is preempted to update the attitude data by the `AttitudeController`. The use of semaphores in the methods ensures that this cannot happen.

The next several code listings (Code Listing 4-19 through Code Listing 4-30) give the code for the elements of the Guarded Call example in Figure 4-16. The first gives the specification for the OS calls used for the semaphores. The next four show the code for the complex data types `Position` and `Attitude`, respectively.

Code Listing 4-23 and Code Listing 4-24 give the code for the resource (`KinematicData`) that uses the `OSSemaphore` to guard its operations to set and get that data. The initializer `KinematicData_Init()` creates a semaphore via a call to `OS_create_semaphore()` while the clean-up function `KinematicData_CleanUp()` calls `OS_destroy_semaphore ()` to free up its memory.   Its accessor and mutator methods to get and set attitude and position data all invoke the `OS_lock_semaphore()`, do their processing, and then invoke `OS_release_semaphore()`.

The code listings after that show the code for the clients: the `FlightDataDisplay`, `AttitudeController`, and `Navigator`. These all share the information stored in the `KinematicData` object and so must be protected from data corruption.

```
#ifndef GuardedCallExample_H
#define GuardedCallExample_H


struct Attitude;
struct AttitudeController;
struct FlightDataDisplay;
struct GuardTester;
struct GuardedCallBuilder;
struct KinematicData;
struct Navigator;
struct Position;

/*  OS provided types and functions  */
struct OSSemaphore;
struct OSSemaphore*  OS_create_semaphore(void);
void OS_destroy_semaphore(struct OSSemaphore*  sPtr);
void OS_lock_semaphore(struct OSSemaphore*  sPtr);
void OS_release_semaphore(struct OSSemaphore*  sPtr);

#endif
```

Code Listing 4-7: GuardedCallExample.h

```
#ifndef Position_H
#define Position_H

#include "GuardedCallExample.h"

typedef struct Position Position;
struct Position {
    double altitude;           /*## attribute altitude  */
    double latitude;           /*## attribute latitude  */
    double longitude;          /*## attribute longitude  */
};

/*  Constructors and destructors:*/
void Position_Init(Position*  const me);
void Position_Cleanup(Position*  const me);

/*  Operations  */
double Position_getAltitude(Position*  const me);
double Position_getLatitude(Position*  const me);
double Position_getLongitude(Position*  const me);
void Position_setAltitude(Position*  const me, double x);
void Position_setLatitude(Position*  const me, double x);
void Position_setLongitude(Position*  const me, double x);

Position  *  Position_Create(void);
void Position_Destroy(Position*  const me);

#endif
```

Code Listing 4-19: Position.h

```
#include "Position.h"

void Position_Init(Position*  const me) {
}

void Position_Cleanup(Position*  const me) {
}

double Position_getAltitude(Position*  const me) {
    return me->altitude;
}

double Position_getLatitude(Position*  const me) {
    return me->latitude;
}
```

```c
double Position_getLongitude(Position*  const me) {
    return me->longitude;
}


void Position_setAltitude(Position*  const me, double x) {
    me->altitude  =  x;
}


void Position_setLatitude(Position*  const me, double x) {
    me->latitude  =  x;
}


void Position_setLongitude(Position*  const me, double x) {
    me->longitude  =  x;
}


Position  *  Position_Create(void) {
    Position*  me  =  (Position  *) malloc(sizeof(Position));
    if(me!=NULL) {
            Position_Init(me);
    }
  return me;
}


void Position_Destroy(Position*  const me) {
    if(me!=NULL) {
        Position_Cleanup(me);
    }
    free(me);
}
```

Code Listing 4-8: Position.c

```c
#ifndef Attitude_H
#define Attitude_H

#include "GuardedCallExample.h"

typedef struct Attitude Attitude;
struct Attitude {
    double pitch;          /*## attribute pitch  */
    double roll;           /*## attribute roll   */
    double yaw;            /*## attribute yaw    */
};

/*  Constructors and destructors:*/
void Attitude_Init(Attitude*  const me);
void Attitude_Cleanup(Attitude*  const me);
```

```
/* Operations */
double Attitude_getPitch(Attitude* const me);
double Attitude_getRoll(Attitude* const me);
double Attitude_getYaw(Attitude* const me);
void Attitude_setPitch(Attitude* const me, double x);
void Attitude_setRoll(Attitude* const me, double x);
void Attitude_setYaw(Attitude* const me, double x);

Attitude * Attitude_Create(void);
void Attitude_Destroy(Attitude* const me);

#endif
```

Code Listing 4-9: Attitude.h

```
#include "Attitude.h"

void Attitude_Init(Attitude* const me) {
}

void Attitude_Cleanup(Attitude* const me) {
}

double Attitude_getPitch(Attitude* const me) {
    return me->pitch;
}

double Attitude_getRoll(Attitude* const me) {
    return me->roll;
}

double Attitude_getYaw(Attitude* const me) {
    return me->yaw;
}

void Attitude_setPitch(Attitude* const me, double x) {
    me->pitch = x;
}

void Attitude_setRoll(Attitude* const me, double x) {
    me->roll = x;
}

void Attitude_setYaw(Attitude* const me, double x) {
    me->yaw = x;
}
```

```
Attitude * Attitude_Create(void) {
    Attitude* me = (Attitude *) malloc(sizeof(Attitude));
    if(me!=NULL)
        {
            Attitude_Init(me);
        }
    return me;
}


void Attitude_Destroy(Attitude* const me) {
    if(me!=NULL)
        {
            Attitude_Cleanup(me);
        }
    free(me);
}
```

Code Listing 4-10: Attitude.c

```
#ifndef KinematicData_H
#define KinematicData_H

#include "GuardedCallExample.h"
#include "Attitude.h"
#include "OSSemaphore.h"
#include "Position.h"

typedef struct KinematicData KinematicData;
struct KinematicData {
    struct Attitude attitude;
    struct Position position;
    OSSemaphore* sema;        /*## mutex semaphore  */
};

/*  Constructors and destructors:*/
void KinematicData_Init(KinematicData* const me);
void KinematicData_Cleanup(KinematicData* const me);

/*  Operations  */

Attitude KinematicData_getAttitude(KinematicData* const me);
struct Position KinematicData_getPosition(KinematicData* const me);

void KinematicData_setAttitude(KinematicData* const me, Attitude a);

void KinematicData_setPosition(KinematicData* const me, Position p);
```

```
KinematicData  *  KinematicData_Create(void);

void KinematicData_Destroy(KinematicData* const me);

#endif
```

Code Listing 4-11: KinematicData.h

```
#include "KinematicData.h"

void KinematicData_Init(KinematicData*  const me) {
    Attitude_Init(&(me->attitude));
    Position_Init(&(me->position));
    me->sema = OS_create_semaphore();
}

void KinematicData_Cleanup(KinematicData*  const me) {
    OS_destroy_semaphore(me->sema);
}

struct Position KinematicData_getPosition(KinematicData*  const me) {
    Position p;

    /*  engage the lock  */
    OS_lock_semaphore(me->sema);

    p = me->position;

    /*  release the lock  */
    OS_release_semaphore(me->sema);

    return p;
}

void KinematicData_setAttitude(KinematicData*  const me, Attitude a) {
  /*  engage the lock  */
  OS_lock_semaphore(me->sema);

  me->attitude = a;

  /*  release the lock  */
  OS_release_semaphore(me->sema);
}

void KinematicData_setPosition(KinematicData*  const me, Position p) {
  /*  engage the lock  */
  OS_lock_semaphore(me->sema);
```

```
    me->position = p;

    /* release the lock */
    OS_release_semaphore(me->sema);

}

KinematicData * KinematicData_Create(void) {
    KinematicData* me = (KinematicData *)
malloc(sizeof(KinematicData));
    if(me!=NULL) {
        KinematicData_Init(me);
    }
    return me;
}
void KinematicData_Destroy(KinematicData* const me) {
    if(me!=NULL) {
        KinematicData_Cleanup(me);
    }
    free(me);
}
Attitude KinematicData_getAttitude(KinematicData* const me) {
    Attitude a;

    /* engage the lock */
    OS_lock_semaphore(me->sema);

    a = me->attitude;

    /* release the lock */
    OS_release_semaphore(me->sema);

    return a;
}
```

Code Listing 4-24: KinematicData.c

```
#ifndef AttitudeController_H
#define AttitudeController_H

#include "GuardedCallExample.h"
#include "Attitude.h"
struct KinematicData;

typedef struct AttitudeController AttitudeController;
struct AttitudeController {
    struct Attitude ownShipAttitude;
```

```
     struct KinematicData*  itsKinematicData;
};

/* Constructors and destructors:*/

void AttitudeController_Init(AttitudeController* const me);
void AttitudeController_Cleanup(AttitudeController* const me);

/* Operations */

void AttitudeController_manageAttitude(AttitudeController* const me);

struct KinematicData*
AttitudeController_getItsKinematicData(const AttitudeController*
const me);

void AttitudeController_setItsKinematicData(AttitudeController*
const me, struct KinematicData* p_KinematicData);

AttitudeController * AttitudeController_Create(void);

void AttitudeController_Destroy(AttitudeController* const me);

#endif
```

Code Listing 4-12: AttitudeController.h

```
#include "AttitudeController.h"
#include "KinematicData.h"

static void cleanUpRelations(AttitudeController* const me);

void AttitudeController_Init(AttitudeController* const me) {
    Attitude_Init(&(me->ownShipAttitude));
    me->itsKinematicData = NULL;
}

void AttitudeController_Cleanup(AttitudeController* const me) {
    cleanUpRelations(me);
}

void AttitudeController_manageAttitude(AttitudeController* const me) {
    KinematicData_setAttitude(me->itsKinematicData, me->ownShipAttitude);
}

struct KinematicData*
AttitudeController_getItsKinematicData(const AttitudeController* const me) {
```

```
    return (struct KinematicData*)me->itsKinematicData;
}

void AttitudeController_setItsKinematicData(AttitudeController*
const me, struct KinematicData* p_KinematicData) {
    me->itsKinematicData = p_KinematicData;
}

AttitudeController * AttitudeController_Create(void) {
    AttitudeController* me = (AttitudeController *)
malloc(sizeof(AttitudeController));
    if(me!=NULL) {
        AttitudeController_Init(me);
    }
    return me;
}

void AttitudeController_Destroy(AttitudeController* const me) {
    if(me!=NULL) {
        AttitudeController_Cleanup(me);
    }
    free(me);
}

static void cleanUpRelations(AttitudeController* const me) {
    if(me->itsKinematicData != NULL) {
        me->itsKinematicData = NULL;
    }
}
```

Code Listing 4-13: AttitudeController.c

```
#ifndef Navigator_H
#define Navigator_H

#include "GuardedCallExample.h"
#include "Position.h"
struct KinematicData;

typedef struct Navigator Navigator;
struct Navigator {
    struct Position ownShipPosition;
    struct KinematicData* itsKinematicData;
};

/* Constructors and destructors:*/
void Navigator_Init(Navigator* const me);
void Navigator_Cleanup(Navigator* const me);
```

```
/* Operations */
void Navigator_updatePosition(Navigator* const me);

struct KinematicData* Navigator_getItsKinematicData(const Navigator* const me);

void Navigator_setItsKinematicData(Navigator* const me, struct
KinematicData* p_KinematicData);

Navigator * Navigator_Create(void);

void Navigator_Destroy(Navigator* const me);

#endif
```

Code Listing 4-14: Navigator.h

```
#include "Navigator.h"
#include "KinematicData.h"

static void cleanUpRelations(Navigator* const me);

void Navigator_Init(Navigator* const me) {
    Position_Init(&(me->ownShipPosition));
    me->itsKinematicData = NULL;
}

void Navigator_Cleanup(Navigator* const me) {
    cleanUpRelations(me);
}

void Navigator_updatePosition(Navigator* const me) {
    KinematicData_setPosition(me->itsKinematicData, me->ownShipPosition);
}

struct KinematicData* Navigator_getItsKinematicData(const
Navigator* const me) {
    return (struct KinematicData*)me->itsKinematicData;
}

void Navigator_setItsKinematicData(Navigator* const me, struct
KinematicData* p_KinematicData) {
    me->itsKinematicData = p_KinematicData;
}

Navigator * Navigator_Create(void) {
    Navigator* me = (Navigator *)malloc(sizeof(Navigator));
    if(me!=NULL) {
        Navigator_Init(me);
    }
```

```
        return me;
    }

    void Navigator_Destroy(Navigator* const me) {
        if(me!=NULL) {
            Navigator_Cleanup(me);
        }
        free(me);
    }

    static void cleanUpRelations(Navigator* const me) {
        if(me->itsKinematicData != NULL) {
            me->itsKinematicData = NULL;
        }
    }
```

Code Listing 4-15: Navigator.c

```
    #ifndef FlightDataDisplay_H
    #define FlightDataDisplay_H

    #include "GuardedCallExample.h"
    struct KinematicData;

    typedef struct FlightDataDisplay FlightDataDisplay;
    struct FlightDataDisplay {
        struct KinematicData* itsKinematicData;
    };

    /* Constructors and destructors:*/
    void FlightDataDisplay_Init(FlightDataDisplay* const me);
    void FlightDataDisplay_Cleanup(FlightDataDisplay* const me);

    /* Operations */

    void FlightDataDisplay_showFlightData(FlightDataDisplay* const me);

    struct KinematicData* FlightDataDisplay_getItsKinematicData(const
    FlightDataDisplay* const me);

    void FlightDataDisplay_setItsKinematicData(FlightDataDisplay*
    const me, struct KinematicData* p_KinematicData);

    FlightDataDisplay * FlightDataDisplay_Create(void);

    void FlightDataDisplay_Destroy(FlightDataDisplay* const me);

    #endif
```

Code Listing 4-16: FlightDataDisplay.h

```c
#include "FlightDataDisplay.h"
#include "KinematicData.h"
#include <stdio.h>

static void cleanUpRelations(FlightDataDisplay* const me);

void FlightDataDisplay_Init(FlightDataDisplay* const me) {
    me->itsKinematicData = NULL;
}

void FlightDataDisplay_Cleanup(FlightDataDisplay* const me) {
    cleanUpRelations(me);
}

void FlightDataDisplay_showFlightData(FlightDataDisplay* const me) {
    Attitude a;
    Position p;
    a = KinematicData_getAttitude(me->itsKinematicData);
    p = KinematicData_getPosition(me->itsKinematicData);
    printf("Roll, pitch, yaw = %f, %f, %f \n", a.roll, a.pitch, a.yaw);
    printf("Lat, Long, Altitude = %f, %f, %f\n", p.latitude, p.longitude, p.altitude);
}

struct KinematicData* FlightDataDisplay_getItsKinematicData(const FlightDataDis-
play* const me) {
    return (struct KinematicData*)me->itsKinematicData;
}

void FlightDataDisplay_setItsKinematicData(FlightDataDisplay* const me, struct Kine-
maticData* p_KinematicData) {
    me->itsKinematicData = p_KinematicData;
}

FlightDataDisplay * FlightDataDisplay_Create(void) {
    FlightDataDisplay* me = (FlightDataDisplay *)malloc(sizeof
(FlightDataDisplay));
    if(me!=NULL) {
    FlightDataDisplay_Init(me);
    }
    return me;
}

void FlightDataDisplay_Destroy(FlightDataDisplay* const me) {
    if(me!=NULL) {
    FlightDataDisplay_Cleanup(me);
    }
    free(me);
}
```

```
static void cleanUpRelations(FlightDataDisplay* const me) {
    if(me->itsKinematicData != NULL) {
    me->itsKinematicData = NULL;
    }
}
```

Code Listing 4-30: FlightDataDisplay.c

4.6 Queuing Pattern

The Queuing Pattern is the most common implementation of task asynchronous communication. It provides a simple means of communication between tasks that are uncoupled in time. The Queuing Pattern accomplishes this communication by storing messages in a queue, a nominally first-in-first-out data structure. The sender deposits a message in the queue and some time later, the receiver withdraws the message from the queue. It also provides a simple means of serializing access to a shared resource. The access messages are queued and handled at a later time and this avoids the mutual exclusion problem common to sharing resources.

4.6.1 Abstract

The Message Queuing Pattern uses asynchronous communications, implemented via queued messages, to synchronize and share information among tasks. This approach has the advantage of simplicity and does not experience mutual exclusion problems because no resource is shared by reference. Any information shared among threads is *passed by value* to the separate thread. While this limits the complexity of the collaboration among the threads, this approach in its pure form is immune to the standard resource corruption problems that plague concurrent systems that share information *passed by reference*. In passed by value sharing, a copy of the information is made and sent to the receiving thread for processing. The receiving thread fully owns the data it receives and so may modify them freely without concern for corrupting data due to multiple writers or due to sharing them among a writer and multiple readers. A downside of the pattern is that the message isn't processed immediately after the sender posts it; the process waits until the receiver task runs and can process the waiting messages.

4.6.2 Problem

In multithreaded systems, tasks must synchronize and share information with others. Two primary things must be accomplished. First, the tasks must synchronize to permit sharing of the information. Secondly, the information must be shared in such a way that there is no chance of corruption or race conditions. This pattern addresses such task interaction.

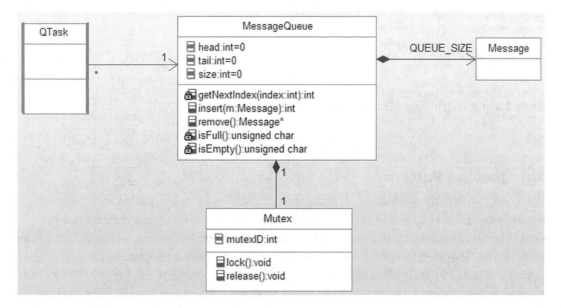

Figure 4-17: Queuing Pattern

4.6.3 Pattern Structure

The Queuing Pattern structure is shown in Figure 4-17. The QUEUE_SIZE declaration deter-mines the maximum number of elements that the queue can retain. Care must be taken to ensure that this is large enough to handle the worst-case situation for your system usage, while not being so large as to waste memory.

4.6.4 Collaboration Roles

This section describes the roles for this pattern.

4.6.4.1 Message

The Message class is an abstraction of a message of relevance between QTasks. It can be anything from a simple data value to an elaborate datagram structure such as used in TCP/IP messaging.

4.6.4.2 MessageQueue

The MessageQueue is the storage of information being exchanged between QTasks. The MessageQueue can store up to a maximum of QUEUE_SIZE message elements. The MessageQueue provides the following methods:

- int MessageQueue_getNextIndex(MessageQueue* me, int index)
  A private (static) function that uses modulo arithmetic to compute the next valid index value.

- `int MessageQueue_insert(MessageQueue*  me, Message m)`
  A public function that, if the `MessageQueue` isn't full, inserts a `Message` into the queue at the head position and updates the `head` index. It returns 0 if the insert succeeds and 1 if the `MessageQueue` was full.

- `Message* MessageQueue_remove(MessageQueue*  me)`
  A public function that, if the `MessageQueue` isn't empty, removes the oldest message, creates a new heap variable to hold the data, updates the `tail` index, and returns that pointer to the caller. It returns a pointer to a `Message` if successful and a `NULL` pointer if not.

- `unsigned char MessageQueue_isFull(MessageQueue*  me)`
  A private function that returns 1 if the `MessageQueue` is full and 0 otherwise.

- `unsigned char MessageQueue_isEmpty(MessageQueue*  me)`
  A private function that returns 1 if the `MessageQueue` is empty and 1 otherwise.

4.6.4.3 Mutex

The `Mutex` is a mutual exclusion semaphore object that serializes access to a `MessageQueue`. The guarded functions of the `MessageQueue` invoke the `lock()` function of the `Mutex` whenever they are called and `release()` once the service is complete. Other client threads that attempt to invoke a service when its associated `Mutex` is locked become blocked until the `Mutex` is unlocked. This element is usually provided by the RTOS. Note this is the same element that appears in the Static Priority Pattern earlier in this chapter.

4.6.4.4 Qtask
The `QTask` class is a client of the `MessageQueue` and invokes either the `insert()` or `remove()` functions to access the data held there.

4.6.5 Consequences

The Queuing Pattern provides good serialization of access to the data as they are passed among tasks. The use of the mutex ensures that the MessageQueue itself isn't corrupted by simultaneous access to the MessageQueue's services. The reception of the data is less timely than the Guarded Call Pattern since the Queuing Pattern implements asynchronous communication.

The number of elements held by the queue can be quite large; this is useful as the products of the elements run in a "bursty" fashion, allowing for the consumer of the elements to catch up later in time. Poorly sized queues can lead to data loss (if too small) or wasted memory (if too large). Very small queues (that can hold one or two elements) can be used for simple asynchronous data exchange when the problem of data loss isn't present.

4.6.6 Implementation Strategies

The simplest implementation of a queue is as an array of message elements. This has the advantage of simplicity but lacks the flexibility of a linked list.

Queues are fairly simple to implement but there are many possible variants. Sometimes some messages are more urgent or important than others and should be processed before waiting for lower priority messages. The MessageQueue can be modified to implement priority by adding either multiple buffers (one for each priority) or by inserting the element in the queue on the basis of the message priority (although this can increase the timing necessary for an element insertion and deletion). In this case, the message should have a priority attribute to enable the MessageQueue to properly insert and remove the element.

In complex systems, it may not be feasible to predict an optimal queue size. In such cases, extensible queues can be implemented. When the MessageQueue is full, more memory is allocated. A linked list implementation of the MessageQueue is more appropriate for an extensible queue. It may be difficult to determine under what circumstances it is appropriate to shrink the queue and release unused memory.

It sometimes happens that the potential number of elements exceeds memory capacity. In this case, a cached queue can be created; in such a queue, there are essentially three storage locations; a local memory newest data buffer, a local memory oldest data buffer, and a file that holds data of intermediate age. When the local new data buffer becomes full, it is emptied to the file system. When the local old data buffer becomes empty, the data from the file system are loaded into the local old data buffer and removed from the disk (or copied from the new data buffer if the file is empty). This allows storage of potentially huge volumes of data, but caching the data can be time intensive.

4.6.7 Related Patterns

The Queuing Pattern serializes access to data by queuing data or commands, allowing the recipient to handle them one at a time. It does this asynchronously, so the time between the sending of the message and its processing is uncoupled. This may not meet the performance needs of the system. The Guarded Call Pattern also serializes access but does so synchronously, so that data and command transfers generally happen closer together in time. However, the Guarded Call Pattern can lead to unbounded priority inversion if not used carefully. In addition, if the receiver isn't ready to receive the message from the sender, the sender must block or take another action.

4.6.8 Example

Figure 4-18 shows an example of producers and consumers that share data via a queue. The key element for this pattern is the GasDataQueue. It implements a small 10-element queue of GasData. GasData is a struct that contains an enumerated GAS_TYPE (with enumerated values

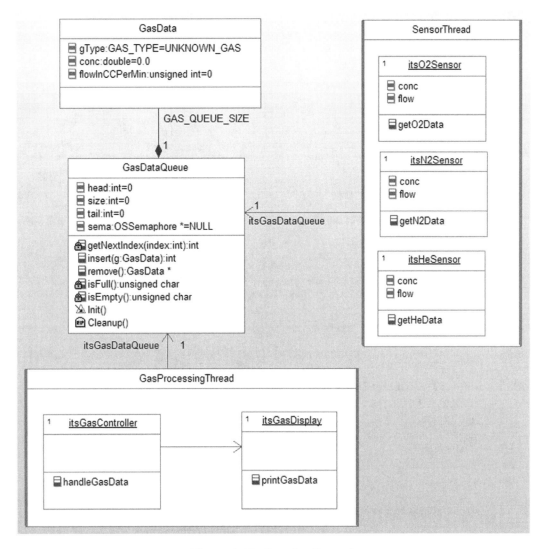

Figure 4-18: Queuing Example

O2_GAS, N2_GAS, HE_GAS, and UNKNOWN_GAS), the gas concentration (a double), and the flow rate in ccs per `minute` (`int`). Code Listing 4-31 gives the `#defines` and general declarations. Code Listing 4-32 and Code Listing 4-33 give the code for the `GasData` struct.

```
#ifndef QueuingExample_H
#define QueuingExample_H

struct GasController;
struct GasData;
struct GasDataQueue;
struct GasDisplay;
```

```
struct GasProcessingThread;
struct HeSensor;
struct N2Sensor;
struct O2Sensor;
struct OSSemaphore;
struct SensorThread;

typedef enum GAS_TYPE {
    O2_GAS,
    N2_GAS,
    HE_GAS,
    UNKNOWN_GAS
} GAS_TYPE;

/* define the size of the queue */
#define GAS_QUEUE_SIZE (10)

/* OS semaphore services */
struct OSSemaphore* OS_create_semaphore(void);
void OS_destroy_semaphore(struct OSSemaphore* sPtr);
void OS_lock_semaphore(struct OSSemaphore* sPtr);
void OS_release_semaphore(struct OSSemaphore* sPtr);

#endif
```

Code Listing 4-17: GasDataExample.h

```
#ifndef GasData_H
#define GasData_H

#include "QueuingExample.h"

typedef struct GasData GasData;
struct GasData {
    double conc;
    unsigned int flowInCCPerMin;
    GAS_TYPE gType;
};

/* Constructors and destructors:*/

void GasData_Init(GasData* const me);
void GasData_Cleanup(GasData* const me);
GasData * GasData_Create(void);
void GasData_Destroy(GasData* const me);

#endif
```

Code Listing 4-32: GasData.h

```
#include "GasData.h"

void GasData_Init(GasData* const me) {
    me->conc = 0.0;
    me->flowInCCPerMin = 0;
    me->gType = UNKNOWN_GAS;
}

void GasData_Cleanup(GasData* const me) {
}

GasData * GasData_Create(void) {
    GasData* me = (GasData *) malloc(sizeof(GasData));
    if(me!=NULL) {
        GasData_Init(me);
    }
    return me;
}

void GasData_Destroy(GasData* const me) {
    if(me!=NULL) {
        GasData_Cleanup(me);
    }
    free(me);
}
```

Code Listing 4-33: GasData.c

The GasDataQueue is the center of the pattern. Its two operations insert() and remove() enqueue new data and dequeue oldest data, respectively. The insert() function returns TRUE if it succeeded and FALSE otherwise. The remove() function returns a pointer to a GasData instance for which the client is responsible for deleting. If the remove() function fails, it returns a NULL pointer.

The class also includes some private (static) helper functions such as getNextIndex() which computes the next index using modulo arithmetic to wrap around the top of the queue, isEmpty(), which returns TRUE if the queue has no elements, and isFull(), which returns TRUE if there is no space to hold additional elements.

Code Listing 4-34 and Code Listing 4-35 give the C code for the GasDataQueue.

```
#ifndef GasDataQueue_H
#define GasDataQueue_H

#include "QueuingExample.h"
#include "GasData.h"
```

```
#include "OSSemaphore.h"

typedef struct GasDataQueue GasDataQueue;
struct GasDataQueue {
    int head;
    OSSemaphore * sema;
    int size;
    int tail;
    struct GasData itsGasData[GAS_QUEUE_SIZE];
};

/* Constructors and destructors:*/

void GasDataQueue_Init(GasDataQueue* const me);
void GasDataQueue_Cleanup(GasDataQueue* const me);

/* Operations */
int GasDataQueue_insert(GasDataQueue* const me, GasData g);

GasData * GasDataQueue_remove(GasDataQueue* const me);

int GasDataQueue_getItsGasData(const GasDataQueue* const me);

GasDataQueue * GasDataQueue_Create(void);

void GasDataQueue_Destroy(GasDataQueue* const me);

#endif
```

Code Listing 4-34: GasDataQueue.h

```
#include "GasDataQueue.h"
#include <stdio.h>

/* private (static) methods */
static void cleanUpRelations(GasDataQueue* const me);
static int getNextIndex(GasDataQueue* const me, int index);
static unsigned char isEmpty(GasDataQueue* const me);
static unsigned char isFull(GasDataQueue* const me);
static void initRelations(GasDataQueue* const me);

void GasDataQueue_Init(GasDataQueue* const me) {
    me->head = 0;
    me->sema = NULL;
    me->size = 0;
    me->tail = 0;
    initRelations(me);
```

```
    me->sema  =  OS_create_semaphore();
}


void GasDataQueue_Cleanup(GasDataQueue*  const me) {
    OS_destroy_semaphore(me->sema);
    cleanUpRelations(me);
}


/*
    Insert puts new gas data elements into the queue
    if possible. It returns 1  if successful, 0  otherwise.
*/
int GasDataQueue_insert(GasDataQueue*  const me, GasData g) {
    OS_lock_semaphore(me->sema);
    if (!isFull(me)) {
        me->itsGasData[me->head]  =  g;
        me->head  =  getNextIndex(me, me->head);
        ++me->size;

        /*  instrumentation  */
        /*  print stuff out, just to visualize the insertions  */
        switch (g.gType) {
        case O2_GAS:
            printf("+++    Oxygen  ");
            break;
        case N2_GAS:
            printf("+++    Nitrogen  ");
            break;
        case HE_GAS:
            printf("+++    Helium  ");
            break;
        default:
                printf("UNKNWON  ");
                break;
        };
        printf("  at conc %f, flow %d\n",g.conc,g.flowInCCPerMin);
        printf("  Number of elements queued %d, head  =  %d, tail  =  %d\n",
            me->size, me->head, me->tail);
        /*  end instrumentation  */

            OS_release_semaphore(me->sema);
        return 1;
    }
    else {  /*  return error indication  */
        OS_release_semaphore(me->sema);
        return 0;
        }
}
/*
        remove creates a  new element on the heap, copies
```

```
        the contents of the oldest data into it, and
        returns the pointer. Returns NULL if the queue
        is empty
    */
GasData * GasDataQueue_remove(GasDataQueue* const me) {
    GasData* gPtr;

    OS_lock_semaphore(me->sema);
    if (!isEmpty(me)) {
        gPtr = (GasData*)malloc(sizeof(GasData));
        gPtr->gType = me->itsGasData[me->tail].gType;
        gPtr->conc = me->itsGasData[me->tail].conc;
        gPtr->flowInCCPerMin = me->itsGasData[me->tail].flowInCCPerMin;
        me->tail = getNextIndex(me, me->tail);
        -me->size;
        /* instrumentation */
        switch (gPtr->gType) {
            case O2_GAS:
                printf("- Oxygen  ");
                break;
            case N2_GAS:
                printf("- Nitrogen  ");
                break;
            case HE_GAS:
                printf("- Helium  ");
                break;
            default:
                    printf("- UNKNWON  ");
                    break;
        };
        printf("  at conc %f, flow %d\n",gPtr->conc,gPtr->flowInCCPerMin);
        printf("  Number of elements queued %d, head = %d, tail = %d\n",
            me->size, me->head, me->tail);
        /* end instrumentation */

        OS_release_semaphore(me->sema);
        return gPtr;
        }
    else {  /* if empty return NULL ptr */
        OS_release_semaphore(me->sema);
        return NULL;
        }
}


static int getNextIndex(GasDataQueue* const me, int index) {
        /* this operation computes the next index from the
        first using modulo arithmetic
        */
    return (index+1) % QUEUE_SIZE;
}
```

```
static unsigned char isEmpty(GasDataQueue*  const me) {
    return (me->size  ==    0);
}

static unsigned char isFull(GasDataQueue*  const me) {
    return (me->size  ==     GAS_QUEUE_SIZE);
}

int GasDataQueue_getItsGasData(const GasDataQueue*  const me) {
    int iter  = 0;
    return iter;
}

GasDataQueue  *  GasDataQueue_Create(void) {
    GasDataQueue*  me  =  (GasDataQueue  *)
malloc(sizeof(GasDataQueue));
    if(me!=NULL) {
        GasDataQueue_Init(me);
    }
    return me;
}

void GasDataQueue_Destroy(GasDataQueue*  const me) {
    if(me!=NULL) {
        GasDataQueue_Cleanup(me);
    }
    free(me);
}

static void initRelations(GasDataQueue*  const me) {
    int iter  = 0;
    while (iter  <  GAS_QUEUE_SIZE){
        GasData_Init(&((me->itsGasData)[iter]));
        iter++;
    }
}

static void cleanUpRelations(GasDataQueue*  const me) {
    int iter  = 0;
    while (iter  <  GAS_QUEUE_SIZE){
        GasData_Cleanup(&((me->itsGasData)[iter]));
        iter++;
    }
}
```

Code Listing 4-35: GasDataQueue.c

In the example (Figure 4-18) two threads run clients. The SensorThread owns three clients itself, an O2Sensor, N2Sensor, and HeSensor. When this thread runs the main function of this thread (`SensorThread_updateData()`), it calls the getXXData() function of each to produce the data, and then marshalls the data up into `GasData` structs and sends them off. The relevant code (less the task initiation and supporting functions) for the `SensorThread` is given in Code Listing 4-36 and Code Listing 4-37.

```
/* ...    initial declaratons stuff above  ...    */

typedef struct SensorThread SensorThread;
struct SensorThread {
    struct GasDataQueue*  itsGasDataQueue;
    struct HeSensor itsHeSensor;
    struct N2Sensor itsN2Sensor;
    struct O2Sensor itsO2Sensor;
};

/* Operations */

void SensorThread_updateData(SensorThread*  const me);

/* ...    other operations declared too  ...    */
#endif
```

Code Listing 4-36: SensorThread.h

```
/*
    updateData runs every period of the sensor thread
    and calls the getXXData() function of each of
    its sensors, then uses a   random number generator
    to decide which data should be published to the
    GasDataQueue.
*/
void SensorThread_updateData(SensorThread*  const me) {
    unsigned char success;
    GasData g;

    GasData_Init(&g);

    O2Sensor_getO2Data(&(me->itsO2Sensor));
    N2Sensor_getN2Data(&(me->itsN2Sensor));
    HeSensor_getHeData(&(me->itsHeSensor));

    if (rand() > RAND_MAX /  3) {
        g.gType =  HE_GAS;
        g.conc = me->itsHeSensor.conc;
        g.flowInCCPerMin = me->itsHeSensor.flow;
```

```
        success  =  GasDataQueue_insert(me->itsGasDataQueue, g);
        if (!success)
            printf("Helium Gas Data queue insertion failed!\n");
    };

    if (rand()  >  RAND_MAX /  3) {
        g.gType  =  N2_GAS;
        g.conc  =  me->itsN2Sensor.conc;
        g.flowInCCPerMin  =  me->itsN2Sensor.flow;
        success  =  GasDataQueue_insert(me->itsGasDataQueue, g);
        if (!success)
            printf("Nitrogen Gas Data queue insertion failed!\n");
    };

    if (rand()  >  RAND_MAX /  3) {
        g.gType  =  O2_GAS;
        g.conc  =  me->itsO2Sensor.conc;
        g.flowInCCPerMin  =  me->itsO2Sensor.flow;
        success  =  GasDataQueue_insert(me->itsGasDataQueue, g);
        if (!success)
            printf("Oxygen Gas Data queue insertion failed!  \n");
    }
}
```

Code Listing 4-37: updateData function of SensorThread.c

The sensors in this example are pretty simple – they just augment their values over time but they show that they just produce data; the `SensorThread_updateData()` function is responsible for pushing the data out to the queue. Code Listing 4-38 and Code Listing 4-39 show the code for one of these sensors.

```
#ifndef O2Sensor_H
#define O2Sensor_H

#include "QueuingExample.h"

typedef struct O2Sensor O2Sensor;
struct O2Sensor {
    double conc;
    unsigned int flow;
};

/*  Constructors and destructors:*/

void O2Sensor_Init(O2Sensor*  const me);
void O2Sensor_Cleanup(O2Sensor*  const me);
```

```
/* Operations */

void O2Sensor_getO2Data(O2Sensor* const me);
O2Sensor * O2Sensor_Create(void);
void O2Sensor_Destroy(O2Sensor* const me);

#endif
```

Code Listing 4-38: O2Sensor.h

```
#include "O2Sensor.h"
#include <stdlib.h>
#include <stdio.h>

void O2Sensor_Init(O2Sensor* const me) {
    me->conc = 0.0;
    me->flow = 0;
}

void O2Sensor_Cleanup(O2Sensor* const me) {
}

/*
    getO2Data() is where the sensor class would
    actually acquire the data. Here it just
    augments it.
*/
void O2Sensor_getO2Data(O2Sensor* const me) {
    me->conc += 20;
    me->flow += 25;
}

O2Sensor * O2Sensor_Create(void) {
    O2Sensor* me = (O2Sensor *) malloc(sizeof(O2Sensor));
    if(me!=NULL) {
        O2Sensor_Init(me);
    }
    return me;
}

void O2Sensor_Destroy(O2Sensor* const me) {
    if(me!=NULL) {
        O2Sensor_Cleanup(me);
    }
    free(me);
}
```

Code Listing 4-39: O2Sensor.c

On the consumer side, the GasProcessingThread_processData() runs periodically and pulls data out. If there are data available, it calls the function GasController_handleGasData () of its nested instance of GasController. This function just prints out the data (by calling the GasDisplay_printGasData() function) but one can imagine that it might find more interesting things to do with the data, such as maintain a constant gas delivery of appropriate concentrations. Code Listing 4-40 through Code Listing 4-43 show the code for these two elements.

```
#ifndef GasController_H
#define GasController_H

#include "QueuingExample.h"
#include "GasData.h"
struct GasDisplay;

typedef struct GasController GasController;
struct GasController {
    struct GasDisplay*  itsGasDisplay;
};

/*  Constructors and destructors:*/

void GasController_Init(GasController*  const me);
void GasController_Cleanup(GasController*  const me);

/*  Operations  */
void GasController_handleGasData(GasController*  const me, GasData
*  gPtr);

struct GasDisplay*  GasController_getItsGasDisplay(const
GasController*  const me);

void GasController_setItsGasDisplay(GasController*  const me,
struct GasDisplay*  p_GasDisplay);

GasController  *  GasController_Create(void);

void GasController_Destroy(GasController*  const me);

#endif
```

Code Listing 4-40: GasController.h

```
#include "GasController.h"
#include "GasDisplay.h"
#include <stdio.h>
static void cleanUpRelations(GasController*  const me);
```

```c
void GasController_Init(GasController* const me) {
    me->itsGasDisplay = NULL;
}


void GasController_Cleanup(GasController* const me) {
    cleanUpRelations(me);
}


void GasController_handleGasData(GasController* const me, GasData * gPtr) {
    if (gPtr) {
        GasDisplay_printGasData(me->itsGasDisplay, gPtr->gType, gPtr->conc,
        gPtr->flowInCCPerMin);
        free(gPtr);
        };
}


struct GasDisplay* GasController_getItsGasDisplay(const GasController* const me) {
    return (struct GasDisplay*)me->itsGasDisplay;
}
void GasController_setItsGasDisplay(GasController* const me, struct GasDisplay*
p_GasDisplay) {
    me->itsGasDisplay = p_GasDisplay;
}


GasController * GasController_Create(void) {
    GasController* me = (GasController *)
malloc(sizeof(GasController));
    if(me!=NULL)
    {
    GasController_Init(me);
    }
    return me;
}


void GasController_Destroy(GasController* const me) {
    if(me!=NULL)
    {
    GasController_Cleanup(me);
    }
    free(me);
}


static void cleanUpRelations(GasController* const me) {
    if(me->itsGasDisplay != NULL)
    {
    me->itsGasDisplay = NULL;
    }
}
```

Code Listing 4-18: GasController.c

```
#include "QueuingExample.h"
#include "GasData.h"

typedef struct GasDisplay GasDisplay;
struct GasDisplay {
    int screenWidth;
};

/* Constructors and destructors:*/
void GasDisplay_Init(GasDisplay* const me);
void GasDisplay_Cleanup(GasDisplay* const me);

/* Operations */
void GasDisplay_printGasData(GasDisplay* const me, const GAS_TYPE gasType, double
gas_conc, unsigned int gas_flow);

GasDisplay * GasDisplay_Create(void);

void GasDisplay_Destroy(GasDisplay* const me);

#endif
```

Code Listing 4-19: GasDisplay.h

```
#include "GasDisplay.h"
#include <stdio.h>

void GasDisplay_Init(GasDisplay* const me) {
    me->screenWidth = 80;
}

void GasDisplay_Cleanup(GasDisplay* const me) {
}

void GasDisplay_printGasData(GasDisplay* const me, const GAS_TYPE gasType, double
gas_conc, unsigned int gas_flow) {
    printf("\n");
    switch (gasType) {
        case O2_GAS:
            printf("Oxygen ");
            break;
        case N2_GAS:
            printf("Nitrogen ");
            break;
        case HE_GAS:
```

```
            printf("Helium  ");
            break;
        default:
                printf("UNKNWON  ");
                break;
    };
    printf("Conc %f, Flow in CC/Min %d\n", gas_conc, gas_flow);
    printf("\n");
}


GasDisplay *  GasDisplay_Create(void) {
    GasDisplay*  me  =  (GasDisplay *) malloc(sizeof(GasDisplay));
    if(me!=NULL) {
    GasDisplay_Init(me);
    }
    return me;
}


void GasDisplay_Destroy(GasDisplay*  const me) {
    if(me!=NULL) {
    GasDisplay_Cleanup(me);
    }
    free(me);
}
```

Code Listing 4-43: GasDisplay.c

Figure 4-19 shows a four-second run of the example in which the GasProcessingThread runs first at a period of 1000 ms and the SensorThread runs second at a period of 500 ms and a queue size of 10. You can see that even though each of the three gas sensors only has a 1/3 chance of producing data during the interval, data are put in faster than they are removed and the queue eventually fills up. The insert operation is instrumented with a printf() statement to show when data are inserted. The GasController's handleGasData() is invoked every period of the GasProcessingThread and it calls the GasDisplay's printGasData() function.

4.7 Rendezvous Pattern

Tasks must synchronize in various ways. The Queuing and Guarded Call Patterns provide mechanisms when the synchronization occurs around a simple function call, sharing a single resource, or passing data. But what if the conditions required for synchronization are more complex? The Rendezvous Pattern reifies the strategy as an object itself and therefore supports even those most complex needs for task synchronization.

```
+++ Nitrogen  at conc 10.000000, flow 15
    Number of elements queued 1, head = 1, tail = 0
+++ Oxygen    at conc 20.000000, flow 25
    Number of elements queued 2, head = 2, tail = 0
--- Nitrogen  at conc 10.000000, flow 15
    Number of elements queued 1, head = 2, tail = 1

Nitrogen Conc 10.000000, Flow in CC/Min 15

+++ Helium    at conc 2.000000, flow 2
    Number of elements queued 2, head = 3, tail = 1
+++ Nitrogen  at conc 20.000000, flow 30
    Number of elements queued 3, head = 4, tail = 1
+++ Oxygen    at conc 40.000000, flow 50
    Number of elements queued 4, head = 5, tail = 1
+++ Oxygen    at conc 60.000000, flow 75
    Number of elements queued 5, head = 6, tail = 1
--- Oxygen    at conc 20.000000, flow 25
    Number of elements queued 4, head = 6, tail = 2

Oxygen   Conc 20.000000, Flow in CC/Min 25

+++ Helium    at conc 4.000000, flow 4
    Number of elements queued 5, head = 7, tail = 2
+++ Nitrogen  at conc 40.000000, flow 60
    Number of elements queued 6, head = 8, tail = 2
+++ Helium    at conc 5.000000, flow 5
    Number of elements queued 7, head = 9, tail = 2
+++ Nitrogen  at conc 50.000000, flow 75
    Number of elements queued 8, head = 0, tail = 2
--- Helium    at conc 2.000000, flow 2
    Number of elements queued 7, head = 0, tail = 3

Helium   Conc 2.000000, Flow in CC/Min 2

+++ Helium    at conc 6.000000, flow 6
    Number of elements queued 8, head = 1, tail = 3
+++ Oxygen    at conc 120.000000, flow 150
    Number of elements queued 9, head = 2, tail = 3
+++ Nitrogen  at conc 70.000000, flow 105
    Number of elements queued 10, head = 3, tail = 3
Oxygen Gas Data queue insertion failed!
--- Nitrogen  at conc 20.000000, flow 30
    Number of elements queued 9, head = 3, tail = 4
+++ Helium    at conc 8.000000, flow 8
Conc 20.000000, Flow in CC/Min 30Conc 20.000000, Flow in CC/Min 30
ued 10, head = 4, tail = 4
Oxygen Gas Data queue insertion failed!
```

Figure 4-19: Sample run of the Queuing example

4.7.1 Abstract

A *precondition* is something that is specified to be true prior to an action or activity. A precondition is a type of constraint that is usually generative, that is, it can be used to generate code either to force the precondition to be true or to check that a precondition is true.

The Rendezvous Pattern is concerned with modeling the preconditions for synchronization or rendezvous of threads as a separate explicit object with its own data and functions. It is a general pattern and easy to apply to ensure that arbitrarily complex sets of preconditions can be met at run-time. The basic behavioral model is that as each thread becomes ready to rendezvous, it registers with the Rendezvous class, and then blocks until the Rendezvous class releases it to run. Once the set of preconditions is met, then the registered tasks are released to run using whatever scheduling policy is currently in force.

4.7.2 Problem

The Rendezvous Pattern solves the problems when a set of tasks must synchronize in a complex fashion.

4.7.3 Pattern Structure

Figure 4-20 shows the structure of the Rendezvous Pattern. Its structural simplicity belies its useful to manage complex task interactions.

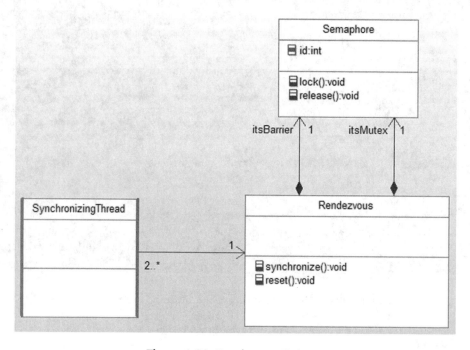

Figure 4-20: Rendezvous Pattern

4.7.4 Collaboration Roles

This section describes the roles for this pattern.

4.7.4.1 Rendezvous

The Rendezvous class manages the synchronization. It does this through its two primary functions:

- void reset(void)
  This function resets the synchronization criteria back to its initial condition.

- void synchronize(void)

  This function is called when a task wants to synchronize. If the criteria are not met, then the task is blocked in some way (see Implementation Strategies, below). This is commonly done either through the use of the Observer Pattern (see Chapter 2) or the Guarded Call Pattern.

The complexity of this class resides primarily in the synchronize() function; this evaluates the synchronization preconditions to see if they are met. These conditions may be internal (such as the Thread Barrier variant of the pattern used for the example) or external, or any combination of the two.

It is important to note that the Rendezvous object itself doesn't normally identify the threads as they rendezvous; as far as the Rendezvous object is concerned the tasks are anonymous. If they are not, such as when the strategy calls for tasks to rendezvous in a specific order, then the synchronize() function can accept that information in the form of one or more parameters.

4.7.4.2 Semaphore

This is a normal counting semaphore. The semaphore provides the following operations:

- Semaphore* create_semaphore(void)
- destroy_semaphore(Semaphore*)
- lock_semaphore(Semaphore*)
- release_semaphore(Semaphore*)

4.7.4.3 SynchronizingThread

This element represents a thread that uses the Rendezvous object for synchronization. It will run until its synchronization point and then call synchronize() function.

There are a number of ways to implement the pattern that result in minor changes to the parameters and return types for the pattern elements.

4.7.5 Consequences

In this pattern, two or more tasks can be synchronized by any arbitrarily complex strategy that can be encoded in the Rendezvous class. The pattern itself is simple and flexible and easily lends itself to specialization and adaptation.

The pattern is most appropriate when the tasks must halt at their synchronization point; that is, the pattern implements a blocking rendezvous. If a nonblocking rendezvous is desired, then this pattern can be mixed with a Queuing Pattern or the Observer Pattern with callbacks to a notify() function within the SynchronizingThread.

4.7.6 Implementation Strategies

This pattern may be implemented with the Observer Pattern from Chapter 3 or with the Guarded Call Pattern from this chapter. If the pattern is implemented with the Observer Pattern, then the Tasks must register with the address of a function to call when the synchronization criteria have been met. If the pattern is implemented with a Guarded Call Pattern, then each Rendezvous object owns a single semaphore that is enabled when the Rendezvous object is created. Each calling task then calls the register() function when it wants to synchronize. When the synchronization criteria is met, the semaphore is released, and the tasks are then all free to run according to the normal scheduling strategy.

4.7.7 Related Patterns

This pattern may be overkill when simpler solutions such as Guarded Call or Queuing Patterns may suffice.

The pattern may be implemented in a number of different ways, as well, using the Guarded Call or Observer Patterns to manage the thread blocking and unblocking.

4.7.8 Example

In this example (Figure 4-21), we have implemented a specific form of the Rendezvous Pattern knows as the Thread Barrier Pattern. In this pattern, the Rendezvous object only knows that a certain number of tasks are expected to rendezvous. When that number of tasks have called the synchronize() function, the set of blocked tasks are released. The example also implemented the pattern with the use of semaphores (Guarded Call Pattern) and lets the operating system block and unlock the tasks.

The algorithm for the Thread Barrier is well understood using counting semaphores. See the code for the synchronize operation in Code Listing 4-45.

```
#ifndef ThreadBarrier_H
#define ThreadBarrier_H

/*## auto_generated */
#include <oxf/Ric.h>
/*## auto_generated */
#include "RendezvousExample.h"
/*## auto_generated */
#include <oxf/RiCTask.h>
/*## package RendezvousPattern::RendezvousExample */

/*## class ThreadBarrier */
typedef struct ThreadBarrier ThreadBarrier;
struct ThreadBarrier {
    int currentCount;
```

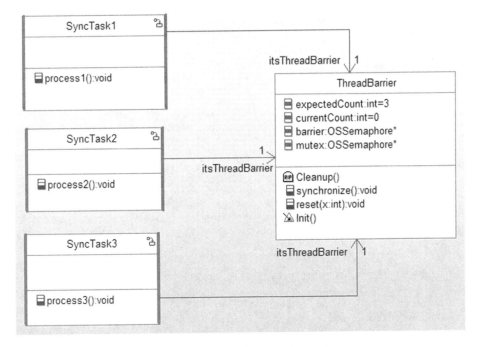

Figure 4-21: Rendezvous Example

```
    int expectedCount;
    OSSemaphore*  barrier;
    OSSemaphore*  mutex;
};

/*  Constructors and destructors:*/
void ThreadBarrier_Init(ThreadBarrier*  const me);
void ThreadBarrier_Cleanup(ThreadBarrier*  const me);

/*  Operations  */
void ThreadBarrier_reset(ThreadBarrier*  const me, int x);
void ThreadBarrier_synchronize(ThreadBarrier*  const me);
ThreadBarrier  *  ThreadBarrier_Create(void);
void ThreadBarrier_Destroy(ThreadBarrier*  const me);
#endif
```

Code Listing 4-20: ThreadBarrier.h

```
#include "ThreadBarrier.h"

void ThreadBarrier_Init(ThreadBarrier*  const me) {
    me->currentCount  =  0;
    me->expectedCount  =  3;
```

```c
    me->mutex = OSSemaphore_Create();
    me->barrier = OSSemaphore_Create();
    if (me->barrier) {
        OSSemaphore_lock(me->barrier);
        printf("BARRIER IS LOCKED FIRST TIME\n");
    };
}

void ThreadBarrier_Cleanup(ThreadBarrier* const me) {
    OSSemaphore_Destroy(me->barrier);
    OSSemaphore_Destroy(me->mutex);
}

void ThreadBarrier_reset(ThreadBarrier* const me, int x) {
    me->expectedCount = x;
    me->currentCount = 0;
}

void ThreadBarrier_synchronize(ThreadBarrier* const me) {
    /*
        protect the critical region around
        the currentCount
    */
    OSSemaphore_lock(me->mutex);
        ++me->currentCount;      /* critical region */
    OSSemaphore_release(me->mutex);

    /*
        are conditions for unblocking met?
        if so, then release the first blocked
        thread or the highest priority blocked
        thread (depending on the OS)
    */
    if (me->currentCount ==    me->expectedCount) {
        printf("Conditions met\n");
        OSSemaphore_release(me->barrier);
    };

    /*
        lock the semaphore and when released
        then release it for the next blocked thread
    */
    OSSemaphore_lock(me->barrier);
    OSSemaphore_release(me->barrier);
}

ThreadBarrier * ThreadBarrier_Create(void) {
    ThreadBarrier* me = (ThreadBarrier *)
  malloc(sizeof(ThreadBarrier));
    if(me!=NULL)
```

```
        ThreadBarrier_Init(me);
        return me;
    }

    void ThreadBarrier_Destroy(ThreadBarrier*  const me) {
        if(me!=NULL)
            ThreadBarrier_Cleanup(me);
        free(me);
    }
```

Code Listing 4-21: ThreadBarrier.c

Deadlock Avoidance Patterns

The last problem I want to address in this chapter is *deadlock*. Deadlock is a situation in which multiple clients are simultaneously waiting for conditions that cannot, in principle, occur.

Figure 4-22 shows an example of deadlock. In this example, the priority of Task 1 is higher than that of Task 2. The figure shows a timing diagram overlayed with a class diagram showing the structure. In addition to the tasks, there are two resources, R1 and R2, shared by both tasks. The problem arises if the tasks lock the resources in reverse order and Task 1 preempts Task 2

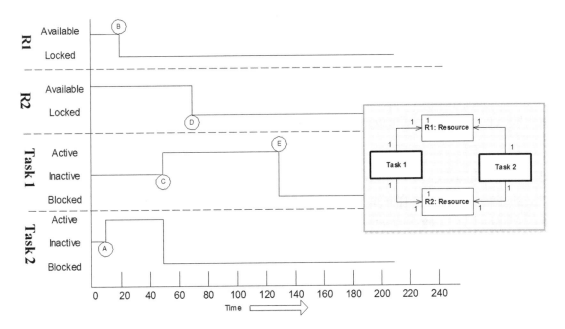

Figure 4-22: Deadlock example

between Task 1 between locking R1 and R2. The following description refers to the named points in Figure 4-22.

A. At this point, Task 2 runs.
B. Task 2 locks R1 and is just about to lock R2.
C. Task 1 runs. Since it is higher priority than Task 2, it preempts Task 2.
D. Task 1 locks R2.
E. Task 1 now attempts to lock R1, however, R1 is currently locked by Task 2. If Task 1 blocks to allow Task 2 to run, Task 2 must block as soon as it tries to lock R1. The system is in deadlock.

It is enough to break any of the four required conditions for deadlock (below) conditions to ensure that deadlock cannot appear. The simultaneous locking, ordered locking, and critical region patterns all avoid the problems of deadlock. For example, had Task 2 enabled a critical region (i.e., disallowed task switching) before it locked R1, then Task 1 would not have been able to preempt it before it locked R2. If both tasks locked the resources in the same order, deadlock would have been prevented. If Task 2 locks both resources at the same time, there would be no problem. Deadlock is easy to prevent, but you must take careful steps to do so.

Leaving aside the common problems of software errors, deadlocks require four conditions to be true:

1. mutual exclusion locking of resources
2. some resources are locked while others are requested
3. preemption while resources are locked is allowed
4. a circular waiting condition exists

Deadlock can be avoided by breaking any of these four conditions. Using the Critical Region Pattern, for example, breaks rules #1 and #3. The Queuing Pattern, because it uses asynchronous message passing avoids #1 (other than the lock on the queue access itself). This chapter finishes up with two other patterns that avoid deadlock.

4.8 Simultaneous Locking Pattern

The Simultaneous Locking Pattern is a pattern solely concerned with deadlock avoidance. It achieves this by locking breaking condition #2 (holding resources while waiting for others). The pattern works in an all-or-none fashion. Either all needed resources are locked at once or none are locked.

4.8.1 Abstract

Deadlock can be solved by breaking any of the four required conditions. This pattern prevents the condition of holding some resources by requesting others by allocating them all at once.

However, this pattern has an advantage over the Critical Region Pattern by allowing higher priority tasks to run if they don't need any of the locked resources.

4.8.2 Problem

The problem of deadlock is a serious enough one in highly reliable computing that many systems design in specific mechanisms to detect it or avoid it. As previously discussed, deadlock occurs when a task is waiting on a condition that can never, in principle, be satisfied. There are four conditions that must be true for deadlock to occur, and it is sufficient to deny the existence of any one of these. The Simultaneous Locking Pattern breaks condition #2, not allowing any task to lock resources while waiting for other resources to be free.

4.8.3 Pattern Structure

The structure of the pattern is shown in Figure 4-23.

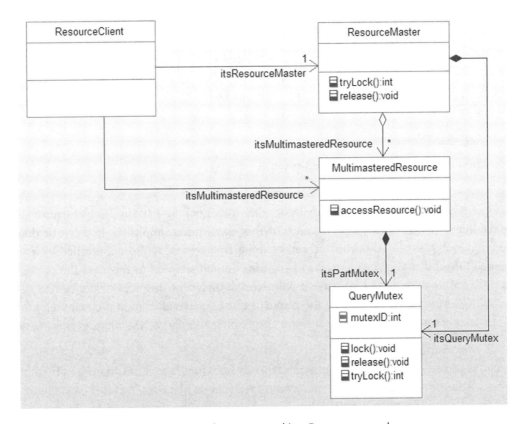

Figure 4-23: Simultaneous Locking Pattern general case

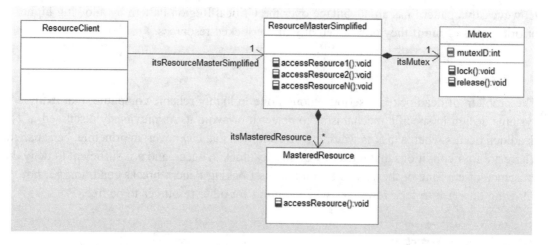

Figure 4-24: Simplified Simultaneous Locking Pattern in which each resource must be assigned to a single Resource Master

In general, a `MultimasteredResource` can be a part of any number of different sets of resources, where a separate instance of `ResourceMaster` manages one such set. This potential sharing of resources across multiple `ResourceMasters` complicates the situation. When this flexibility is not needed, the situation simplifies to that shown in Figure 4-24. While the difference may not appear to be great, it removes the requirement for a `tryLock()` function on the `Mutex` and it is algorithmically much simpler.

4.8.4 Collaboration Roles

This section describes the roles for this pattern.

4.8.4.1 MasteredResource

This class is an element of the Simplified Simultaneous Locking Pattern. It represents a shared resource that is owned by a single instance of ResourceMasterSimplified. If it can be determined that each `MasteredResource` can be uniquely assigned to a set controlled by a single instance of `ResourceMasterSimplified`, life is much simpler; in this case the resource does not need its own `Mutex` because it will execute under the auspices of the `Mutex` of the `ResourceMasterSimplified`. If the resource must be shared in multiple resource groups, then its becomes a `MultimasteredResource`, which is a part of the more general pattern.

4.8.4.2 MultimasteredResource

This class is an element of the general Simultaneous Locking Pattern. It is managed by multiple `ResourceMasters`, then it requires its own mutex (via its like `itsPartMutex`). The `Mutex` it uses must provide the `tryLock()` function so that it can be determined by the `ResourceMaster` whether all the attempted locks will succeed or fail.

4.8.4.3 Mutex

This element is a normal mutex semaphore as used in previous patterns; if offers `lock()` and `release()` functions that work in the way standard for a mutex semaphore.

4.8.4.4 QueryMutex

This element is a normal mutex semaphore except that it also provides a `tryLock()` function for the general pattern case. This function works like the normal `lock()` function except that it returns with an error code if the lock fails rather than blocking the current thread. In the specialized form of the pattern (in which a single resource is owned by a single `ResourceMaster`), then the `tryLock()` function isn't needed.

4.8.4.5 ResourceClient

This element is a client that wants to access a set of resources all at once to avoid potential deadlock. Although it has direct access (via its `itsMultimasteredResource` link) to the clients, it doesn't access them until it has successfully received a lock on the `ResourceMaster`. This requires discipline on the part of the developers of the `ResourceClients` to ensure that they never attempt to access any of the client services without first getting a lock, and once the client is done with the resources, it frees the `ResourceMaster`.

4.8.4.6 Resource Master

The `ResourceMaster` controls the lock of the entire set of resources. It is used in the general case of the pattern and it must return a lock if and only if it can lock all the resources. While it can be implemented with mutexes solely on the parts (instances of `ResourceMaster`) because it may be blocked multiple times while it attempts to lock the set of individual `MultimasteredResources`, it may have unacceptable execution times in practice. A solution, implemented here, is to use the `tryLock()` function of the mutex to test. If the `tryLock()` calls for the entire set of `MultimasteredResources` to all succeed, then the `ResourceMaster` itself is locked and the `ResourceClient` is allowed to use its link to the `MultimasteredResources`. If the `ResourceClient` cannot lock all of the `MultimasteredResources`, then the `ResourceMaster` unlocks all of the `MultimasteredResources` it locked (before it found one where it failed), and returns with an error code to the `ResourceClient`.

4.8.4.7 ResourceMasterSimplified

The Simplified Simultaneous Locking Pattern doesn't require the additional developer discipline required for the general case. The `ResourceClient` no longer has direct links to the `MasterResources` *per se*; instead, the `ResourceMasterSimplified` republishes all the services of its `MasteredResources` and delegates the functions directly to them. All of the functions share the same mutex.

4.8.5 Consequences

The Simultaneous Locking Pattern prevents deadlock by removing required condition #2 – some resources are locked while others are requested – by locking either all of the resources needed at once, or none of them. However, while the pattern removes the possibility of deadlock, it may increase the delay of the execution of other tasks. This delay may occur even in situations in which there is no actual resource conflict. The problem is far more pronounced for resources that are widely shared than for resources that are shared more narrowly. In addition, the pattern does not address priority inversion and in fact may make it worse, unless some priority inversion bounding pattern is deployed, such as the Guarded Call Pattern with priority inheritance. The general case for the pattern, shown in Figure 4-24, requires developer discipline in that they must avoid using the direct links to the `MultimasteredResources` until they have successfully been given a lock. In large project teams, this discipline can be ignored resulting in subtle and hard-to-identify errors. The simplified model, shown in Figure 4-24, doesn't have this problem because the `ResourceMasterSimplified` republishes the resource access methods.

Of course this pattern only applies when deadlock is a possibility, such as when multiple resources are shared with at least two in common by at least two threads.

4.8.6 Implementation Strategies

The implementation of this pattern should pose no difficulties other than the need to use, in the general case, a mutex semaphore that supports a `tryLock()` operation, and the need to ensure that the precondition of access (successful locking of the `MultimasteredResource`) is achieved before accessing the resource directly.

4.8.7 Related Patterns

Deadlock can be prevented by avoiding any of the required four conditions. The Critical Regions Pattern avoids by breaking condition #1 – mutual exclusion locking of resources. The Ordered Locking Pattern, described later in this chapter, prevents deadlock by breaking condition #4 – circular waiting condition.

Because this pattern doesn't address the concern of unbounded priority inversion, it can be mixed with the Guarded Call Pattern with priority inheritance.

4.8.8 Example

Figure 4-25 shows an example of the simplified form of Figure 4-24. In this case, the `SensorMaster` replicates the functions of the sensors (although the names are munged with the

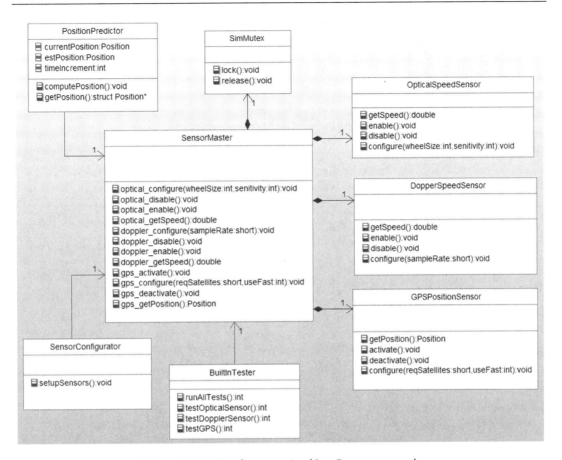

Figure 4-25: Simultaneous Locking Pattern example

names of the servers to disambiguate services). When a client (such as the `PositionPredictor`) invokes one of the services, the `SensorMaster` locks the mutex and releases it before returning to the client. The `SensorMaster` owns a mutex semaphore for this purpose.

```
#ifndef SensorMaster_H
#define SensorMaster_H
#include "Position.h"
struct DopplerSpeedSensor;
struct GPSPositionSensor;
struct OpticalSpeedSensor;
struct SimMutex;

typedef struct SensorMaster SensorMaster;
struct SensorMaster {
    struct DopplerSpeedSensor*  itsDopplerSpeedSensor;
    struct GPSPositionSensor*  itsGPSPositionSensor;
```

```
      struct OpticalSpeedSensor* itsOpticalSpeedSensor;
      struct SimMutex* itsSimMutex;
};

/* Constructors and destructors:*/

void SensorMaster_Init(SensorMaster* const me);
void SensorMaster_Cleanup(SensorMaster* const me);

/* Operations */
void SensorMaster_doppler_configure(SensorMaster* const me, short sampleRate);
void SensorMaster_doppler_disable(SensorMaster* const me);
void SensorMaster_doppler_enable(SensorMaster* const me);
double SensorMaster_doppler_getSpeed(SensorMaster* const me);
void SensorMaster_gps_activate(SensorMaster* const me);
void SensorMaster_gps_configure(SensorMaster* const me, short
reqSatellites, int useFast);

void SensorMaster_gps_deactivate(SensorMaster* const me);
struct Position SensorMaster_gps_getPosition(SensorMaster* const me);
void SensorMaster_optical_configure(SensorMaster* const me, int
wheelSize, int sensitivity);
void SensorMaster_optical_disable(SensorMaster* const me);
void SensorMaster_optical_enable(SensorMaster* const me);
double SensorMaster_optical_getSpeed(SensorMaster* const me);

struct DopplerSpeedSensor*
SensorMaster_getItsDopplerSpeedSensor(const SensorMaster* const me);
void SensorMaster_setItsDopplerSpeedSensor(SensorMaster* const
me, struct DopplerSpeedSensor* p_DopplerSpeedSensor);

struct GPSPositionSensor*
SensorMaster_getItsGPSPositionSensor(const SensorMaster* const me);

void SensorMaster_setItsGPSPositionSensor(SensorMaster* const me,
struct GPSPositionSensor* p_GPSPositionSensor);

struct OpticalSpeedSensor*
SensorMaster_getItsOpticalSpeedSensor(const SensorMaster* const me);
void SensorMaster_setItsOpticalSpeedSensor(SensorMaster* const me,
struct OpticalSpeedSensor* p_OpticalSpeedSensor);

struct SimMutex* SensorMaster_getItsSimMutex(const SensorMaster* const me);

void SensorMaster_setItsSimMutex(SensorMaster* const me, struct
SimMutex* p_SimMutex);

SensorMaster * SensorMaster_Create(void);
```

```
void SensorMaster_Destroy(SensorMaster* const me);
```

```
#endif
```

Code Listing 4-22: SensorMaster.h

```
#include "SensorMaster.h"
#include "DopplerSpeedSensor.h"
#include "GPSPositionSensor.h"
#include "OpticalSpeedSensor.h"
#include "SimMutex.h"

static void cleanUpRelations(SensorMaster* const me);

void SensorMaster_Init(SensorMaster* const me) {
    me->itsDopplerSpeedSensor = NULL;
    me->itsGPSPositionSensor = NULL;
    me->itsOpticalSpeedSensor = NULL;
    me->itsSimMutex = NULL;
}

void SensorMaster_Cleanup(SensorMaster* const me) {
    cleanUpRelations(me);
}

void SensorMaster_doppler_configure(SensorMaster* const me, short sampleRate) {
    SimMutex_lock(me->itsSimMutex);
    DopplerSpeedSensor_configure(me->itsDopplerSpeedSensor,sampleRate);
    SimMutex_release(me->itsSimMutex);
}

void SensorMaster_doppler_disable(SensorMaster* const me) {
    SimMutex_lock(me->itsSimMutex);
    DopplerSpeedSensor_disable(me->itsDopplerSpeedSensor);
    SimMutex_release(me->itsSimMutex);
}

void SensorMaster_doppler_enable(SensorMaster* const me) {
    SimMutex_lock(me->itsSimMutex);
    DopplerSpeedSensor_enable(me->itsDopplerSpeedSensor);
    SimMutex_release(me->itsSimMutex);
}

double SensorMaster_doppler_getSpeed(SensorMaster* const me) {
    double speed;
    SimMutex_lock(me->itsSimMutex);
    speed = DopplerSpeedSensor_getSpeed(me->itsDopplerSpeedSensor);
```

```
    SimMutex_release(me->itsSimMutex);
    return speed;
}

void SensorMaster_gps_activate(SensorMaster*  const me) {
    SimMutex_lock(me->itsSimMutex);
    GPSPositionSensor_activate(me->itsGPSPositionSensor);
    SimMutex_release(me->itsSimMutex);
}

void SensorMaster_gps_configure(SensorMaster*  const me, short reqSatellites, int use-
Fast) {
    SimMutex_lock(me->itsSimMutex);
    GPSPositionSensor_configure(me->itsGPSPositionSensor,reqSatellites,useFast);
    SimMutex_release(me->itsSimMutex);
}

void SensorMaster_gps_deactivate(SensorMaster*  const me) {
    SimMutex_lock(me->itsSimMutex);
    GPSPositionSensor_deactivate(me->itsGPSPositionSensor);
    SimMutex_release(me->itsSimMutex);
}

struct Position SensorMaster_gps_getPosition(SensorMaster*  const me) {
    Position p;
    SimMutex_lock(me->itsSimMutex);
    p = GPSPositionSensor_getPosition(me->itsGPSPositionSensor);
    SimMutex_release(me->itsSimMutex);
    return p;
}

void SensorMaster_optical_configure(SensorMaster*  const me, int wheelSize,
int sensitivity) {
    SimMutex_lock(me->itsSimMutex);
    OpticalSpeedSensor_configure(me->itsOpticalSpeedSensor,wheelSize, sensitivity);
    SimMutex_release(me->itsSimMutex);
}

void SensorMaster_optical_disable(SensorMaster*  const me) {
    SimMutex_lock(me->itsSimMutex);
    OpticalSpeedSensor_disable(me->itsOpticalSpeedSensor);
    SimMutex_release(me->itsSimMutex);
}

void SensorMaster_optical_enable(SensorMaster*  const me) {
    SimMutex_lock(me->itsSimMutex);
    OpticalSpeedSensor_enable(me->itsOpticalSpeedSensor);
    SimMutex_release(me->itsSimMutex);
}
```

```
double SensorMaster_optical_getSpeed(SensorMaster* const me) {
    double speed;
    SimMutex_lock(me->itsSimMutex);
    speed = OpticalSpeedSensor_getSpeed(me->itsOpticalSpeedSensor);
    SimMutex_release(me->itsSimMutex);
    return speed;
}

struct DopplerSpeedSensor* SensorMaster_getItsDopplerSpeedSensor(const
SensorMaster* const me) {
    return (struct DopplerSpeedSensor*)me->itsDopplerSpeedSensor;
}

void SensorMaster_setItsDopplerSpeedSensor(SensorMaster* const me, struct Doppler-
SpeedSensor* p_DopplerSpeedSensor) {
    me->itsDopplerSpeedSensor = p_DopplerSpeedSensor;
}

struct GPSPositionSensor* SensorMaster_getItsGPSPositionSensor(const
SensorMaster* const me) {
    return (struct GPSPositionSensor*)me->itsGPSPositionSensor;
}

void SensorMaster_setItsGPSPositionSensor(SensorMaster* const me,
struct GPSPositionSensor* p_GPSPositionSensor) {
    me->itsGPSPositionSensor = p_GPSPositionSensor;
}

struct OpticalSpeedSensor* SensorMaster_getItsOpticalSpeedSensor(const
SensorMaster* const me) {
    return (struct OpticalSpeedSensor*)me->itsOpticalSpeedSensor;
}

void SensorMaster_setItsOpticalSpeedSensor(SensorMaster* const me, struct Optical-
SpeedSensor* p_OpticalSpeedSensor) {
    me->itsOpticalSpeedSensor = p_OpticalSpeedSensor;
}

struct SimMutex* SensorMaster_getItsSimMutex(const SensorMaster* const me) {
    return (struct SimMutex*)me->itsSimMutex;
}

void SensorMaster_setItsSimMutex(SensorMaster* const me, struct
SimMutex* p_SimMutex) {
    me->itsSimMutex = p_SimMutex;
}

SensorMaster * SensorMaster_Create(void) {
```

```
    SensorMaster* me = (SensorMaster *)
malloc(sizeof(SensorMaster));
    if(me!=NULL)
      SensorMaster_Init(me);
    return me;
}

void SensorMaster_Destroy(SensorMaster* const me) {
    if(me!=NULL)
        SensorMaster_Cleanup(me);
    free(me);
}

static void cleanUpRelations(SensorMaster* const me) {
    if(me->itsDopplerSpeedSensor != NULL)
        me->itsDopplerSpeedSensor = NULL;
    if(me->itsGPSPositionSensor != NULL)
        me->itsGPSPositionSensor = NULL;
    if(me->itsOpticalSpeedSensor != NULL)
        me->itsOpticalSpeedSensor = NULL;
    if(me->itsSimMutex != NULL)
        me->itsSimMutex = NULL;
}
```

Code Listing 4-23: SensorMaster.c

4.9 Ordered Locking

The Ordered Locking Pattern is another way to ensure that deadlock cannot occur, this time by preventing condition 4 (circular waiting) from occurring. It does this by ordering the resources and requiring that they always be locked by a client in that specified order. If this is religiously enforced, then no circular waiting condition can ever occur.

4.9.1 Abstract

The Ordered Locking Pattern eliminates deadlock by ordering resources and enforcing a policy in which resources must be allocated only in a specific order. For monadic (that is, functions that can be used independently of others) operations, the client can invoke them without special consideration and the called function will lock and release the resource internally. For holding locks over the long term – accessing dyadic (that is, functions that require the caller to explicitly lock and unlock them) functions – the client must explicitly lock and release the resources, rather than doing it implicitly by merely invoking a service on a resource. This means that the potential for neglecting to unlock the resource exists.

4.9.2 Problem

The Ordered Locking Pattern solely addresses the problem of deadlock elimination, as does the previous Simultaneous Locking Pattern.

4.9.3 Pattern Structure

Figure 4-26 shows the pattern structure for the Ordered Locking Pattern.

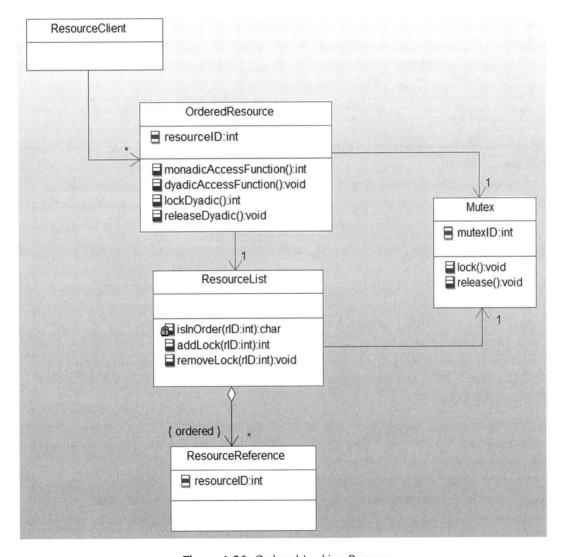

Figure 4-26: Ordered Locking Pattern

4.9.4 Collaboration Roles

This section describes the roles for this pattern.

4.9.4.1 Mutex

The Mutex is used by two other pattern elements: OrderedResource and ResourceList. Of course, this means that each of these elements will use different instances of the Mutex class. As with the other patterns in this chapter, the Mutex class provides two basic services. The lock() function locks the resource (if currently unlocked) or blocks the caller (if already locked). The release() function removes the lock and allows any blocked callers to continue.

4.9.4.2 OrderedResource

The OrderedResource is really the heart of the pattern. It has a resourceID attribute. This is a unique ID associated with each resource instance and, along with the ResourceList, this class enforces the ordered locking rule:

A resource may only be locked if its resourceID *is greater than the largest* resourceID *of any locked resource.*

The orderedResource may have different kinds of functions. The first kind, called monadic functions, internally lock the mutex, perform their magic, unlock the Mutex, and then return. The magic inside the function may call functions of other resources, hence the need to use the ResourceList to prevent possible deadlocks. Figure 4-27 shows the flow of control for such a monadicAccessFunction.

The other kind of access function is dyadic, called so because it requires a separate pair of calls to manage the lock on the resource. That is, before a dyadic access function is called, the ResourceClient should first call lockDyadic() and after the ResourceClient is done with the resource, it should call releaseDyadic(). Because of the ordered locking enforcement, the ResourceClient can hold any number of resources in the locked state while it performs its work without fear of causing deadlock. However, the ResourceClient must lock them in the order of their resourceIDs or errors will be returned and access will be denied. Figure 4-28 shows the flowchart for locking the resource and Figure 4-29 shows the flowchart for its release.

An interesting question arises as to how to assign the resourceIDs. In many ways, the best approach is to assign them at design time so that the constructor OrderedResource_Init() simply initializes the value. This approach requires design-time analysis to ensure that the resourceIDs are unique but also provides an opportunity to use *a priori* information to identify an optimal assignment of resourceIDs depending on the usage patterns of the various resources. Design-time analysis may discover resourceID assignments that lessen blocking due to deadlock prevention mechanisms (that is, if we assign the resourceIDs in *this* order, we minimize the occurrence of resource lock rejection due to potential deadlocking).

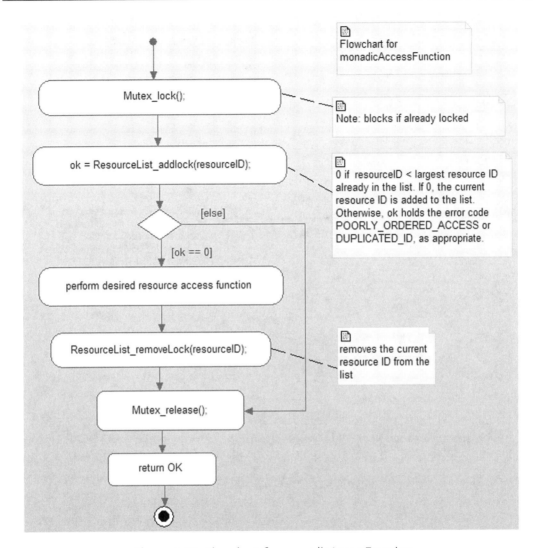

Figure 4-27: Flowchart for monadicAccessFunction

On the other hand, this analysis may be very difficult for larger systems. However, dynamically assigning the resourceIDs with a resourceID server is problematic as it may lead to pathological conditions under which the order in which a set of resources was called in one case leads to the permanent failure of another ResourceClient. For this reason, I recommend design-time assignment of resourceIDs.

4.9.4.3 ResourceClient

The ResourceClient represents the set of software elements that want to invoke the services of the OrderedResources. The ResourceClient doesn't need to know

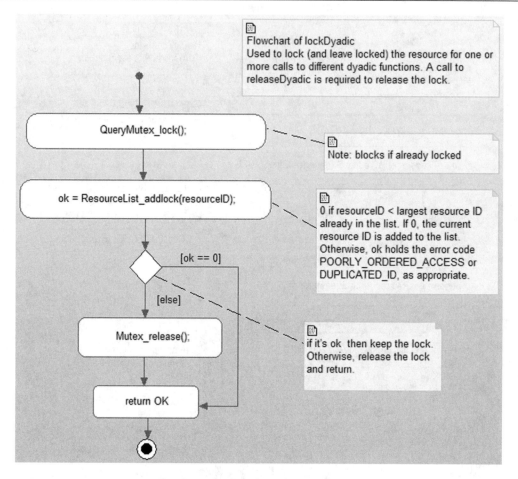

Figure 4-28: Flowchart for lockDyadic

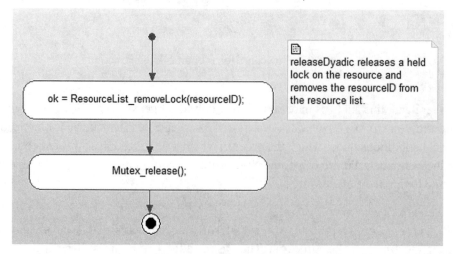

Figure 4-29: Flowchart for releaseDyadic

anything about the resourceIDs *per se*, but as a part of testing, they need to be designed so that they only lock the resources in the correct order. For monadic functions, they don't need to do anything special – they will just invoke the monadicAccessFunction(). For dyadic access functions, the ResourceClient must explicitly lock the resource first, invoke the needed services, and then call unlock. The use of the dyadic form is more flexible – since the ResourceClient can hold the locks on any number of resources as long as it needs them – but requires additional discipline on the part of the developer.

4.9.4.4 ResourceList

This element provides two important functions. addLock( )returns 0 ("ok") if the passed resourceID is greater than that of the largest currently locked resource. If the passed referenceID is *less than* the largest resource ID, then POORLY_ORDERED_ACCESS is returned; if the passed referenceID equals it, then DUPLICATE_IDS is returned. (Note that if the same resource is accessed while it is locked, its mutex will block before the call is made to addLock( ), so the only way this can occur is if the resourceID is nonunique). It is also important to note that there must be only one ResourceList for the application, giving the resourceIDs an effective global scope.

4.9.4.5 ResourceReference

This is just the resourceID held in an ordered list; most likely an array. It is insufficient to just keep the largest value only since many resources might be locked at any one time.

4.9.5 Consequences

This pattern effectively removes the possibility of resource-based deadlocks by removing the possibility of condition 4 – circular waiting. For the algorithm to work, any ordering of OrderedResources will do provided that this ordering is global. However, some orderings are better than others and will result is less blocking overall. This may take some analysis at design time to identify the best ordering of the OrderedResources. As mentioned above, if OrderedResources are themselves ResourceClients (a reasonable possibility), then they should *only* invoke services of OrderedResources that have higher valued IDs than they do. If they invoke a lower-valued OrderedResources, then they are in effect violating the ordered locking protocol by the transitive property of locking (if A locks B and then B locks C, then A is in effect locking C). This results in a returned error code.

While draconian, one solution to the potential problem of transitive violation of the ordering policy is to enforce the rule that a OrderedResources may never invoke services or lock other OrderedResources. If your system design does allow such transitive locking, then each transitive path must be examined to ensure that the ordering policy is not violated.

4.9.6 Implementation Strategies

The implementation of the pattern requires the addition of a `resourceID` to each `OrderedResource` and logic in the `ResourceList` to ensure that the `resourceID` of each `OrderedResource` is larger than any currently locked `resourceID`. In addition, a list of locked `resourceIDs` must be maintained so that as `OrderedResources` are released, additional ones can be locked appropriately.

The most common implementation for the `ResourceList` is as an array of integers representing the `resourceIDs` in the order in which they are locked.

4.9.7 Related Patterns

This pattern is specifically designed to remove the possibilities of deadlocks. The Simultaneous Locking Pattern also does this, although through the different mechanism of preventing condition 2 (some resources are locked while others are requested). The Critical Region Pattern also removes the possibility of deadlock by preventing condition 3 (preemption while resources are locked is allowed). The Cyclic Executive Pattern avoids deadlock by preventing condition 1 (mutual exclusion locking of resources).

4.9.8 Example

The example in Figure 4-30 shows an application of the Ordered Locking Pattern. In this (simplified) aircraft example, three servers hold related, but distinct information; `AttitudeSensor` has information about the attitude (roll, pitch, and yaw) of the aircraft, `PositionSensor` has information about where the aircraft is in the airspace (latitude, longitude, and altitude), and the `VelocitySensor` has information about how fast the aircraft is moving in three dimensions.

The example has two different clients. The `KinematicModelClient` needs to get time-coherent information about the state of the aircraft for the purpose of precise and detailed control of actuators and engines. The `RoutePlanningClient` needs the same information to determine updates to the route depending on fuel usage, threat assessments, mission criticality, and so on.

Let's suppose, for the purpose of this discussion that for both the clients it is crucial to get data in the same sample interval. A way to do that is to lock all the resources, gather and manipulate the data, and then release all the resources. If the KinematicModel locks them in the order 1) `PositionSensor`, 2) `VelocitySensor`, and 3) `AttitudeSensor` but the `RoutePlanningClient` (done by a different subcontractor) locks them in the reverse order, there is an opportunity for deadlock. Ergo, we've added the Ordered Locking Pattern to specifically prevent deadlock. We've also used the dyadic form of the pattern so the

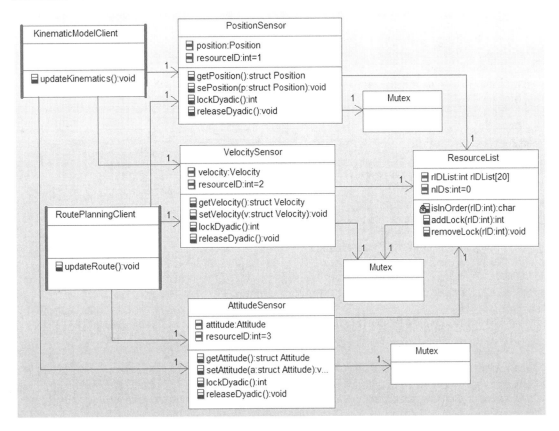

Figure 4-30: Ordered Locking Pattern Example

clients are expected to manually lock the resources before use and release the resources once done.

Note that the mutex class shows up multiple times in the diagram. This is done to minimize line crossing in the diagram. There is a single mutex class, but four distinct instances of that class, each providing a separate locking semaphore for its owner.

I've included the code for two of the classes, the `VelocitySensor` and the `ResourceList` to illustrate how this pattern may be implemented. The error codes are simply implemented in the OrdseredLockingExample.h file as

```
#define POORLY_ORDERED_ACCESS (1)
#define DUPLICATED_IDS (2)
```

Code Listing 4-48 has the VelocitySensor.h code and Code Listing 4-49 has the corresponding C code.

```
#ifndef VelocitySensor_H
#define VelocitySensor_H
/*  brings in the error codes  */
#include "OrderedLockingExample.h"
#include "Velocity.h"
struct Mutex;

struct ResourceList;

typedef struct VelocitySensor VelocitySensor;
struct VelocitySensor {
    int resourceID;
    struct Velocity velocity;
    struct Mutex*  itsMutex;
    struct ResourceList*  itsResourceList;
};

void VelocitySensor_Init(VelocitySensor*  const me);

void VelocitySensor_Cleanup(VelocitySensor*  const me);

/*  Operations  */
struct Velocity VelocitySensor_getVelocity(VelocitySensor*  const me);

int VelocitySensor_lockDyadic(VelocitySensor*  const me);

void VelocitySensor_releaseDyadic(VelocitySensor*  const me);

void VelocitySensor_setVelocity(VelocitySensor*  const me, struct Velocity v);

struct Mutex*  VelocitySensor_getItsMutex(const VelocitySensor*  const me);

void VelocitySensor_setItsMutex(VelocitySensor*  const me, struct Mutex*  p_Mutex);

struct ResourceList*  VelocitySensor_getItsResourceList(const
VelocitySensor* const me);

void VelocitySensor_setItsResourceList(VelocitySensor*  const me,
struct ResourceList*  p_ResourceList);

VelocitySensor  *  VelocitySensor_Create(void);

void VelocitySensor_Destroy(VelocitySensor*  const me);

#endif
```

Code Listing 4-24: VelocitySensor.h

```c
#include "VelocitySensor.h"
#include "Mutex.h"
#include "ResourceList.h"

static void cleanUpRelations(VelocitySensor* const me);

void VelocitySensor_Init(VelocitySensor* const me) {
    me->resourceID = 2;
    {
        Velocity_Init(&(me->velocity));
    }
    me->itsMutex = NULL;
    me->itsResourceList = NULL;
}

void VelocitySensor_Cleanup(VelocitySensor* const me) {
    cleanUpRelations(me);
}

struct Velocity VelocitySensor_getVelocity(VelocitySensor* const me) {
    return me->velocity;
}

int VelocitySensor_lockDyadic(VelocitySensor* const me) {
    int ok;

    Mutex_lock(me->itsMutex);
    ok = ResourceList_addLock(me->itsResourceList, me->resourceID);
    if (ok != 0)
        Mutex_release(me->itsMutex);
    return ok;
}

void VelocitySensor_releaseDyadic(VelocitySensor* const me) {
    ResourceList_removeLock(me->itsResourceList, me->resourceID);
    Mutex_release(me->itsMutex);
}

void VelocitySensor_setVelocity(VelocitySensor* const me, struct Velocity v) {
    me->velocity = v;
}

struct Mutex* VelocitySensor_getItsMutex(const VelocitySensor* const me) {
    return (struct Mutex*)me->itsMutex;
}

void VelocitySensor_setItsMutex(VelocitySensor* const me, struct Mutex* p_Mutex) {
    me->itsMutex = p_Mutex;
}
```

```
struct ResourceList*  VelocitySensor_getItsResourceList(const
VelocitySensor*  const me) {
    return (struct ResourceList*)me->itsResourceList;
}


void VelocitySensor_setItsResourceList(VelocitySensor*  const me,
struct ResourceList*  p_ResourceList) {
    me->itsResourceList  =  p_ResourceList;
}


VelocitySensor  *  VelocitySensor_Create(void) {
    VelocitySensor*  me  =  (VelocitySensor  *)
malloc(sizeof(VelocitySensor));
    if(me!=NULL)
            VelocitySensor_Init(me);
    return me;
}


void VelocitySensor_Destroy(VelocitySensor*  const me) {
    if(me!=NULL)
    VelocitySensor_Cleanup(me);
    free(me);
}


static void cleanUpRelations(VelocitySensor*  const me) {
    if(me->itsMutex !=  NULL)
            me->itsMutex  =  NULL;
    if(me->itsResourceList !=  NULL)
            me->itsResourceList = NULL;
}
```

Code Listing 4-49: VelocitySensor.c

Of course, the key class for this pattern is the ResourceList. ResourceList.h is in Code Listing 4-50 with the corresponding C file shown in Code Listing 4-51.


```
#ifndef ResourceList_H
#define ResourceList_H

#include "OrderedLockingExample.h"
struct Mutex;

typedef struct ResourceList ResourceList;
struct ResourceList {
    int nIDs;
    int rIDList[20];
    struct Mutex*  itsMutex;
};
```

```
/*  Constructors and destructors:*/
void ResourceList_Init(ResourceList*  const me);
void ResourceList_Cleanup(ResourceList*  const me);

/*  Operations  */
int ResourceList_addLock(ResourceList*  const me, int rID);

/*  The elements are added in order but  */
/*  can be removed in any order. Therefore,  */
/*  the stored resoiurceIDs above the current  */
/*  one in the list must be moved down to  */
/*  lower in the list.  */
void ResourceList_removeLock(ResourceList*  const me, int rID);

struct Mutex*  ResourceList_getItsMutex(const ResourceList*  const me);

void ResourceList_setItsMutex(ResourceList*  const me, struct Mutex*  p_Mutex);

ResourceList  *  ResourceList_Create(void);

void ResourceList_Destroy(ResourceList*  const me);

#endif
```

Code Listing 4-50: ResourceList.h

```
#include "ResourceList.h"
#include "Mutex.h"

static char isInOrder(ResourceList*  const me, int rID);

static void cleanUpRelations(ResourceList*  const me);

void ResourceList_Init(ResourceList*  const me) {
    me->nIDs  =  0;
    me->itsMutex  =  NULL;
}

void ResourceList_Cleanup(ResourceList*  const me) {
    cleanUpRelations(me);
}

int ResourceList_addLock(ResourceList*  const me, int rID) {
    int retVal;
    Mutex_lock(me->itsMutex);
    if (isInOrder(me,rID)) {
        me->rIDList[me->nIDs++]  =  rID;
        retVal  =  0;
        }
```

```c
        else
            if (rID ==    me->rIDList[me->nIDs])
                retVal =  DUPLICATED_IDS;
            else
                retVal =  POORLY_ORDERED_ACCESS;
    Mutex_release(me->itsMutex);
    return retVal;
}


void ResourceList_removeLock(ResourceList* const me, int rID) {
    int j,k;

    if (me->nIDs) {
        for (j=0; j<me->nIDs; j++) {
            if (rID ==    me->rIDList[j]) {
                for (k=j; k<me->nIDs-1; k++)
                    me->rIDList[k] = me->rIDList[k+1];
                -me->nIDs;
                break;
                };
            };
        };
}


static char isInOrder(ResourceList* const me, int rID) {
    if (me->nIDs)
        return rID > me->rIDList[me->nIDs-1];
    else
        return 1;
}


struct Mutex* ResourceList_getItsMutex(const ResourceList* const me) {
    return (struct Mutex*)me->itsMutex;
}


void ResourceList_setItsMutex(ResourceList* const me, struct Mutex* p_Mutex) {
    me->itsMutex = p_Mutex;
}


ResourceList * ResourceList_Create(void) {
    ResourceList* me = (ResourceList *)
malloc(sizeof(ResourceList));
        if(me!=NULL)
            ResourceList_Init(me);
    return me;
}


void ResourceList_Destroy(ResourceList* const me) {
    if(me!=NULL)
    ResourceList_Cleanup(me);
```

```
      free(me);
}


static void cleanUpRelations(ResourceList*  const me) {
    if(me->itsMutex !=  NULL)
          me->itsMutex  =  NULL;
}
```

Code Listing 4-51: ResourceList.c

4.10 So, What Have We Learned?

The chapter started off with a basic introduction to concurrency concepts, such as what do we mean by the terms concurrency, urgency, criticality, resource, deadlock, and so on. Concurrency is something that we deal with constantly in our daily lives; many many things happen in parallel and we must account for when to start, stop, and synchronize. It is no different in programming embedded systems, except that we must be more precise in our definition and usage of the terms to ensure our systems' predictably to the right thing all the time.

We also talked about how to represent concurrency in the Unified Modeling Language (UML) so that we could draw pictures of the patterns and understand what they mean. Primarily we use «active» classes to represent threads and nest (part) objects inside these structured classes to indicate the parts running within the thread context of the enclosing class. In the same thread, to use a pun, we talked about the structure and services of a typical Real-Time Operating System (RTOS).

This chapter then presented a number of patterns that dealt with concurrency and resource management that are in common use in embedded systems. The first two – the Cyclic Executive and the Static Priority Patterns – provided two different ways for scheduling threads. The Cyclic Executive Pattern is very simple and easy to implement but is demonstrably suboptimal in terms of responsiveness to incoming events. The Static Priority Pattern relies on scheduling services of an RTOS kernel and gives very good responsiveness to events, but it is less predictable and more complex than a Cyclic Executive.

The next three patterns – the Critical Region, the Guarded Call, and the Queuing Patterns – addressed the concern of serializing access to resources in a multitasking environment. The Critical Region Pattern turns off task switching during resource access, thus preventing the possibility of resource data corruption. This approach works but blocks higher-priority tasks that may never access the resource. The Guarded Call Pattern accomplishes the same resource protection goal by using mutex semaphore. The naïve implementation of this pattern can lead to unbounded priority inversion, so the pattern can be augmented with others, such as the Priority

Inheritance Pattern, to limit priority inversion to a single level. The last pattern in this group, the Queuing Pattern, serializes access to a resource by putting the requests in a queue and the resource removes them from the queue in a first-in-first-out (FIFO) order. The Queuing is conceptually very simple but can lead to delays in response to a request made of the resource.

The last two patterns – the Simultaneous Locking and the Ordered Locking Patterns – concerned themselves with deadlock prevention. Deadlock is a very real problem that occurs when there are conditions waited for that can never actually occur, and this happens far more often than most people would suspect. Four conditions are required for deadlock to occur and deadlock can be eliminated by violating any of these four conditions. The Simultaneous Locking Pattern violates the requirement of holding one resource while requesting others through the simple expedient of locking all resource it needs at once. The Ordered Locking Pattern breaks another requirement of deadlocks, that a circular waiting condition occurs; by ensuring that all clients locking the resource in the same order, circular waiting (and deadlock) can be eliminated.

Design Patterns for State Machines

Design Patterns for Embedded Systems in C
DOI:10.1016/B978-1-85617-707-8.00005-4
Copyright © 2011 Elsevier Inc.

Patterns in this chapter

Single Event Receptor Pattern – Implement state machines with a single event receptor

Multiple Event Receptor Pattern – Implement state machines with many event receptors

State Table Pattern – Implement a table-driven state machine

State Design Pattern – Implement state machines with creation of state objects

Decomposed And-State Pattern – Implement and-states through the decomposition of the composite state

---

5.1 Oh Behave!

Behavior can be defined as a change in condition or value over time. It is often a response to an external event or request, but autonomous systems decide for themselves when and how to behave. I like to further characterize behavior into four basic kinds[1]. *Simple behavior* is behavior independent of any history of the object or system. For example, computing $\cos(\frac{\pi}{4})$ returns the same result regardless of what value the cos() function was invoked with before or when it was computed. Similarly computing the maximum of two values is self-contained in the sense that it doesn't rely on the history of the system. *Continuous behavior*, on the other hand, does rely upon the history of the object or system. For example, a moving average digital filter computes an output that depends not only on current data but also on the history of the data (see Equation 5-1).

$$f_j = \frac{d_j + d_{j-1} + d_{j-2} + d_{j-3}}{4}$$

Equation 5-1: Moving average digital filter

Similarly, a Proportional Integral-Differential (PID) control loop uses feedback to update its output, as we see in Figure 5-1. In both cases, the output depends not only on current information

---

[1] See my previous book *Real-Time UML Third Edition: Advances in the UML for Real-Time Systems* (Addison-Wesley, 2004) for more detailed information. We won't deal with all of the rich semantics available in UML state machines in this book.

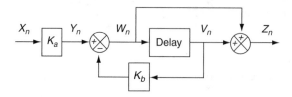

Figure 5-1: Simple PID control loop

but also on previous information, but it does so in a smooth, continuous way. Put another way, the output varies in value but not in kind as a function of new data and the system history.

The third kind of behavior, to which this is dedicated, is *discontinuous* or *stateful behavior*. In stateful systems. the behavior depends on the new data (typically incoming events of interest) and the current state of the system, which are determined by that system's history. However, the difference is the kind of actions the system performs not the value of the outputs. In Figure 5-2, we see a state machine for a microwave oven. The history of the system is represented in the states and the behaviors are the actions, such as `initialize()` (executed on entry into the state `Ready`) and `timeTocook -= currentTime()` (executed on the transition from `Microwaving` to

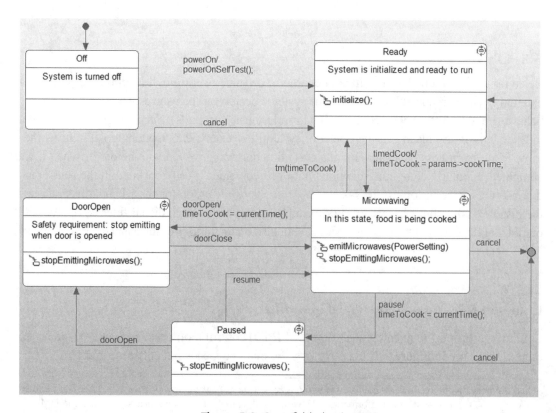

Figure 5-2: Stateful behavior

Table 5-1: State transition table

	Initial	powerOn	timedCook	doorOpen	doorClose	cancel	pause	resime	resume	Null
ROOT	Off									
Off		Ready								
Ready			Microwaving							
Microwaving				DoorOpen		junctionconnector_11	Paused			Ready
DoorOpen					Microwaving	Ready				
Paused				DoorOpen		junctionconnector_11			Microwaving	
junctionconnector_11										Ready

the DoorOpen or Paused states). The important thing to note here is that the kind of actions the system performs varies on the basis of its state (and therefore on its history). Microwaves are only emitted in the Microwaving state; the system is only initialized in the Ready state, and so on.

The state machine may also be visualized as a state x transition table, such as Table 5-1. Many people prefer this notation. It has some advantages but, as you can see, the actual behaviors are not shown in the table. Instead, the rows indicate the current state and the columns are the transitions; in the cells of the table are the next states after the events are processed by the state machine.

State machines are an extremely powerful means by which the behavior of a complex, embedded system can be controlled and orchestrated. As mentioned, it will be the subject of this and the next chapters.

Lastly, the fourth type of behavior is a combination of the previous two, known as *piece-wise continuous behavior* (see Figure 5.3). In PC behavior, within a state, continuous (and, for that matter, simple) the behaviors that occur are of the same kind, but between states, different kinds of behaviors occur. A simple example of this is using different sets of differential equations depending on the state of the system.

5.2 Basic State Machine Concepts

A Finite State Machine (FSM) is a directed graph composed of three primary elements: states, transitions, and actions. A *state* is a condition of a system or element (usually a class in an object-oriented system). Looking back to , the element is the microwave oven (or at least the collection of elements involved in cooking) and the states are the conditions of that system – Off, Ready, Microwaving, DoorOpen, and Paused. Usual names for states are verbs, adverbs, and verb phrases.

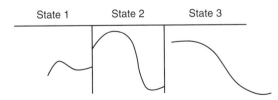

State 1   State 2   State 3

Figure 5-3: Piece-wise continuous behavior

A *transition* is a path from one state to another, usually initiated by an event of interest; it connects a predecessor state with a subsequent state when the element is in the predecessor state and receives the triggering event. The events of interest in Figure 5-2 include the `powerOn`, `timedCook`, `doorOpen`, `cancel`, `pause`, `resume`, and `tm` (timeout) events. If an event occurs while the element is in a state that doesn't respond to that particular event, the event is "quietly discarded" with no effect – that is, state machines only make affirmative statements "When I'm in this state and this occurs, I'll do this") not negative statements ("If I'm not in this state or this doesn't happen, then I'll do that"). Events not explicitly handled within a state are discarded should they occur when the element is in that state.

The actual behaviors executed by the element are represented in *actions*. In UML state machine, actions may be associated with entry into a state, exit from a state, a transition *per se*, or what is called an "internal transition" (we will get into this in more detail shortly). Actions are specified in an *action language*, which for us will be C. Actions shown on Figure 5-2 include `initialize()`, `emitMicrowaves()`, `stopEmittingMicrowaves()`, and `timeToCook -= currentTime()`.

Figure 5-4 shows the basic syntax for UML state machines. The rounded rectangles represent the states of the element. The element remains in a state until a transition causes it to change state. Within a state, three annotations may be made. Entry actions are actions the element executes when the state is entered regardless of which transition path was taken (it is possible to

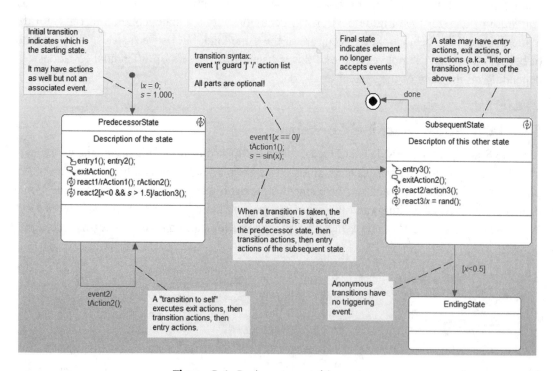

Figure 5-4: Basic state machine syntax

make many different transitions terminate on any given state). The exit actions are actions taken whenever the state is left, regardless of which path is taken. Reactions ("internal transitions") are taken when the element is in the specified state and the triggering event occurs; in this case, the actions are taken but the state is not exited. Contrast the `PredecessorState` response to the event `react1` with the transition taken with `event2`. In both cases, the element ends up in the same state (`PredecessorState`), but in the former case, only the actions `rAction1()` and `rAction2()` are executed. In the latter case, the exit actions (in this case `exitAction()`) are taken, then the transition actions (in this case `tAction2()`), and finally the entry actions are taken (in this case `entry1()` and `entry2()`).

Transitions are shown as directed arrows. They are normally initiated by the element receiving a triggering event. Transitions have three syntactic fields with three special characters delimiting them. The first field is the *event name*; this is the name of the event type that causes this transition to be taken (provided, of course, that the element is in the predecessor state when the event occurs). The second field is known as the *guard*. The guard is a Boolean condition (without side effects!); if the event occurs AND the guard evaluates to true, then the transition is taken. If the event occurs and the guard is false, then the event is "quietly discarded" and no actions are taken, including the exit actions of the predecessor state. This means that the guard, if present, *must* be evaluated before any actions are executed. The guard, when present, is delimited by square brackets. Lastly, there may be a list of actions that are executed when the transition is taken. The *action list* is preceded by a slash mark.

It should be noted that all of these fields are optional. When the triggering event is missing (known sometimes as an "anonymous transition"), then the transition is triggered as soon as the element enters the predecessor state, that is, as soon as the entry actions for the predecessor state are completed. If the anonymous transition has a guard that evaluates to false, then the anonymous event is discarded and the element remains in the predecessor state even if the guard condition should later become true.

There is a special kind of transition, known as the *initial transition* (or "default pseudostate" if you want to be fancy) that indicates the starting state and actions when the element is created. This transition may not have a triggering event.

There is a special state, known as the *final pseudostate*, that indicates the element no longer accepts events. Usually, this means that the element is about to be deleted. The icon for the final state is a filled circle within an unfilled circle. In Figure 5-4, the pseudostate is entered when the element is in the `SubsequentState` and the element receives the `done` event.

5.2.1 OR-States

In simple state machines, an element must always be in *exactly* one of its states, no more and no less. For this reason, such states are called *OR-states*. UML state machines extend the notion of OR-states by adding the notion of nesting within states. Nested states add a level of abstraction

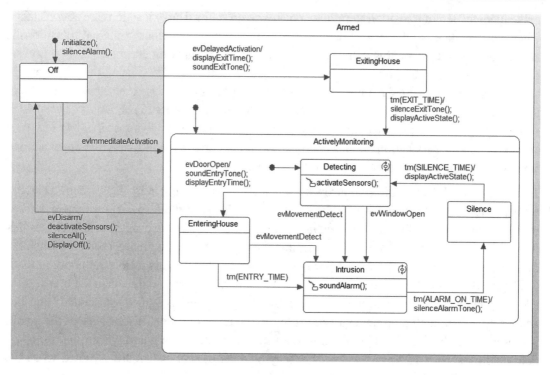

Figure 5-5: OR-states[2]

to the FSM world and this requires a minor modification to the basic rule: the element must be in exactly one state *at each level of abstraction*.

Figure 5-5 shows some nested states for the monitoring part of a home alarm system. The system must either be Off or Armed – no other conditions are possible. If it is Armed, then it must either be in the ExitingHouse or ActivelyMonitoring states. And, if it is ActivelyMonitoring, then it must be in exactly one of the following states: Detecting, EnteringHouse, Intrusion, or Silence.

There are a few interesting things to note. First, there must be a default pseudostate at every level of abstraction and it indicates the default state when that level of abstraction is entered. This default can be bypassed when desired, however. If you take the "default" transition path from the Off state via the evImmediateActivation transition, the system ends up in the Armed::ActivelyMonitoring::Detecting state. However, if the user activates the alarm from inside the house, the system gives them a couple of minutes to get out of the house before beginning detecting – this is the evDelayedActivation transition path. Also note that regardless of which deeply nested state of Armed the system is in, when the system is deactivated

---

[2] Thanks to Mark Richardson of LDRA for the idea of this state machine example.

Table 5-2: OR-state table

	Initial	evDelayedActivation	evImmediateActivation	evDisarm	evDoorOpen	evMovementDetect	evWindowOpen	Null
ROOT	Off							
Off		ExitingHouse	Armed					
Armed	ActivelyMonitoring			Off				
ExitingHouse								ActivelyMonitoring
ActivelyMonitoring	Detecting							
Detecting					EnteringHouse	Intrusion	Intrusion	
EnteringHouse						Intrusion		Intrusion
Intrusion								Silence
Silence								Detecting

via the evDisarm transition, the system transitions to the Off state. This transition can be activated whenever the system is in the Armed state, regardless of which nested state it is currently in.

As a side note, the state diagram in Figure 5-5 can also be shown in tabular form, as it is in Table 5-2.

5.2.2 AND-States

Sometimes state machines are complex in the sense that they have independent (or at least mostly so) aspects. Consider the state machine for a light that has two independent aspects; color (Off, Red, Yellow, and Green) and display style (Off, Steady, Flashing Slowly, and Flashing Quickly). How many states of this light are there? If you look at Figure 5-6,

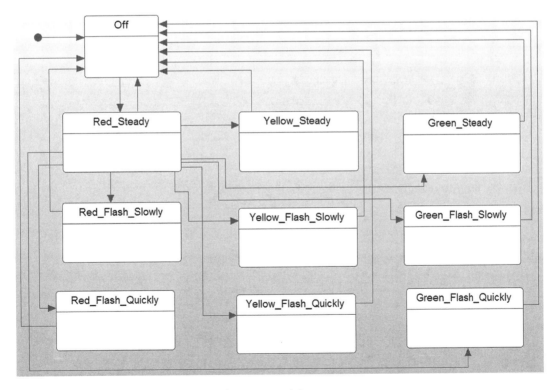

Figure 5-6: Light states

you see that there are 10. Further, in this light, it is possible to go from any of these states to any other. The figure shows one source state going to all other states (Red_Steady) and one target state receiving transitions from all other states (Off); you can imagine the rat's nest that would result from actually drawing all transition paths.

You can see that to consider all the states of the independent aspects, you must multiply the independent state spaces together, resulting in a possibly enormous set of states (imagine now adding a third independent aspect, brightness – low, medium, and high – into the mix). This problem is so common it even has a name – *state explosion*. UML state machines have a solution to the state explosion problem – AND-states.

AND-states are special independent states that are always nested inside other states. The special rule for AND-states is that the element must be in exactly one of each active AND-state at each level of abstraction. In the case of the light, we'll have one AND-state for color and another for display rate. You can see that this simplifies the problem so much that we have no problem representing all the transitions in the state machine even though we couldn't before (and maintain any semblance of readability). This is shown in Figure 5-7. The dashed line in the middle of the On state is an AND-line and separates the two AND-states.

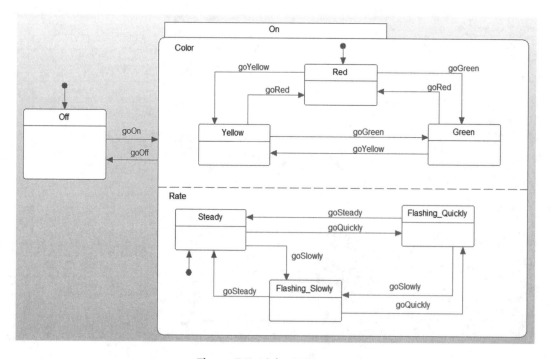

Figure 5-7: Light AND-states

Table 5-3: Light AND-state table

	Initial	goSteady	goQuickly	goSlowly	goYellow	goGreen	goRed	goOn	goOff
ROOT	Off								
Off								On	
On									Off
Color	Red								
Green					Yellow		Red		
Yellow						Green	Red		
Red					Yellow	Green			
Rate	Steady								
Flashing_Slowly		Steady	Flashing_Quickly						
Steady			Flashing_Quickly	Flashing_Slowly					
Flashing_Quickly		Steady		Flashing_Slowly					

And, yes, the state machine in Figure 5-7 can also be shown in a table. See Table 5-3. You just have to pay special attention to the icon of the On state to notice that the nested states are AND-states and not OR-states to correctly interpret the table.

5.2.3 Special Stuff: Timeouts, Conditionals, and More

All FSMs can be ultimately reduced to the core elements discussed above. However, the UML standard adds a number of syntactic elements that simplify representing complex situations. Most of these fall into the category of *pseudostates*, an unfortunate name for a set of annotations that are in *no way* states, but are nonetheless useful.

First, let's discuss timeouts semantics. The UML has the notion that elapsed time is a kind of event that can cause a change of state in a state machine. The notation for this is normally either tm(<time interval>) or after(<time interval>). The way to think about a timeout transition is as follows. When entering a state that has an exiting timeout transition, a logical timer begins as soon as the state is entered (i.e., as soon as the state's exit actions complete). If the element is still in the state when the timeout expires, then the transition triggered by the timeout fires. If the element leaves that state and reenters it later, the timer starts over from zero.

The state machine in Figure 5-8 shows how timeout transitions might be used. When the system is created, it begins in the Off state. Once it receives the start event, the system transitions to the Waiting state. In WAIT_TIME milliseconds later[3], the system transitions to the DoneWaiting state. Let's suppose that WAIT_TIME is set to 1000 milliseconds. If the system remains in the Waiting state for a full second, then it will transition to the DoneWaiting state. However, let's suppose that 950 ms after entering the Waiting state, the system receives the pause event; in this case, the timer (logically) disappears and the system enters the Paused state. If, two days later, the system receives the resume event, it will transition back to the Waiting state and resume timing, but it will again wait a full second before timing out. Similarly, let's set QUIT_TIME to

---

[3] Milliseconds are the most common units for state implementation and will be used here, but you are free to use whatever timing resolution makes sense for you.

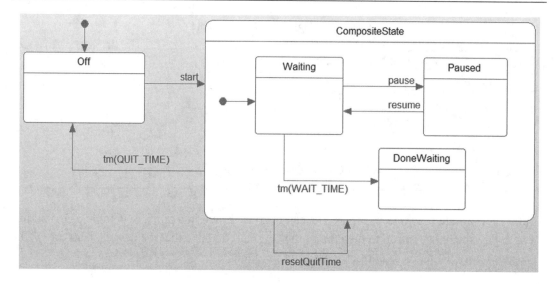

Figure 5-8: Timeouts

20,000 ms. We can effectively reset the system by issuing the resetQuitTime event. Let's suppose the system is in the Paused state and the QUIT_TIME timer has counted down 10,000 ms. What happens then?

The transition triggered by the resetQuitTime causes the system to reenter the WaitingState with the WAIT_TIME counter starting over and the QUIT_TIME counter also restarted. What if we want to reenter the Paused state? UML has a solution for that concern too – the history pseudostate.

The UML provides a number of pseudostates but we will only discuss the most common of these:

- Default pseudostate – identifies the starting state for a system or within a level of abstraction
- Conditional pseudostate – shows branch points where a transition may terminate on different target states and/or have different actions depending on the results of guards
- Fork pseudostate – allows bypassing the defaults when going into a composite with AND nested states
- Join pseudostate – allows specific nested AND-states as source states for an exit from a composite state with AND nested states
- History pseudostate – sets the default nested state to be the "last active" nested state
- Final pseudostate – indicates that the state machine no longer receives events, usually because it is about to be destroyed

Figure 5-9 shows these different annotations. The default and final pseudostates we've seen before. The conditional pseudostate is activated while the system is in the Ready state and the go event is received. If the guard useDefaults() returns TRUE, then the default nest AND-states

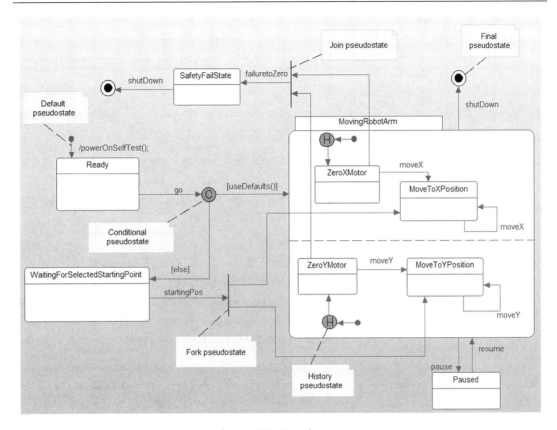

Figure 5-9: Pseudostates

are entered. Alternatively, if it is FALSE, the (optional) else clause routes the transition to the WaitingForSelectedStartingPoint state. If used, the else clause path is taken if and only if all other branching guards are FALSE. If all the guards evaluate to FALSE (impossible if the else clause is included), then the event is discarded without taking a transition.

The fork is used between the WaitingForSelectedStartingPoint and the MovingRobotArm states. In this case, the system bypasses the defaults (taken in the go [useDefaults()] transition) and instead transitions to two nondefault states inside the MovingRobotArm composite state.

Similarly, a join occurs between the ZeroXMotor and ZeroYMotor nested AND-states on one side and the SafetyFailState on the other. That means that the system must be in *both* the predecessor states AND the event failuretoZero must occur; if the system is in any other condition, the event is discarded.

Lastly, the history pseudostate is used to return to the last active nested state. In the example, when the system is in the MovingRobotArm state and receives the event pause, it transitions to the Paused state. A subsequent resume event will return to the same nested states the system was in when it was in the MovingRobotArm state. For example, imagine the system is in the ZeroXMotor and MoveToYPosition states when the pause occurs. With the resume event, the system will return to those two states. Without the history state, it would return to the ZeroXMotor and ZeroYMotor nested states, since they are the defaults.

5.2.4 Synchronous versus Asynchronous?

A synchronous event is activated by a call to an operation on the target state machine and the caller waits (blocked) until the state machine completes its response to the event. In contrast, an asynchronous event is sent via a send-and-forget mechanism by the sender and the sender may continue on regardless of the actions of the target state machine.

The state machine itself doesn't care how the event arrives, but must handle the event in either case. This means that to avoid race conditions, synchronous state machines must block callers (using the Guarded Call or Critical Region Patterns of the previous chapter) and asynchronous state machines must use queuing to store their events until they can be handled. While a state machine can handle both synchronous and asynchronous events, a given state machine usually handles one or the other but not both to eliminate potential race conditions.

We really want to be able to have the system respond in the same way to the same sequence of events; such a system is said to have *deterministic semantics*. There are a couple of conditions under which determinism is threatened with which any good implementation of a state machine must deal.

What should happen, for example, if a state machine is in the middle of executing an action and an incoming event occurs? Should the state machine stop immediately and handle the event? The answer, of course, is *no*. The state machine must be idle and waiting for events before handling a new one. Mealy-Moore state machines "handle" this by assuming that actions take zero time, a proposition that is not only clearly false but also complicates timing analysis. The UML has a different answer – run-to-completion semantics. That is, actions may take time, but any pending events must wait until the state machine has completed the entire exit action-transition action-entry action sequence. The problem is simplified if asynchronous events are used, since the delivery mechanism for the event can be triggered when the state machine enters a state. For synchronous events in the presence of multiple threads, though, the action sequence must block any attempt to invoke an event receptor operation until the action executions are complete.

Let's consider the basic elements of both synchronous and asynchronous state machines.

For a synchronous state machine, the owning element of the state machine will provide one or more event handler functions. The most obvious way to handle the deterministic semantics problem is to lock the semaphore when the event receptor is called and unlock it when the event receptor returns.

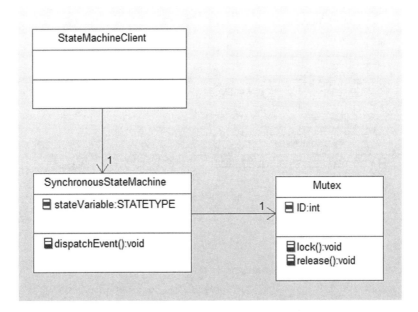

Figure 5-10: Synchronous state machine basic interaction model

Figure 5-10 shows this basic model. In the figure, the `SynchronousStateMachine` has a single `stateVariable` (of an enumerated type `STATETYPE`, although most people just encode this with integers) to track the current state of the FSM. It also has a single function `dispatchEvent` that receives the event and any data that it passes. The basic flow of logic for the function would be:

```
Mutex_lock(me->itsMutex);
/* perform list of actions */
/* update stateVariable */
Mutex_release(me->itsMutex);
```

Code Listing 5-1: Basic dispatchEvent Flow

Figure 5-11 shows the basic form for an asynchronous state machine. The first difference you see is that the client actually sends events to the queue via the `post()` operation and then queues them. This locks the queue's mutex, stores it in the queue, unlocks its `Mutex`, calls the `dispatchEvent` on the `AsynchronousStateMachine` and returns. Code Listing 5-2 shows the basic flow for the `post()` function.

```
Mutex_lock(me->itsMutex);
insert(ev); /* insert into the queue */
```

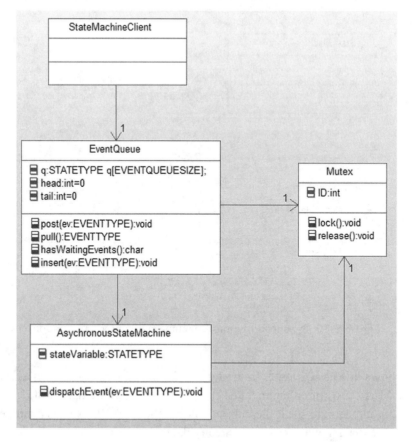

Figure 5-11: Asynchronous state machine basic interaction model

```
Mutex_release(me->itsMutex);
AsynchronousStateMachine_dispatchEvent(ev);
```

Code Listing 5-2: Basic post flow

This ensures that the AsynchronousStateMachine is notified when an event is posted (note that the Mutex associated with the AsynchronousStateMachine protects any active action sequence from interruption). Note that we expect the mutex in Figure 5-11 to be a recursive mutex (see the discussion in the previous chapter) so that the EventQueue can invoke (and be blocked by) the Mutex associated with the AsynchronousStateMachine as many times as necessary.

The EventType referenced in Figure 5-11 may be a simple enumerated type or something more complex. If you want to pass data along with the event, then it needs to be a struct (usually with a union clause) to hold the data. That is because the type must be stored within an event queue regardless of the size and subcomponents of the data being passed. In the case of a synchronous state machine, this can be encoded directly in the various event receptors' argument lists. For example, the follow code snippet shows the definition of three event receptors for a synchronous state machine that each take different arguments:

```
void event1(FSM* me); /* no arguments */
void event2(FMS* me, int n, double torque); /* two scalar arguments */
void event3(FSM* me, struct NavData* nd); /* pointer to struct */
```

Code Listing 5-3: Event receptor data passing

Note that because the synchronous state machine can call a specific function for each different event separately, there is no need to encode the event type explicitly, unlike the case in which events must be queued for later consumption.

The above discussion is applicable regardless of exactly how we implement the state machine. Let us now turn our attention to different implementation patterns for finite state machines. In all cases, we will implement the state machine shown in Figure 5-12. This state machine implements a simple parser for input characters and computes a numeric result. This works for such character strings as "12.34" and ".1234").

It is assumed that the client has already determined, on a character-by-character basis, whether the character is a digit ('0' to '9'), a dot ('.') or whitespace (' ', or '\t') and has generated one of

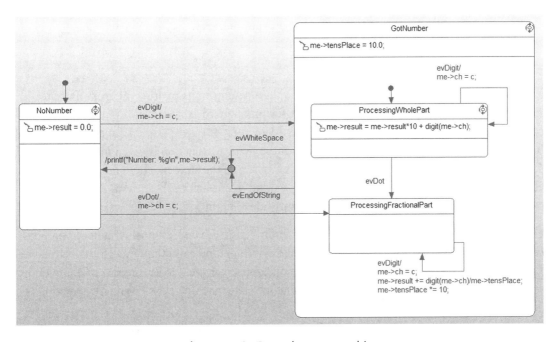

Figure 5-12: Example state machine

the events accordingly. For multiple receptors, the client will generate one of the following events: evDigit (passing the character as a parameter), evDot, evWhiteSpace, or evEndOfString. For a single event receptor, there will be a single event evChar and it will have to carry data of the character itself. It is easy enough to make the character type identification within the state machine but it is left as an exercise to the reader.

5.3 Single Event Receptor Pattern

Single event receptor state machines (henceforth known as SERSMs) can be used for both synchronous and asynchronous events.

5.3.1 Abstract

The Single Event Receptor Pattern relies on a single event receptor to provide the interface between the clients and the state machine. Internally, this single event receptor must accept an event data type that not only identifies which event has occurred but also any data that accompanies the event. This pattern may be used in synchronous state machines – receiving events directly from the clients – or asynchronous, via an intermediate event queue.

5.3.2 Problem

This pattern addresses the problem of providing simple implementation of a state machine that applies to both synchronous and asynchronous event delivery.

5.3.3 Pattern Structure

This pattern can be used in two ways – for either asynchronous or for synchronous state machines. The asynchronous version context for this pattern is shown in Figure 5-13. In this case, it should be noted that at the end of the TSREventQueue_post() function, the OS function postSignal() is called. The code for the post() function is given in Code Listing 5-4. On the state machine side, the asynchronous version of the state machine receptor class has a task loop that waits for the signal, indicating that an event has been queued, and then iterates over all queued events. That code is shown in Code Listing 5-5.

The synchronous version of the pattern is shown in Figure 5-14. The most notable differences between the two is the absence of the event queue and that the class that owns the state machine TokenizeSyncSingleReceptor has an association to a mutex to ensure run-to-completion semantics for the state machine actions. Note that the state machine class has many operations but the only one called by the client is TokenizerSyncSingleReceptor_eventDispatch(). Internally, this function will determine which event is being passed and process the incoming event.

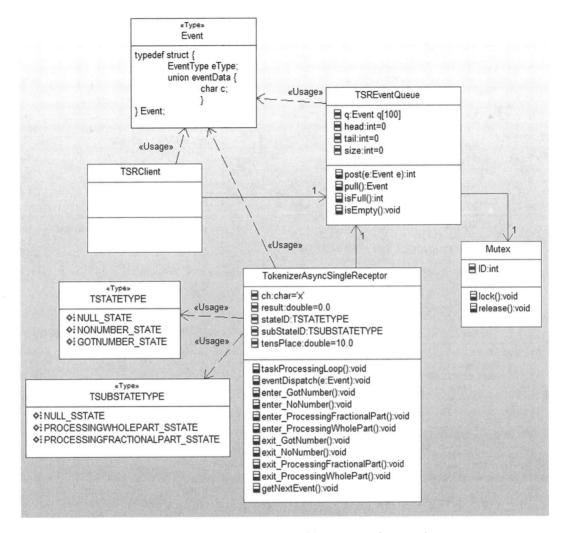

Figure 5-13: Single event receptor state machine context for asynchronous events

5.3.4 Collaboration Roles

This section describes the roles for this pattern.

5.3.4.1 Event

The event is a struct that contains an attribute to identify the type of the event and other data that is passed along with the different events. This is normally implemented as a union in C.

5.3.4.2 EventType

The EventType is an enumerated type for the events. Alternatively, ints can be used but I recommend enums as being more informative.

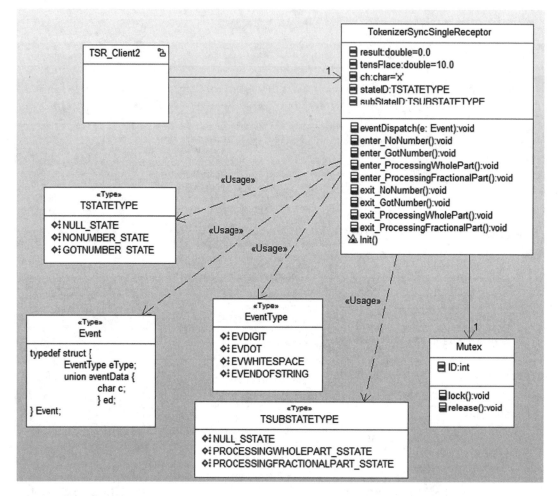

Figure 5-14: Single event receptor state machine context for synchronous events

5.3.4.3 Mutex

For the asynchronous version of the state machine, the `Mutex` ensures the integrity of the `TSREventQueue` in the case that multiple clients are simultaneously trying to post events or if a client tries to post an event while the state machine class is removing one.

For the synchronous version of the pattern, the mutex ensures the deterministic run-to-completion semantics of transitions by preventing other clients from interrupting the state machine execution of an action sequence. This is accomplished by locking the semaphore before the event is processed and releasing it once the transition processing is complete.

5.3.4.4 TokenizerAsyncSingleReceptor

This element is the owner and implementer of the desired state machine (see Figure 5-12). It has two kinds of attributes: those involved with the intent of the state machine (such as `result`)

and those involved with the management of the state machine (such as stateID). Of the former type, we find:

- result which holds the current value of the number being parsed
- tensPlace which is used in computing fractional parts of numbers, as in 0.345 of the number 10.345
- ch which holds the current character being processed

Of the latter type, we see:

- stateID which stores the current high level state (NULL_STATE, NONUMBER_STATE, or GOTNUMBER_STATE)
- subStateID which stores the current substate (NULL_SSTATE, PROCESSINGWHOLEPART_SSTATE, or PROCESSINGFRACTIONALPART_SSTATE)

These values are defined in the enumerated types TSTATETYPE and TSUBSTATETYPE, respectively.

Functions of this state machine are likewise of three types. The first is the task loop that pends on the signal (which is posted when an event is enqueued). The second is the eventDispatch that identifies and processes the event. The third type of operations are for the entry and exit actions of the states. Note that the transitions actions are all embedded within the event-receptors since they only occur once, but there are often many paths into and out of a state, so we explicitly create entry and exit functions for each state, making them reusable.

5.3.4.5 *TokenizerSyncSingleReceptor*
This class is exactly like the previously described TokenizerAsyncSingleReceptor except that 1) it does not have a task loop because it is invoked synchronously, and 2) it uses a mutex to enforce run-to-completion semantics of its state machine. The only function directly invoked by the client is TokenizerSyncSingleReceptor_eventDispatch().

5.3.4.6 *TSRClient*
This is the client of the state machine. For the asynchronous version, it queues the event by calling TSREventQueue_post() function using an instance of the event type as the passed parameter. For the synchronous case, it invokes the single event receptor TokenizerSyncSingleReceptor and passes the appropriate event (and associated data) as typed with the event type.

5.3.4.7 *TSREventQueue*
This is a queue that holds events. It is only used in the asynchronous event processing case, to hold the incoming events until the state machine gets around to processing them. It provides two primary functions. The TSREventQueue_post() function enqueues an event and signals

the OS that an event has arrived. The TSREventQueue_pull() function is invoked by the TSRAsyncSingleReceptor class to get the next pending event.

5.3.4.8 TSTATE

TSTATE is an enumerated type of the high-level states.

5.3.4.9 TSUBSTATE

TSUBSTATE is an enumerated type of the nested states. In this case, there are only two levels of nesting. If there were more, then more enumerated types could be added.

5.3.5 Consequences

This pattern provides a straightforward state machine implementation easily adapted to either synchronous or asynchronous state machine event passing. The fact that all the state logic is encapsulated within a single event receptor limits the scalability of the approach.

5.3.6 Implementation Strategies

It should be noted that *one way* of implementing this pattern is by simply taking a Multiple Event Receptor Pattern and adding a single publicly available event receptor (to be invoked by clients) that calls the multiple event receptors already present. You'll still need to create the abstract event type to be passed into the single event receptor. This is one way to take a synchronous state machine done using that pattern and enabling it to be used in an asynchronous way – add an event queue, a polling mechanism to look for new events, a single event receptor and you can use it without other modification.

A more common implementation of the Single Event Receptor Pattern is to encode all the logic in a big switch-case statement within the single event receptor.

5.3.7 Related Patterns

The synchronous version of the pattern uses the Guarded Call Pattern (see Chapter 4) to ensure deterministic run-to-completion of transitions while the asynchronous version uses the Queuing Pattern. The Critical Region Pattern can be used instead for either synchronous or asynchronous state machines to ensure the run-to-completion semantics.

5.3.8 Example

The asynchronous version differs in a couple of ways from the synchronous version. First, let's consider what is special about the asynchronous version.

The most obvious difference is, of course, the presence of the TSREventQueue. The clients don't invoke the event receptor on the state machine directly; instead they send the events to the event queue, which then posts the signal to the OS on which the state machine pends (see Code

Listing 5-4). The state machine class has its own task processing loop that waits for the signal to be posted, and when it is, loops to remove and process all events that are currently queued (Code Listing 5-5).

```
int TSREventQueue_post(Event e) {
    Mutex_lock(me->itsMutex);
    if (!TSRQueue_isFull(me)) {
        me->q[me->head] = e;
        me->head = (me->head + 1) % QSIZE;
        me->size += 1;
        Mutex_release(me->itsMutex);
        postSignal(); /* signal that an event is present */
        return 1;
    }
    else {
        Mutex_release(me->itsMutex);
        return 0;
        };
};
```

Code Listing 5-4: TSREventQueue_post() function

```
void TokenizerAsyncSingleReceptor_taskLoop() {
    while (1) {
        waitOnSignal(); /* wait until event occurs */
        while (! TSREventQueue_isEmpty(me->itsTSREventQueue)) {
            TokenizerAsyncSingleReceptor_eventDispatch(
                TSREventQueue_pull(me->itsTSREventQueue));
        }
    }
}
```

Code Listing 5-5: TokenizeAsyncSingleReceptor_taskLoop() function

Code Listing 5-1 shows the enumerated types and other definitions used by the elements of the pattern.

```
/* this is the size of the event queue */
#define QSIZE 100

typedef enum EventType {
  EVDIGIT,
  EVDOT,
  EVWHITESPACE,
  EVENDOFSTRING
} EventType;

typedef struct {
    EventType eType;
    union eventData {
        char c;
        } ed;
```

```
} Event;

typedef enum TSTATETYPE {
  NULL_STATE,
  NONUMBER_STATE,
  GOTNUMBER_STATE
} TSTATETYPE;

typedef enum TSUBSTATETYPE {
  NULL_SSTATE,
  PROCESSINGWHOLEPART_SSTATE,
  PROCESSINGFRACTIONALPART_SSTATE
} TSUBSTATETYPE;

/* helper function returns the digit */
/* held by a char */
int digit(char c) {
  return c-'0';
}

/*
  OS signal and wait functions for
  task synchronization
*/

void postSignal(void);
void waitOnSignal(void);
```

Code Listing 5-6: TokenizeAsyncSinglePkg.h

Code Listing 5-7 and Code Listing 5-8 show the code for the event queue used in the pattern.

```
#include "TokenizerAsyncSinglePkg.h"

struct Mutex;

typedef struct TSREventQueue TSREventQueue;
struct TSREventQueue {
  Event q[100];
  int size;
  int head;
  int tail;
  struct Mutex* itsMutex;
};

/* Constructors and destructors:*/
void TSREventQueue_Init(TSREventQueue* const me);

void TSREventQueue_Cleanup(TSREventQueue* const me);

/* Operations */

void TSREventQueue_isEmpty(TSREventQueue* const me);
int TSREventQueue_isFull(TSREventQueue* const me);
/* puts the passed event in */
/* the event queue and then */
```

```
/* calls the event receptor of */
/* the state machine. Note */
/* if the queue overflows an */
/* error code (1) is returned. */
int TSREventQueue_post(TSREventQueue* const me, Event e);

/* It is assumed that the caller has */
/* ensured that there is at least one */
/* event in the queue (via the isEmpty() */
/* funtion) prior to calling this function. */
/* Otherwise they get a default event. */
Event TSREventQueue_pull(TSREventQueue* const me);

struct Mutex* TSREventQueue_getItsMutex(const TSREventQueue* const me);

void TSREventQueue_setItsMutex(TSREventQueue* const me, struct Mutex* p_Mutex);

TSREventQueue * TSREventQueue_Create(void);

void TSREventQueue_Destroy(TSREventQueue* const me);

#endif
```

Code Listing 5-7: TSREventQueue.h

```
#include "TSREventQueue.h"
#include "Mutex.h"

static void cleanUpRelations(TSREventQueue* const me);

void TSREventQueue_Init(TSREventQueue* const me) {
  me->head = 0;
  me->size = 0;
  me->tail = 0;
  me->itsMutex = NULL;
}

void TSREventQueue_Cleanup(TSREventQueue* const me) {
  cleanUpRelations(me);
}

void TSREventQueue_isEmpty(TSREventQueue* const me) {
  return me->size <= 0;
}

int TSREventQueue_isFull(TSREventQueue* const me) {
  return me->size >= QSIZE-1;
}

/* post enqueues an event and signals that fact */
int TSREventQueue_post(TSREventQueue* const me, Event e) {
  Mutex_lock(me->itsMutex);
  if (!TSRQueue_isFull(me)) {
    me->q[me->head] = e;
      me->head = (me->head + 1) % QSIZE;
      me->size += 1;
      Mutex_release(me->itsMutex);
```

```
          postSignal(); /* signal that an event is present */
          return 1;
    }
    else {
       Mutex_release(me->itsMutex);
       return 0;
       };
}

/* pulls removes the oldest event from the queue */
/* pull should only be called when there is an event waiting */
Event TSREventQueue_pull(TSREventQueue* const me) {
   Event e;
   Mutex_lock(me->itsMutex);
   if (!TSREventQueue_isEmpty(me)) {
      e = me->q[me->tail];
      me->tail = (me->tail + 1) % QSIZE
      size -= 1;
   };
   Mutex_release(me->itsMutex);
   return e;
}

struct Mutex* TSREventQueue_getItsMutex(const TSREventQueue* const me) {
   return (struct Mutex*)me->itsMutex;
}

void TSREventQueue_setItsMutex(TSREventQueue* const me, struct Mutex* p_Mutex) {
   me->itsMutex = p_Mutex;
}

TSREventQueue * TSREventQueue_Create(void) {
   TSREventQueue* me = (TSREventQueue *) malloc(sizeof(TSREventQueue));
   if(me!=NULL)
   TSREventQueue_Init(me);
   return me;
}

void TSREventQueue_Destroy(TSREventQueue* const me) {
   if(me!=NULL)
   TSREventQueue_Cleanup(me);
   free(me);
}

static void cleanUpRelations(TSREventQueue* const me) {
   if(me->itsMutex != NULL)
   me->itsMutex = NULL;
}
```

Code Listing 5-8: TSREventQueue.c

Of course, the real action here takes place in the class with the state machine. In this case, we'll show just the code for the asynchronous version in Code Listing 5-9 and Code Listing 5-10.

```
#include <stdio.h>
#include "TokenizerSyncSingleReceptorPkg.h"

struct Mutex;

typedef struct TokenizerSyncSingleReceptor TokenizerSyncSingleReceptor;
struct TokenizerSyncSingleReceptor {
  char ch;
  double result;
  TSTATETYPE stateID;
  TSUBSTATETYPE subStateID;
  double tensPlace;
  struct Mutex* itsMutex;
};

void
TokenizerSyncSingleReceptor_enter_GotNumber(TokenizerSyncSingleReceptor* const me);

void
TokenizerSyncSingleReceptor_enter_NoNumber(TokenizerSyncSingleReceptor* const me);

void
TokenizerSyncSingleReceptor_enter_ProcessingFractionalPart(TokenizerSyncSingleRe-
ceptor* const me);

void
TokenizerSyncSingleReceptor_enter_ProcessingWholePart(TokenizerSyncSingleReceptor*
const me);

void
TokenizerSyncSingleReceptor_exit_GotNumber(TokenizerSyncSingleReceptor* const me);

void
TokenizerSyncSingleReceptor_exit_NoNumber(TokenizerSyncSingleReceptor* const me);

void
TokenizerSyncSingleReceptor_exit_ProcessingFractionalPart(TokenizerSyncSingleRecep-
tor* const me);

void
TokenizerSyncSingleReceptor_exit_ProcessingWholePart(TokenizerSyncSingleReceptor*
const me);

void
TokenizerSyncSingleReceptor_Init(TokenizerSyncSingleReceptor* const me);

void
TokenizerSyncSingleReceptor_Cleanup(TokenizerSyncSingleReceptor* const me);

void
TokenizerSyncSingleReceptor_eventDispatch(TokenizerSyncSingleReceptor*   const   me,
Event e);

struct Mutex* TokenizerSyncSingleReceptor_getItsMutex(const TokenizerSyncSingleRecep-
tor* const me);

void
TokenizerSyncSingleReceptor_setItsMutex(TokenizerSyncSingleReceptor* const me, struct
Mutex* p_Mutex);
```

```
TokenizerSyncSingleReceptor *
TokenizerSyncSingleReceptor_Create(void);

void
TokenizerSyncSingleReceptor_Destroy(TokenizerSyncSingleReceptor* const me);

#endif
```

Code Listing 5-9: TSRSyncSingleReceptor.h

```
#include "TSRSyncSingleReceptor.h"
#include "Mutex.h"

void
TokenizerSyncSingleReceptor_enter_GotNumber(TokenizerSyncSingleReceptor* const me) {
  me->tensPlace = 10.0;
}

void
TokenizerSyncSingleReceptor_enter_NoNumber(TokenizerSyncSingleReceptor* const me) {
  me->result = 0.0;
}

void
TokenizerSyncSingleReceptor_enter_ProcessingFractionalPart(TokenizerSyncSingle
Receptor* const me) {
    /* enter appropriate actions, if any here */
}

void
TokenizerSyncSingleReceptor_enter_ProcessingWholePart(TokenizerSyncSingleReceptor*
const me) {
  me->result = me->result*10 + digit(me->ch);
}

void
TokenizerSyncSingleReceptor_exit_GotNumber(TokenizerSyncSingleReceptor* const me) {
  me->subStateID = NULL_SSTATE;
}

void
TokenizerSyncSingleReceptor_exit_NoNumber(TokenizerSyncSingleReceptor* const me) {
    /* enter appropriate actions, if any here */
}

void
TokenizerSyncSingleReceptor_exit_ProcessingFractionalPart(TokenizerSyncSingle
Receptor* const me) {
  /* enter appropriate actions, if any here */
}

void
TokenizerSyncSingleReceptor_exit_ProcessingWholePart(TokenizerSyncSingleReceptor*
const me) {
  /* enter appropriate actions, if any here */
}
```

```
void
TokenizerSyncSingleReceptor_Init(TokenizerSyncSingleReceptor* const me) {
  me->ch = 'x';
  me->result = 0.0;
  me->tensPlace = 10.0;
  me->itsMutex = NULL;

  me->stateID = NONUMBER_STATE;
  me->subStateID = NULL_SSTATE;
}

void
TokenizerSyncSingleReceptor_eventDispatch(TokenizerSyncSingleReceptor*   const   me,
Event e) {
  Mutex_lock(me->itsMutex);
  switch (e.eType) {
    case EVDIGIT:
        switch (me->stateID) {
            case NONUMBER_STATE:
                /* transition to GotNumber state, default substate */

            TokenizerSyncSingleReceptor_exit_NoNumber(me);
                me->ch = e.ed.c;

            TokenizerSyncSingleReceptor_enter_GotNumber(me);
                me->stateID = GOTNUMBER_STATE;

            TokenizerSyncSingleReceptor_enter_ProcessingWholePart(me);
                me->subStateID = PROCESSINGWHOLEPART_SSTATE;
                printf("Current value of result: %g\n", me->result);
                break;
            case GOTNUMBER_STATE:
                switch (me->subStateID) {
                    case PROCESSINGWHOLEPART_SSTATE:

                    TokenizerSyncSingleReceptor_exit_ProcessingWholePart(me);
                        me->ch = e.ed.c;
                    TokenizerSyncSingleReceptor_enter_ProcessingWholePart(me);
                        printf("Current value of result: %g\n", me->result);
                        break;
                    case PROCESSINGFRACTIONALPART_SSTATE:

                    TokenizerSyncSingleReceptor_exit_ProcessingFractionalPart(me);
                        me->ch = e.ed.c;
                        me->result += digit(me->ch) / me->tensPlace;
                        me->tensPlace *= 10.0;

                    TokenizerSyncSingleReceptor_enter_ProcessingFractionalPart(me);
                        printf("Current value of result: %g\n", me->result);
                        break;
                };
            };
        break;
    case EVDOT:
        me->ch = '.';
```

```
        switch (me->stateID) {
            case NONUMBER_STATE:
                /* transition to GotNumber state, default substate */
            TokenizerSyncSingleReceptor_exit_NoNumber(me);

            TokenizerSyncSingleReceptor_enter_GotNumber(me);
                    me->stateID = GOTNUMBER_STATE;

            TokenizerSyncSingleReceptor_enter_ProcessingFractionalPart(me);
                    me->subStateID = PROCESSINGFRACTIONALPART_SSTATE;
                break;
            case GOTNUMBER_STATE:
                switch (me->subStateID) {
                    case PROCESSINGWHOLEPART_SSTATE:

                    TokenizerSyncSingleReceptor_exit_ProcessingWholePart(me);

                    TokenizerSyncSingleReceptor_enter_ProcessingFractionalPart(me);
                        me->subStateID = PROCESSINGFRACTIONALPART_SSTATE;
                        break;
                };
            };
        break;
    case EVWHITESPACE:
    case EVENDOFSTRING:
        switch (me->stateID) {
            case GOTNUMBER_STATE:
                switch (me->subStateID) {
                    case PROCESSINGWHOLEPART_SSTATE:

                    TokenizerSyncSingleReceptor_exit_ProcessingWholePart(me);
                        break;

                    case PROCESSINGFRACTIONALPART_SSTATE:

                    TokenizerSyncSingleReceptor_exit_ProcessingFractionalPart(me);
                        break;
                    };
            TokenizerSyncSingleReceptor_exit_GotNumber(me);
                printf("Number: %g\n", me->result);
            TokenizerSyncSingleReceptor_enter_NoNumber(me);
                me->stateID = NONUMBER_STATE;
                break;
        };
        break;
    }; // end switch e.eType
  Mutex_release(me->itsMutex);
}

struct Mutex* TokenizerSyncSingleReceptor_getItsMutex(const TokenizerSyncSingleReceptor*
const me) {
  return (struct Mutex*)me->itsMutex;
}

void
TokenizerSyncSingleReceptor_setItsMutex(TokenizerSyncSingleReceptor* const me, struct
Mutex* p_Mutex) {
```

```
   me->itsMutex = p_Mutex;
}

TokenizerSyncSingleReceptor *
TokenizerSyncSingleReceptor_Create(void) {
  TokenizerSyncSingleReceptor* me = (TokenizerSyncSingleReceptor *) malloc(sizeof
(TokenizerSyncSingleReceptor));
  if(me!=NULL)
  TokenizerSyncSingleReceptor_Init(me);
  return me;
}
void
TokenizerSyncSingleReceptor_Destroy(TokenizerSyncSingleReceptor* const me) {
  if(me!=NULL)
  TokenizerSyncSingleReceptor_Cleanup(me);
  free(me);
}
static void cleanUpRelations(TokenizerSyncSingleReceptor* const me) {
  if(me->itsMutex != NULL)
  me->itsMutex = NULL;
}
void
TokenizerSyncSingleReceptor_Cleanup(TokenizerSyncSingleReceptor* const me) {
  cleanUpRelations(me);
}
```

Code Listing 5-10: TSRSyncSingleReceptor.c

5.4 Multiple Event Receptor Pattern

Multiple event receptor finite state machines (MERSMs) are generally only used for synchronous state machines because the client is often aware of the set of events that it might want to send to the server state machine. In this pattern, there is a single event receptor for each event sent from the client.

5.4.1 Abstract

The MERSM approach for state machine implementation is probably the most common implementation pattern for synchronous state machines. It is usually the simplest because each event receptor concerns itself only with the handling of a single event and the execution of the associated actions.

5.4.2 Problem

This pattern addresses the problem of providing a simple robust implementation for synchronous state machines.

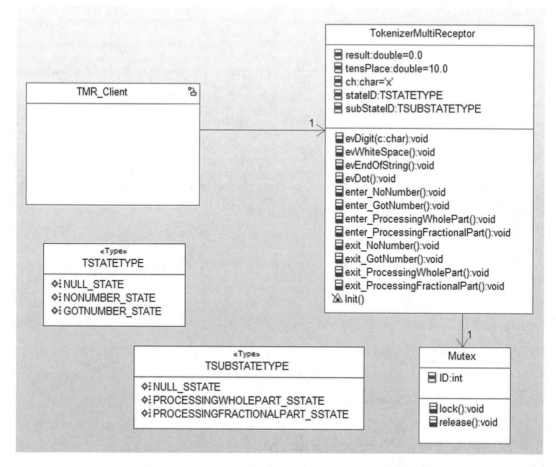

Figure 5-15: Multiple event receptor state machine context

5.4.3 Pattern Structure

The context for the state machine implementation is shown in Figure 5-15. You can see that the TokenizerMultiReceptor has an event receptor for each event passed to the state machine plus additional functions representing the entry and exit actions of each of the states. In addition, the figure explicitly shows the enumerated types used for the state and substates (as mentioned, it is also common to simply use integer values). The event type need not be explicitly represented because the client implicitly indicates it by invoking the appropriate event receptor.

5.4.4 Collaboration Roles

This section describes the roles for this pattern.

5.4.4.1 Mutex

The mutex ensures the deterministic run-to-completion semantics of transitions by preventing other clients from interrupting the state machine execution of an action sequence. This is accomplished by locking the semaphore before the event is processed and releasing it once the transition processing is complete.

5.4.4.2 TMRClient

This is the client of the state machine. It invokes the event receptors (in this case, those functions with the "ev" prefix, such as evDigit and evDot.

5.4.4.3 TokenizeMultiReceptor

This element is the owner and implementer of the desired state machine (see Figure 5-12). It has two kinds of attributes: those involved with the intent of the state machine (such as result) and those involved with the management of the state machine (such as stateID). Of the former type, we find:

- result, which holds the current value of the number being parsed
- tensPlace, which is used in computing fractional parts of numbers, as in 0.345 of the number 10.345
- ch, which holds the current character being processed

Of the latter type, we see:

- stateID, which stores the current high level state (NULL_STATE, NONUMBER_STATE, or GOTNUMBER_STATE)
- subStateID, which stores the current substate (NULL_SSTATE, PROCESSINGWHOLEPART_SSTATE, or PROCESSINGFRACTIONALPART_SSTATE)

These values are defined in the enumerated types TSTATETYPE and TSUBSTATETYPE, respectively.

Functions of this state machine are likewise of two types. The first type is the event receptors. These are invoked by the client to "send" the event and pass data as needed. These include:

- evDigit(c: char) sends the evDigit event and passes the character of the digit as a parameter
- evDot sends the dot event, indicating that the decimal point was the next character in the string
- evEndOfString indicates that there are no more characters in the string
- evWhiteSpace indicates that whitespace, such as a blank or tab character, was found

The other operations are for the entry and exit actions of the states. Note that the transitions actions are all embedded within the event receptors since they only occur once, but there are often many paths into and out of a state so we explicitly create entry and exit functions for each state. These are not called directly by the clients.

5.4.4.4 TSTATE

TSTATE is an enumerated type of the high level states.

5.4.4.5 TSUBSTATE

TSUBSTATE is an enumerated type of the nested states. In this case, there are only two levels.

5.4.5 Consequences

This pattern simplifies the implementation by dividing up the state logic into a set of event receptors. Each of which provides a switch-case statement to implement the appropriate behavior dependent on states. The pattern is almost exclusively applied to synchronous state machines because the event queues do not, generally, parse the internal content of the data they manage, and so do not know which event receptor to call, if there is more than one.

5.4.6 Implementation Strategies

The primary variation points for the implementation of this pattern are to use If-then statements instead of switch-case statements and to use raw integers for state variables rather than enumerated types.

5.4.7 Related Patterns

This pattern uses the Guarded Call Pattern (see Chapter 4) to ensure deterministic run-to-completion of transitions. It is also possible to use the Critical Region Pattern to simply disable task switching or even disable interrupts instead. The single event receptor provides a similar approach that can be easily adapted for asynchronous event processing.

5.4.8 Example

The implementation of the sample state machine with this pattern is shown the following code lists. Code Listing 5-11 shows the enumerated types and the definition of the digit() function. The next two code listings provide the header and implementation files for the TokenizeMultiReceptor class. Note that this class defines functions for entry and exit of each state even if empty. You may choose to omit the empty ones if desired. The init() function operates as a constructor and puts the object into its starting state and invokes the relevant entry action.

```
typedef enum TSTATETYPE {
  NULL_STATE,
  NONUMBER_STATE,
  GOTNUMBER_STATE
} TSTATETYPE;

typedef enum TSUBSTATETYPE {
  NULL_SSTATE,
  PROCESSINGWHOLEPART_SSTATE,
  PROCESSINGFRACTIONALPART_SSTATE
} TSUBSTATETYPE;
```

```
int digit(char c) {
  return c-'0';
}
```

Code Listing 5-11: Basic types used by TokenizeMultiReceptor

```
#include "TokenizeMultiReceptor.h"
#include <stdio.h>

struct Mutex;

typedef struct TokenizerMultiReceptor TokenizerMultiReceptor;
struct TokenizerMultiReceptor {
  char ch;
  double result;
  TSTATETYPE stateID;
  TSUBSTATETYPE subStateID;
  double tensPlace;
  struct Mutex* itsMutex;
};
/* Constructors and destructors:*/
void TokenizerMultiReceptor_Cleanup(TokenizerMultiReceptor* const me);
/* Operations */

void TokenizerMultiReceptor_evDigit(TokenizerMultiReceptor* const me, char c);

void TokenizerMultiReceptor_evDot(TokenizerMultiReceptor* const me);

void TokenizerMultiReceptor_evEndOfString(TokenizerMultiReceptor* const me);

void TokenizerMultiReceptor_evWhiteSpace(TokenizerMultiReceptor* const me);

void
TokenizerMultiReceptor_enter_GotNumber(TokenizerMultiReceptor* const me);

void
TokenizerMultiReceptor_enter_NoNumber(TokenizerMultiReceptor* const me);

void
TokenizerMultiReceptor_enter_ProcessingFractionalPart(TokenizerMultiReceptor*  const
me);

void
TokenizerMultiReceptor_enter_ProcessingWholePart(TokenizerMultiReceptor* const me);

void
TokenizerMultiReceptor_exit_GotNumber(TokenizerMultiReceptor* const me);

void
TokenizerMultiReceptor_exit_NoNumber(TokenizerMultiReceptor* const me);

void
TokenizerMultiReceptor_exit_ProcessingFractionalPart(TokenizerMultiReceptor* const me);

void
TokenizerMultiReceptor_exit_ProcessingWholePart(TokenizerMultiReceptor* const me);
TokenizerMultiReceptor * TokenizerMultiReceptor_Create(void);
```

```
void TokenizerMultiReceptor_Destroy(TokenizerMultiReceptor* const me);

void TokenizerMultiReceptor_Init(TokenizerMultiReceptor* const me);

struct Mutex* TokenizerMultiReceptor_getItsMutex(const TokenizerMultiReceptor* const
me);

void    TokenizerMultiReceptor_setItsMutex(TokenizerMultiReceptor*    const    me,    struct
Mutex* p_Mutex);

#endif
```

Code Listing 5-12: TokenizeMultiReceptor.h

```c
#include "TokenizerMultiReceptor.h"
#include "Mutex.h"

static void cleanUpRelations(TokenizerMultiReceptor* const me);

void TokenizerMultiReceptor_Cleanup(TokenizerMultiReceptor* const me) {
  cleanUpRelations(me);
}

/*
    process the evDigit event, passing the character
    c as a parameter
*/
void TokenizerMultiReceptor_evDigit(TokenizerMultiReceptor* const me, char c) {
  Mutex_lock(me->itsMutex);
  switch (me->stateID) {
    case NONUMBER_STATE:
        /* transition to GotNumber state, default substate */
        TokenizerMultiReceptor_exit_NoNumber(me);
        me->ch = c;
        TokenizerMultiReceptor_enter_GotNumber(me);
        me->stateID = GOTNUMBER_STATE;
        TokenizerMultiReceptor_enter_ProcessingWholePart(me);
        me->subStateID = PROCESSINGWHOLEPART_SSTATE;
        printf("Current value of result: %g\n", me->result);
        break;
    case GOTNUMBER_STATE:
        switch (me->subStateID) {
           case PROCESSINGWHOLEPART_SSTATE:

        TokenizerMultiReceptor_exit_ProcessingWholePart(me);
                me->ch = c;

        TokenizerMultiReceptor_enter_ProcessingWholePart(me);
                printf("Current value of result: %g\n", me->result);
                break;
           case PROCESSINGFRACTIONALPART_SSTATE:

        TokenizerMultiReceptor_exit_ProcessingFractionalPart(me);
                me->ch = c;
                me->result += digit(me->ch) / me->tensPlace;
                me->tensPlace *= 10.0;
```

```
                    TokenizerMultiReceptor_enter_ProcessingFractionalPart(me);
                        printf("Current value of result: %g\n", me->result);
                        break;
                };
            };
    Mutex_release(me->itsMutex);
}
/*
    process the evDot
*/
void TokenizerMultiReceptor_evDot(TokenizerMultiReceptor* const me) {
    Mutex_lock(me->itsMutex);
    me->ch = '.';
    switch (me->stateID) {
        case NONUMBER_STATE:
            /* transition to GotNumber state, default substate */
            TokenizerMultiReceptor_exit_NoNumber(me);
            TokenizerMultiReceptor_enter_GotNumber(me);
            me->stateID = GOTNUMBER_STATE;

    TokenizerMultiReceptor_enter_ProcessingFractionalPart(me);
            me->subStateID = PROCESSINGFRACTIONALPART_SSTATE;
            break;
        case GOTNUMBER_STATE:
            switch (me->subStateID) {
                case PROCESSINGWHOLEPART_SSTATE:

    TokenizerMultiReceptor_exit_ProcessingWholePart(me);

                TokenizerMultiReceptor_enter_ProcessingFractionalPart(me);
                    me->subStateID = PROCESSINGFRACTIONALPART_SSTATE;
                    break;
                };
        };
    Mutex_release(me->itsMutex);
}
/*
    process the evEndOfString event
*/
void TokenizerMultiReceptor_evEndOfString(TokenizerMultiReceptor* const me) {
    Mutex_lock(me->itsMutex);
    switch (me->stateID) {
        case GOTNUMBER_STATE:
            switch (me->subStateID) {
                case PROCESSINGWHOLEPART_SSTATE:

    TokenizerMultiReceptor_exit_ProcessingWholePart(me);
                    break;
                case PROCESSINGFRACTIONALPART_SSTATE:

    TokenizerMultiReceptor_exit_ProcessingFractionalPart(me);
                    break;
                };
```

```
        TokenizerMultiReceptor_exit_GotNumber(me);
        printf("Number: %g\n", me->result);

        TokenizerMultiReceptor_enter_NoNumber(me);
        me->stateID = NONUMBER_STATE;
        break;
    };
  Mutex_release(me->itsMutex);
}

/*
    process the evWhiteSpace event
*/
void TokenizerMultiReceptor_evWhiteSpace(TokenizerMultiReceptor* const me) {
  Mutex_lock(me->itsMutex);
  switch (me->stateID) {
    case GOTNUMBER_STATE:
        switch (me->subStateID) {
            case PROCESSINGWHOLEPART_SSTATE:

    TokenizerMultiReceptor_exit_ProcessingWholePart(me);
                break;
            case PROCESSINGFRACTIONALPART_SSTATE:

    TokenizerMultiReceptor_exit_ProcessingFractionalPart(me);
                break;
            };
            TokenizerMultiReceptor_exit_GotNumber(me);
            printf("Number: %g\n", me->result);
            TokenizerMultiReceptor_enter_NoNumber(me);
            me->stateID = NONUMBER_STATE;
            break;
    };
  Mutex_release(me->itsMutex);
}

/*
    entry and exit actions for each state
*/
void
TokenizerMultiReceptor_enter_GotNumber(TokenizerMultiReceptor* const me) {
  me->tensPlace = 10.0;
}

void
TokenizerMultiReceptor_enter_NoNumber(TokenizerMultiReceptor* const me) {
  me->result = 0.0;
}

void
TokenizerMultiReceptor_enter_ProcessingFractionalPart(TokenizerMultiReceptor*  const
me) {
}
```

```c
void
TokenizerMultiReceptor_enter_ProcessingWholePart(TokenizerMultiReceptor* const me) {
  me->result = me->result*10 + digit(me->ch);
}

void
TokenizerMultiReceptor_exit_GotNumber(TokenizerMultiReceptor* const me) {
  me->subStateID = NULL_SSTATE;
}

void TokenizerMultiReceptor_exit_NoNumber(TokenizerMultiReceptor* const me) {
}

void
TokenizerMultiReceptor_exit_ProcessingFractionalPart(TokenizerMultiReceptor*    const
me) {
}

void
TokenizerMultiReceptor_exit_ProcessingWholePart(TokenizerMultiReceptor* const me) {
}

/* helper functions */
TokenizerMultiReceptor * TokenizerMultiReceptor_Create(void) {
  TokenizerMultiReceptor* me = (TokenizerMultiReceptor *) malloc(sizeof
  (TokenizerMultiReceptor));
  if(me!=NULL)
    TokenizerMultiReceptor_Init(me);
  return me;
}

void TokenizerMultiReceptor_Destroy(TokenizerMultiReceptor* const me) {
  if(me!=NULL)
  TokenizerMultiReceptor_Cleanup(me);
  free(me);
}

void TokenizerMultiReceptor_Init(TokenizerMultiReceptor* const me) {
  me->ch = 'x';
  me->result = 0.0;
  me->tensPlace = 10.0;
  me->itsMutex = NULL;

  /* initialize state variables */
  me->stateID = NONUMBER_STATE;
  me->subStateID = NULL_SSTATE;
}

struct Mutex* TokenizerMultiReceptor_getItsMutex(const TokenizerMultiReceptor* const me) {
  return (struct Mutex*)me->itsMutex;
}

void  TokenizerMultiReceptor_setItsMutex(TokenizerMultiReceptor*  const  me,  struct
Mutex* p_Mutex) {
  me->itsMutex = p_Mutex;
}
```

```
static void cleanUpRelations(TokenizerMultiReceptor* const me) {
   if(me->itsMutex != NULL)
   me->itsMutex = NULL;
}
```

Code Listing 5-13:TokenizeMultiReceptor.c

5.5 State Table Pattern

The State Table Pattern is a create pattern for large, flat (i.e., without nesting or AND-states) state machines. It gives good performance (O(constant)) in use although it requires more initialization time before it becomes available. They are relatively easy to extend to more states but don't handle nesting very easily.

5.5.1 Abstract

The State Table Pattern uses a two-dimensional array to hold the state transition information. This table is usually constructed as a state x event table; the state variable is the first index and the received event ID is the second index. The contents of the table is a structure containing links to the appropriate actions and guards and the new state, should the transition be taken. This requires a somewhat more elaborate data structuring than the other state implementation patterns.

5.5.2 Problem

The problem addressed by the State Table Pattern is to implement a state machine for potentially large state spaces that are flat or can easily be made so. Also, this pattern is applicable when you want the performance of the state machine, in terms of the response time to a given event, independent of the size of the state space but you are unconcerned about the time necessary to initialize the state machine.

5.5.3 Pattern Structure

Figure 5-16 shows the structure for the State Table Pattern. It shows two important aspects of the data typing. First, the table itself is set up as a two-dimensional array of which the first index is the state (held within the state variable `stateID`) and the second index is the event ID received. The event ID, as before, is a field within the struct `Event`. To add levels of nesting requires the nesting of such tables and most developers feel that it is easier to flatten the state space (that is always possible – proof is left to the interested reader) than to nest the tables.

The other notable complexity of the data structuring is in the `TableEntry` type. This is a struct containing a set of `ActionTypes`, a `GuardType`, and the state to enter when the transition is complete. The `GuardType` is a parameterless function (except for the ubiquitous

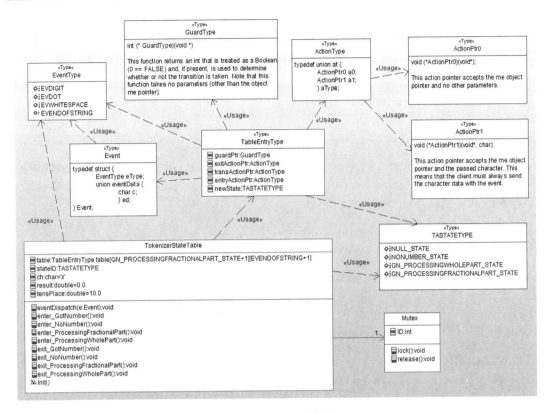

Figure 5-16: State Table Pattern

object me pointer) that returns an int indication `TRUE` (non-zero) or `FALSE` (zero). The `ActionType`, however, is a union with either a parameterless function pointer (`Action0`) or a pointer to a function that has a `char` parameter (`Action1`). The `TableEntry` struct has separate entries for entry, exit, and transition actions. If an action field doesn't apply, the pointer should be set to `NULL`. This means that if there are multiple actions to perform within one of those fields, they need to be composed into a single action that calls them in sequence. The initialization of all the table entries is done in the `init()` function of the `TokenizerStateTable` class.

The `eventDispatch()` function now has a straightforward task. Given an incoming event, invoke the guard (if any) in the cell `table[stateID][eType]` and if `TRUE` (or missing) then invoke the exit actions, transition actions, and entry actions and set the `stateID` equal to the value of the `newState` field found within the cell.

5.5.4 Collaboration Roles

This section describes the roles for this pattern.

5.5.4.1 ActionPtr0

This type is a pointer to a function that has 0 parameters, other than the obligatory me object pointer.

5.5.4.2 ActionPtr1

This type is a pointer to a function that takes 1 parameter of type char in addition to the obligatory me object pointer.

5.5.4.3 ActionType

This type is a union of the two ActionPtr types; as a union it only has a single pointer, but it is either of type ActionPtr0 or ActionPtr1. The ActionType defines the TableEntryType fields for entry, transition, and exit actions.

5.5.4.4 Event

This type is a struct that holds the event type and the associated data passed within some of the events – in this case, a char. This is normally implemented as a union in C.

5.5.4.5 EventType

The EventType is an enumerated type for the events. Alternatively, ints can be used but I recommend enums as being more informative.

5.5.4.6 GuardType

This type implements a function pointer that takes no arguments but returns an int that will be treated as a Boolean and used to determine whether a triggered transition will be taken or not. Note that in this example, no guards are used so all these elements for this example will be NULL.

5.5.4.7 Mutex

As with the other implementation patterns, the Mutex is used to ensure the run-to-completion semantics required for deterministic state behavior.

5.5.4.8 TableEntryType

This composite type specifies the elements within the cells of the state table. It consists of a set of function pointers for exit, transition, entry, and guard actions plus the new state of the state machine should the transition be taken.

5.5.4.9 TokenizerStateTable

This class implements the state machine. Internally it contains the state table *per se* as one of its data members (along with the stateID and other attributes used for the computation required of the state machine). It also includes the functions necessary to implement the actions called.

5.5.4.10 TSTATETYPE

TSTATE is an enumerated type of the high level states. Note that the nested state machine is flattened to remove nesting. This greatly simplifies the state table. The enumerated values for this type are:

- NULL_STATE
- NONUMBER_STATE
- GN_PROCESSINGWHOLEPART_STATE
- GN_PROCESSINGFRACTIONALPART_STATE

Where the GN_ prefix is a reminder that the last two states were nested states of the GotNumber state.

5.5.5 Consequences

This pattern has excellent execution performance that is independent of the size of the state space (i.e., the number of cells in the state table). This is because it is directly indexed with a combination of the current stateID and the incoming eType. On the other hand, the initialization of the table, while not particularly complex, is involved, since the fields for every StateTableEntry must be initialized separately in the init() function. This means that the initialization of this pattern is much slower than for the other patterns, with the possible exception of the State Pattern, discussed in the next section. Thus, this implementation pattern is a good choice for circumstances in which the start-up time of the system is relatively unimportant but the state space is potentially large.

The State Table Pattern is also easier to extend than the other patterns since extension is mostly a matter of adding new elements to the state table and defining the StateTableEntries for the relevant transitions.

It should be noted that this pattern will waste memory if the state space is sparse, that is, when many of the cells are vacant, because the table allocates StateTableEntries for all cells. This can be addressed by storing pointers to StateTableEntries rather than the elements themselves if memory is tight.

This pattern, as presented, supports only synchronous state machines. However, it can be mixed with the elements of the asynchronous version of the State Single Receptor Pattern (with suitable enhancement of the TokenizerStateTable class with a task processing loop to pull events from the event queue), it can likewise handle asynchronous events.

5.5.6 Implementation Strategies

Normally, the table is implemented as a two-dimensional array of StateTableEntries. This can waste memory if the table is sparsely populated – memory can be saved by implementing the table as a two-dimensional array of pointers to StateTableEntries instead (at the cost of an additional pointer dereference during execution).

5.5.7 Related Patterns

This pattern employs the Mutex of the Guarded Call Pattern (see Chapter 4). It can also use a critical region within the eventDispatch() function but that would not permit higher priority tasks to run during the state processing even when they could not possibly interfere with the execution of the state machine. As noted above, the Queuing Pattern could be mixed in as well, to support asynchronous event handling.

5.5.8 Example

This pattern has some interesting programming aspects with respect to the data typing, event dispatching, and initialization. Code Listing 5-14 shows the definitions of the types used by the TokenizerStateTable class. Code Listing 5-15 and Code Listing 5-16 show the header and implementation files for the TokenizerStateTable class respectively.

```c
#ifndef StateTablePattern_H
#define StateTablePattern_H

typedef enum TSTATETYPE {
  NULL_STATE,
  NONUMBER_STATE,
  GN_PROCESSINGWHOLEPART_STATE,
  GN_PROCESSINGFRACTIONALPART_STATE
} TSTATETYPE;

/* This action pointer accepts the me object pointer and no other parameters. */
typedef void (*ActionPtr0)(void*);

/* This action pointer accepts the me object pointer and the passed character. This means
that the client must always send the character data with the event. */
typedef void (*ActionPtr1)(void*, char);

typedef enum EventType {
  EVDIGIT,
  EVDOT,
  EVWHITESPACE,
  EVENDOFSTRING
} EventType;
typedef struct {
    EventType eType;
    union eventData {
        char c;
        } ed;
} Event;

typedef struct ap {
    int nParams; /* 0 or 1 in this case */
    union {
        ActionPtr0 a0;
        ActionPtr1 a1;
    } aPtr;
} ActionType;

typedef ActionType* ActionPtr;
```

```
/* This function returns an int that is treated as a Boolean (0 == FALSE) and, if present, is
used to determine whether or not the transition is taken. Note that this function takes no
parameters (other than the object me pointer). */
typedef int (*GuardType)(void *);

typedef struct TableEntryType {
  ActionPtr entryActionPtr;
  ActionPtr exitActionPtr;
  GuardType guardPtr;
  TSTATETYPE newState;
  ActionPtr transActionPtr;
} TableEntryType;

/* digit returns an int value of the char
  That is: return c-'0' */
int digit(char c); */

#endif
```

Code Listing 5-14: StateTablePattern.h

```
#ifndef TokenizerStateTable_H
#define TokenizerStateTable_H

#include "StateTablePattern.h"
struct Mutex;

typedef struct TokenizerStateTable TokenizerStateTable;
struct TokenizerStateTable {
  char ch;
  double result;
  TSTATETYPE stateID;
  TableEntryType
table[GN_PROCESSINGFRACTIONALPART_STATE+1][EVENDOFSTRING+1];
  double tensPlace;
  struct Mutex* itsMutex;
};
/* Constructors and destructors:*/
void TokenizerStateTable_Init(TokenizerStateTable* const me);

void TokenizerStateTable_Cleanup(TokenizerStateTable* const me);

/* Operations */
void TokenizerStateTable_eventDispatch(TokenizerStateTable* const me, Event e);

void TokenizerStateTable_enter_GotNumber(TokenizerStateTable* const me);

void TokenizerStateTable_enter_NoNumber(TokenizerStateTable* const me);

void
TokenizerStateTable_enter_ProcessingFractionalPart(TokenizerStateTable* const me);

void
TokenizerStateTable_enter_ProcessingWholePart(TokenizerStateTable* const me);

void TokenizerStateTable_exit_GotNumber(TokenizerStateTable* const me);
```

```
void TokenizerStateTable_exit_NoNumber(TokenizerStateTable* const me);

void
TokenizerStateTable_exit_ProcessingFractionalPart(TokenizerStateTable* const me);

void
TokenizerStateTable_exit_ProcessingWholePart(TokenizerStateTable* const me);

struct Mutex* TokenizerStateTable_getItsMutex(const TokenizerStateTable* const me);

void  TokenizerStateTable_setItsMutex(TokenizerStateTable*  const  me,  struct  Mutex*
p_Mutex);

TokenizerStateTable * TokenizerStateTable_Create(void);

void TokenizerStateTable_Destroy(TokenizerStateTable* const me);

void TokenizerStateTable_assignCh(TokenizerStateTable* const me, char c);

void TokenizerStateTable_NoNum2GN(TokenizerStateTable* const me, char c);

void TokenizerStateTable_Frac2Frac(TokenizerStateTable* const me, char c);

void TokenizerStateTable_printResult(TokenizerStateTable* const me);

#endif
```

Code Listing 5-15: TokenizerStateTable.h

```
#include "TokenizerStateTable.h"
#include "Mutex.h"
static void cleanUpRelations(TokenizerStateTable* const me);

void TokenizerStateTable_Init(TokenizerStateTable* const me) {
  me->ch = 'x';
  me->result = 0.0;
  me->tensPlace = 10.0;
  me->itsMutex = NULL;
  TSTATETYPE st;
  EventType ev;
  TableEntryType te;

  /* initialize each cell of the table to null condition */
  for (st=NULL_STATE; st<=GN_PROCESSINGFRACTIONALPART_STATE; st++) {
    for (ev=EVDIGIT; ev<=EVENDOFSTRING; ev++) {
      me->table[st][ev].newState = NULL_STATE;
      me->table[st][ev].guardPtr = (GuardType)NULL;
      me->table[st][ev].exitActionPtr = NULL;
      me->table[st][ev].transActionPtr = NULL;
      me->table[st][ev].entryActionPtr = NULL;
      }
  }

  /*                              */
  /* now add the specific transitions */
  /*                              */
  /* evDigit from NoNumber to ProcessingWholePart */
```

```
  te.guardPtr = (GuardType)NULL;
  te.exitActionPtr = (ActionPtr)malloc(sizeof(ActionPtr));
  te.exitActionPtr->nParams = 0;
  te.exitActionPtr->aPtr.a0 = (ActionPtr0)TokenizerStateTable_exit_NoNumber;
  te.transActionPtr = (ActionPtr)malloc(sizeof(ActionPtr));
  te.transActionPtr->nParams = 1;
  te.transActionPtr->aPtr.a1 = (ActionPtr1)TokenizerStateTable_NoNum2GN;
  te.entryActionPtr = (ActionPtr)malloc(sizeof(ActionPtr));
  te.entryActionPtr->nParams = 0;
  te.entryActionPtr->aPtr.a0 = (ActionPtr0)
TokenizerStateTable_enter_ProcessingWholePart;
  te.newState = GN_PROCESSINGWHOLEPART_STATE;
  me->table[NONUMBER_STATE][EVDIGIT] = te;

  /* evDigit from Proessing Whole Part to itself */
  te.guardPtr = (GuardType)NULL;
  te.exitActionPtr = (ActionPtr)malloc(sizeof(ActionPtr));
  te.exitActionPtr->nParams = 0;
  te.exitActionPtr->aPtr.a0 = (ActionPtr0)TokenizerStateTable_exit_NoNumber;
  te.transActionPtr = (ActionPtr)malloc(sizeof(ActionPtr));
  te.transActionPtr->nParams = 1;
  te.transActionPtr->aPtr.a1 = (ActionPtr1)TokenizerStateTable_assignCh;
  te.entryActionPtr = (ActionPtr)malloc(sizeof(ActionPtr));
  te.entryActionPtr->nParams = 0;
  te.entryActionPtr->aPtr.a0 = (ActionPtr0)
TokenizerStateTable_enter_ProcessingWholePart;
  te.newState = GN_PROCESSINGWHOLEPART_STATE;
  me->table[GN_PROCESSINGWHOLEPART_STATE][EVDIGIT] = te;

  /* evDot from NoNumber to ProcessingFractionalPart */
  te.guardPtr = (GuardType)NULL;
  te.exitActionPtr = (ActionPtr)malloc(sizeof(ActionPtr));
  te.exitActionPtr->nParams = 0;
  te.exitActionPtr->aPtr.a0 = (ActionPtr0)TokenizerStateTable_exit_NoNumber;
  te.transActionPtr = (ActionPtr)malloc(sizeof(ActionPtr));
  te.transActionPtr->nParams = 1;
  te.transActionPtr->aPtr.a1 = (ActionPtr1)TokenizerStateTable_NoNum2GN;
  te.entryActionPtr = (ActionPtr)malloc(sizeof(ActionPtr));
  te.entryActionPtr->nParams = 0;
  te.entryActionPtr->aPtr.a0 = (ActionPtr0)
TokenizerStateTable_enter_ProcessingFractionalPart;
  te.newState = GN_PROCESSINGFRACTIONALPART_STATE;
  me->table[NONUMBER_STATE][EVDOT] = te;

  /* evDot from ProcessingWholePart to ProcessingFractionalPart */
  te.guardPtr = (GuardType)NULL;
  te.exitActionPtr = (ActionPtr)malloc(sizeof(ActionPtr));
  te.exitActionPtr->nParams = 0;
  te.exitActionPtr->aPtr.a0 = (ActionPtr0)
TokenizerStateTable_exit_ProcessingWholePart;
  te.transActionPtr = (ActionPtr)NULL;
```

```
  te.entryActionPtr = (ActionPtr)malloc(sizeof(ActionPtr));
  te.entryActionPtr->nParams = 0;
  te.entryActionPtr->aPtr.a0 = (ActionPtr0)
TokenizerStateTable_enter_ProcessingFractionalPart;
  te.newState = GN_PROCESSINGFRACTIONALPART_STATE;
  me->table[GN_PROCESSINGWHOLEPART_STATE][EVDOT] = te;


  /* evDigit from ProcessingFractionalPart to ProcessingFractionalPart */
  te.guardPtr = (GuardType)NULL;
  te.exitActionPtr = (ActionPtr)malloc(sizeof(ActionPtr));
  te.exitActionPtr->nParams = 0;
  te.exitActionPtr->aPtr.a0 = (ActionPtr0)
TokenizerStateTable_exit_ProcessingFractionalPart;
  te.transActionPtr = (ActionPtr)malloc(sizeof(ActionPtr));
  te.transActionPtr->nParams = 1;
  te.transActionPtr->aPtr.a1 = (ActionPtr1)TokenizerStateTable_Frac2Frac;
  te.entryActionPtr = (ActionPtr)malloc(sizeof(ActionPtr));
  te.entryActionPtr->nParams = 0;
  te.entryActionPtr->aPtr.a0 = (ActionPtr0)
TokenizerStateTable_enter_ProcessingFractionalPart;
  te.newState = GN_PROCESSINGFRACTIONALPART_STATE;
  me->table[GN_PROCESSINGFRACTIONALPART_STATE][EVDIGIT] = te;


  /* evWhiteSpace from ProcessingWholePart to NoNumber */
  te.guardPtr = (GuardType)NULL;
  te.exitActionPtr = (ActionPtr)malloc(sizeof(ActionPtr));
  te.exitActionPtr->nParams = 0;
  te.exitActionPtr->aPtr.a0 = (ActionPtr0)
TokenizerStateTable_exit_ProcessingWholePart;
  te.transActionPtr = (ActionPtr)malloc(sizeof(ActionPtr));
  te.transActionPtr->nParams = 0;
  te.transActionPtr->aPtr.a0 = (ActionPtr0)TokenizerStateTable_printResult;
  te.entryActionPtr = (ActionPtr)malloc(sizeof(ActionPtr));
  te.entryActionPtr->nParams = 0;
  te.entryActionPtr->aPtr.a0 = (ActionPtr0)TokenizerStateTable_enter_NoNumber;
  te.newState = NONUMBER_STATE;
  me->table[GN_PROCESSINGWHOLEPART_STATE][EVWHITESPACE] = te;


  /* evWhiteSpace from ProcessingFractionalPart to NoNumber */
  te.guardPtr = (GuardType)NULL;
  te.exitActionPtr = (ActionPtr)malloc(sizeof(ActionPtr));
  te.exitActionPtr->nParams = 0;
  te.exitActionPtr->aPtr.a0 = (ActionPtr0)
TokenizerStateTable_exit_ProcessingFractionalPart;
  te.transActionPtr = (ActionPtr)malloc(sizeof(ActionPtr));
  te.transActionPtr->nParams = 0;
  te.transActionPtr->aPtr.a0 = (ActionPtr0)TokenizerStateTable_printResult;
  te.entryActionPtr = (ActionPtr)malloc(sizeof(ActionPtr));
  te.entryActionPtr->nParams = 0;
  te.entryActionPtr->aPtr.a0 = (ActionPtr0)TokenizerStateTable_enter_NoNumber;
```

```
      te.newState = NONUMBER_STATE;
      me->table[GN_PROCESSINGFRACTIONALPART_STATE][EVWHITESPACE] = te;

      /* evEndOfString from ProcessingWholePart to NoNumber */
      te.guardPtr = (GuardType)NULL;
      te.exitActionPtr = (ActionPtr)malloc(sizeof(ActionPtr));
      te.exitActionPtr->nParams = 0;
      te.exitActionPtr->aPtr.a0 = (ActionPtr0)
TokenizerStateTable_exit_ProcessingWholePart;
      te.transActionPtr = (ActionPtr)malloc(sizeof(ActionPtr));
      te.transActionPtr->nParams = 0;
      te.transActionPtr->aPtr.a0 = (ActionPtr0)TokenizerStateTable_printResult;
      te.entryActionPtr = (ActionPtr)malloc(sizeof(ActionPtr));
      te.entryActionPtr->nParams = 0;
      te.entryActionPtr->aPtr.a0 = (ActionPtr0)TokenizerStateTable_enter_NoNumber;
      te.newState = NONUMBER_STATE;
      me->table[GN_PROCESSINGWHOLEPART_STATE][EVENDOFSTRING] = te;

      /* evEndOfString from ProcessingFractionalPart to NoNumber */
      te.guardPtr = (GuardType)NULL;
      te.exitActionPtr = (ActionPtr)malloc(sizeof(ActionPtr));
      te.exitActionPtr->nParams = 0;
      te.exitActionPtr->aPtr.a0 = (ActionPtr0)
TokenizerStateTable_exit_ProcessingFractionalPart;
      te.transActionPtr = (ActionPtr)malloc(sizeof(ActionPtr));
      te.transActionPtr->nParams = 0;
      te.transActionPtr->aPtr.a0 = (ActionPtr0)TokenizerStateTable_printResult;
      te.entryActionPtr = (ActionPtr)malloc(sizeof(ActionPtr));
      te.entryActionPtr->nParams = 0;
      te.entryActionPtr->aPtr.a0 = (ActionPtr0)TokenizerStateTable_enter_NoNumber;
      te.newState = NONUMBER_STATE;
      me->table[GN_PROCESSINGFRACTIONALPART_STATE][EVENDOFSTRING] = te;

      /* Last initialization activity - enter the initial state */
      me->stateID = NONUMBER_STATE;
      TokenizerStateTable_enter_NoNumber(me);
}

void TokenizerStateTable_Cleanup(TokenizerStateTable* const me) {
   cleanUpRelations(me);
}

void TokenizerStateTable_eventDispatch(TokenizerStateTable* const me, Event e) {
   int takeTransition = 0;
   Mutex_lock(me->itsMutex);
   /* first ensure the entry is within the table boundaries */
   if (me->stateID >= NULL_STATE && me->stateID <= GN_PROCESSINGFRACTIONALPART_STATE) {
      if (e.eType >= EVDIGIT && e.eType <= EVENDOFSTRING) {
         /* is there a valid transition for the current state and event? */
       if (me->table[me->stateID][e.eType].newState != NULL_STATE) {
           /* is there a guard? */
```

```
                if (me->table[me->stateID][e.eType].guardPtr == NULL)
                    /* is the guard TRUE? */
                    takeTransition = TRUE; /* if no guard, then it "evaluates" to TRUE */
                else
                    takeTransition =(me->table[me->stateID][e.eType].guardPtr(me));
            if (takeTransition) {
                if (me->table[me->stateID][e.eType].exitActionPtr != NULL)
                    if (me->table[me->stateID][e.eType].exitActionPtr->nParams == 0)
                        me->table[me->stateID][e.eType].exitActionPtr->aPtr.a0(me);
                    else
                      me->table[me->stateID][e.eType].exitActionPtr->aPtr.a1(me, e.ed.c);
                if (me->table[me->stateID][e.eType].transActionPtr != NULL)
                    if (me->table[me->stateID][e.eType].transActionPtr->nParams == 0)
                        me->table[me->stateID][e.eType].transActionPtr->aPtr.a0(me);
                    else
                      me->table[me->stateID][e.eType].transActionPtr->aPtr.a1(me, e.ed.c);
                if (me->table[me->stateID][e.eType].entryActionPtr != NULL)
                    if (me->table[me->stateID][e.eType].entryActionPtr->nParams == 0)
                        me->table[me->stateID][e.eType].entryActionPtr->aPtr.a0(me);
                    else
                      me->table[me->stateID][e.eType].entryActionPtr->aPtr.a1(me, e.ed.c);
                me->stateID = me->table[me->stateID][e.eType].newState;
            }
        }
      }
    }
  Mutex_release(me->itsMutex);
}

void TokenizerStateTable_enter_GotNumber(TokenizerStateTable* const me) {
  me->tensPlace = 10.0;
}

void TokenizerStateTable_enter_NoNumber(TokenizerStateTable* const me) {
  me->result = 0.0;
}

void
TokenizerStateTable_enter_ProcessingFractionalPart(TokenizerStateTable*
const me) {
}

void
TokenizerStateTable_enter_ProcessingWholePart(TokenizerStateTable* const me) {
  me->result = me->result*10 + digit(me->ch);
}

void TokenizerStateTable_exit_GotNumber(TokenizerStateTable* const me) {
}

void TokenizerStateTable_exit_NoNumber(TokenizerStateTable* const me) {
}
```

```
void
TokenizerStateTable_exit_ProcessingFractionalPart(TokenizerStateTable*
const me) {
}

void
TokenizerStateTable_exit_ProcessingWholePart(TokenizerStateTable* const me) {
}

struct Mutex* TokenizerStateTable_getItsMutex(const TokenizerStateTable* const me) {
    return (struct Mutex*)me->itsMutex;
}

void TokenizerStateTable_setItsMutex(TokenizerStateTable* const me, struct Mutex*
p_Mutex) {
    me->itsMutex = p_Mutex;
}

TokenizerStateTable * TokenizerStateTable_Create(void) {
    TokenizerStateTable* me = (TokenizerStateTable *) malloc(sizeof
(TokenizerStateTable));
    if(me!=NULL)
    TokenizerStateTable_Init(me);
    return me;
}

void TokenizerStateTable_Destroy(TokenizerStateTable* const me) {
    if(me!=NULL)
    TokenizerStateTable_Cleanup(me);
    free(me);
}

static void cleanUpRelations(TokenizerStateTable* const me) {
    if(me->itsMutex != NULL)
    me->itsMutex = NULL;
}

void TokenizerStateTable_assignCh(TokenizerStateTable* const me, char c) {
    me->ch = c;
}

void TokenizerStateTable_NoNum2GN(TokenizerStateTable* const me, char c) {
    TokenizerStateTable_assignCh(me, c);
    me->tensPlace = 10.0;
}

void TokenizerStateTable_Frac2Frac(TokenizerStateTable* const me, char c) {
    me->ch = c;
    me->result += digit(me->ch)/me->tensPlace;
    me->tensPlace *= 10;
}

void TokenizerStateTable_printResult(TokenizerStateTable* const me) {
    printf("Number: %g\n", me->result);
}
```

Code Listing 5-16: TokenizerStateTable.c

Figure 5-17 shows an example execution of the system in which a client sends the characters of the string "1.3" to the `TokenizerStateTable`.

5.6 State Pattern

This, perhaps poorly named, pattern comes from *Design Patterns: Elements of Reusable Object Oriented Software* by Gamma, et. al.[4] It provides an implementation strategy that can be optimized either for data memory at the expense of run-time performance or vice versa. It is applicable for sparse and dense state spaces alike and has relatively low initial set-up costs. However, it makes rather heavy use of dynamic memory. It is simple to implement in C++ because of its use of polymorphism and therefore somewhat more complex in C.

5.6.1 Abstract

The State Pattern implements a state machine through the creation of state objects, one per state. The class owning the state machine, known as the `Context`, maintains a list of these objects with an internal variable that identifies which of these is the current state. All event receptors are passed to the currently active state object.

5.6.2 Problem

This pattern provides a means for implementing state machines that simplifies the stateful class by creating a set of objects, each of which implements the behavior of a single state. It makes the states explicit, rather easy to modify, and enables the extension of the state space by adding new state objects as needed.

5.6.3 Pattern Structure

Figure 5-18 shows the structure for the pattern for the parsing example. Note that the states are all implemented as different classes, and each of these classes provides a function for each of the events (even if the implementation is empty). The `Context` passes the control off to the current state (implemented as the `currentState` index into the `stateList` array). Because the actions are almost always implemented in terms of manipulating variables and invoking functions of the `Context`, each state class requires a pointer back to the `Context`. In a fashion similar to the State Table Pattern, the pattern is made much easier if you flatten the state machine (i.e., eliminate nesting).

---

[4] Gamma, E., Helm, R., Johnson, R., Vlissides, J., 1995. *Design Patterns: Elements of Reusable Object-Oriented Software*. Addison-Wesley.

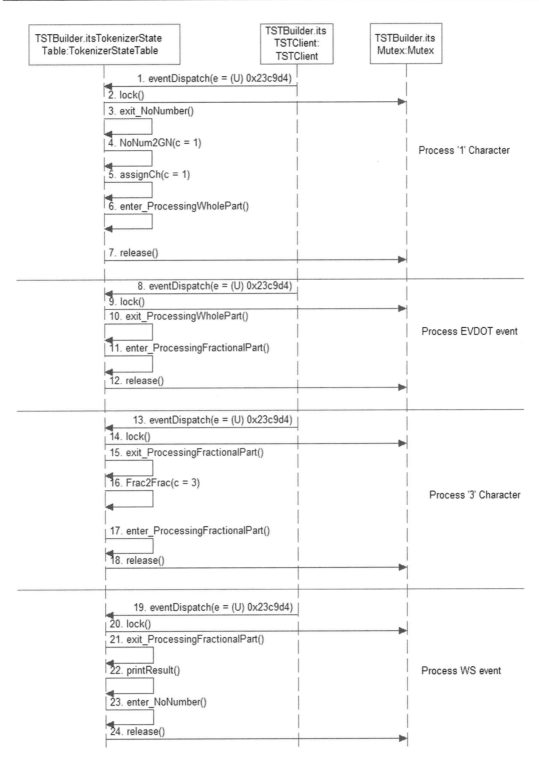

Figure 5-17: State Table Pattern sequence for execution of "1.3" string

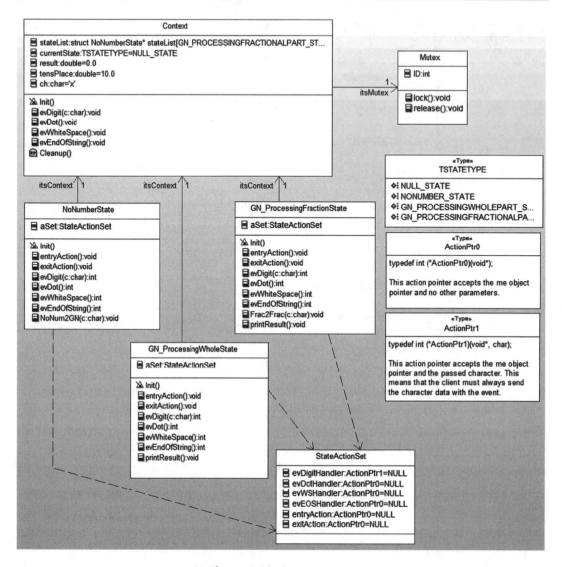

Figure 5-18: State Pattern

5.6.4 Collaboration Roles

5.6.4.1 ActionPtr0

This type is a pointer to a function that has 0 parameters, other than the obligatory me object pointer, and returns an int. The value of the return is either 0 (event discarded) or 1 (event handled).

5.6.4.2 ActionPtr1

This type is a pointer to a function that takes 1 parameter of type `char` in addition to the obligatory `me` object pointer and returns an `int`. The value of the return is either 0 (event discarded) or 1 (event handled).

5.6.4.3 Context

The `Context` is the class that owns the state machine and has the data manipulated by the actions. It implements a `stateList` as an array of pointers to the state classes and a `currentState` variable that indicates which of the current states is active. Since the structure of the state classes is all the same (in terms of the struct), the elements of the stateList array can be typed as pointers to one of them.

The `Context` provides a set of event handlers similar to the Multiple Event Receptor Pattern. The implementation of each event handler delegates the event handling to the current state object, which is free to act on, or discard the event as appropriate. If it handles the event, the state object will update the `Context`'s `currentState` variable; once the event handler returns (and if it handled the event), the `Context` will execute the entry actions of the new state.

During initialization, the `Context` creates each of the state objects, puts it in the `stateList`, and then enters the initial state.

5.6.4.4 Mutex

The `Mutex` ensures the deterministic run-to-completion semantics of transitions by preventing other clients from interrupting the state machine execution of an action sequence. This is accomplished by locking the semaphore before the event is processed and releasing it once the transition processing is complete.

5.6.4.5 State Classes

The state classes each implement one of the states of the `Context` and implements all of the event handlers. The event handlers return 1 if they handled the event and 0 if they discarded the event. If it handles the event, the event handler is expected to execute the exit actions of the current state, then the transition actions, and finally set the `Context`'s `currentState` to the new state. The entry actions of the new state will be executed by the `Context` after the return of the state object's event handler. The state classes are:

- `NoNumberState`
- `GN_ProcessingFractionState`
- `GN_ProcessingWholeState`

Each event handler may have additional helper functions invoked by its event handlers, particularly for the transition actions, in addition to the entry and exit actions.

Each state object has an instance of the `StateActionSet` class that implements function pointers to each of the event handlers. During the state object's initialization, these function pointers are set to point to the appropriate event handler functions of the state class. In this way, polymorphism is addressed to support general event handling regardless of which is the current state.

The actions of each state class is normally a manipulation of the values owned by the `Context`. For this reason, each state class has an association (pointer) back to the `Context` to access those values. An alternative implementation, discussed later, is to pass the `Context`'s me pointer to each action to make the state classes truly context-free.

5.6.4.6 *StateActionSet*

This class is the key to implementing the required polymorphism. Each state class owns its own instance of this class in a variable called `aSet`. Each event handler (and the entry and exit actions) have a function pointer that will be dereferenced to invoke the event handler for the current state. In this case, the handlers are either of type `ActionPtr0` (no parameters other than the me pointer) and `ActionPtr1` (one char parameter plus the me pointer). To implement events with other parameter lists, new action pointer types must be created.

This class has no behavior other than an initializer that sets all the pointers to `NULL` initially. It is expected that the owner of each instance of the `StateActionSet` will set the function pointers to access the correct functions.

5.6.4.7 *TSTATETYPE*

`TSTATE` is an enumerated type of the high level states. Note that the nested state machine is flattened to remove nesting. This greatly simplifies the state pattern implementation. The enumerated values for this type are:

- `NULL_STATE`
- `NONUMBER_STATE`
- `GN_PROCESSINGWHOLEPART_STATE`
- `GN_PROCESSINGFRACTIONALPART_STATE`

Where the `GN_` prefix is a reminder that the last two states were nested states of the `GotNumber` state.

5.6.5 *Consequences*

This pattern localizes the behavior specific to a state to a single object, simplifying and encapsulating state-specific behavior. This means new states can be added by adding a new state class. New events are more work to add, since they must be at least added to the `StateActionSet` and usually to each state class as well. The State Pattern makes the event transitions explicit by providing an event handler for each incoming event in each state class.

This pattern uses more memory than a state table but simplifies the implementation by distributing the complexity among the various state objects.

Another advantage of this pattern is that with a small change in implementation, it is possible to share state objects among multiple instances of the Context class – for example, if the system has multiple sensors all typed by the same class. This makes it very lightweight in terms of memory usage in those cases.

5.6.6 Implementation Strategies

This pattern is straightforward in C++ but rather more complex in C because it uses inheritance and polymorphism. In the implementation shown here, polymorphism is implemented with a combination of two techniques. First, all the state classes implement the same event handling interface; that is each state implements an event handler for all events even if the event is discarded. The event handlers return a 1 if the event was handled and 0 if the event was discarded. This is necessary because the Context needs to know if it should execute the entry actions of the new state.

The StateActionSet class then provided the polymorphic dispatch of the functions by providing a set of pointers for the event handlers and the entry and exit handlers. Each state class implemented the event handlers appropriately and then put a function pointer in its StateActionSet instance. The Context class uses the currentState variable as the index to dereference the pointer to the state that is "current," and further dereferences the state's StateActionSet to invoke the appropriate event handler.

Two primary implementation variants occur around whether to create all the state objects at once and simply switch among them based on the currentState variable of the Context, or to dynamically create and destroy the instances of the different state. In general, if the system changes state only infrequently, then memory can be saved by only creating the state as it is needed and destroying it when the Context is no longer in that state. If the state changes are frequent, then this approach will likely impose a significant performance penalty, so at the cost of some additional memory, it is best to keep all the state instances in memory.

An interesting property of the pattern is that the state objects themselves have no state context other than the itsContext pointer. This means that the state objects can be shared by multiple instances of the Context class. This might occur if, for example, the system has many sensor instances all typed by the same Context class. In this case the State Pattern can implement shared state objects, resulting in significant memory savings. To decontextualize the state classes, remove the class's itsContext pointer and add a Context pointer (that's the Context's me pointer) to each event handler parameter list (and also to the entry and exit actions). When this is done, each state class is completely independent of its Context.

5.6.7 Related Patterns

The State Pattern implements a separate event receptor for each event, similar to the Single Event Receptor Pattern. The State Pattern is useful for more complex state machines because the event handling is decomposed across different state objects. On the other hand, implementation of nested states is somewhat simpler in the case of the Single Event Receptor Pattern.

The State Table Pattern is similar in usage to the State Pattern, in that both are greatly simplified by flattening the state hierarchy. However, the former cannot easily share implementation so when the system has multiple instances of the Context, the latter may be more memory efficient. Both patterns have good performance.

5.6.8 Example

This section shows the implementation of the various elements of this pattern for the tokenizer example in Figure 5-12. Code Listing 5-17 shows the common types used by the other elements.

```
#ifndef StatePattern_H
#define StatePattern_H

struct Context;
struct GN_ProcessingFractionState;
struct GN_ProcessingWholeState;
struct Mutex;
struct NoNumberState;
struct SPClient;
struct StateActionSet;
struct StatePatternBulder;

typedef enum TSTATETYPE {
  NULL_STATE,
  NONUMBER_STATE,
  GN_PROCESSINGWHOLEPART_STATE,
  GN_PROCESSINGFRACTIONALPART_STATE
} TSTATETYPE;

typedef int (*ActionPtr0)(void*);
typedef int (*ActionPtr1)(void*, char);

int digit(char c); /* returns c-'0' */

#endif
```

Code Listing 5-17: StatePattern.h

Next we have the Context code in Code Listing 5-18 and Code Listing 5-19.

```
#ifndef Context_H
#define Context_H

#include "StatePattern.h"
#include "GN_ProcessingFractionState.h"
```

```
#include "GN_ProcessingWholeState.h"
#include "NoNumberState.h"

struct Mutex;

typedef struct Context Context;
struct Context {
  char ch;
  TSTATETYPE currentState;
  double result;
  struct NoNumberState*
stateList[GN_PROCESSINGFRACTIONALPART_STATE+1];
  double tensPlace;
  struct Mutex* itsMutex;
};

/* Constructors and destructors:*/
void Context_Init(Context* const me);
void Context_Cleanup(Context* const me);

/* Operations */
void Context_evDigit(Context* const me, char c);
void Context_evDot(Context* const me);
void Context_evEndOfString(Context* const me);
void Context_evWhiteSpace(Context* const me);

struct Mutex* Context_getItsMutex(const Context* const me);
void Context_setItsMutex(Context* const me, struct Mutex* p_Mutex);

Context * Context_Create(void);
void Context_Destroy(Context* const me);

#endif
```

Code Listing 5-18: Context.h

```
#include "Context.h"
#include "Mutex.h"

static void cleanUpRelations(Context* const me);

void Context_Init(Context* const me) {
  me->ch = 'x';
  me->currentState = NULL_STATE;
  me->result = 0.0;
  me->tensPlace = 10.0;
  me->itsMutex = NULL;

  me->stateList[NULL_STATE] = NULL;

  /* set up state list */
  me->stateList[NONUMBER_STATE] = NoNumberState_Create();
  me->stateList[NONUMBER_STATE]->itsContext = me;
  me->stateList[GN_PROCESSINGWHOLEPART_STATE] = (NoNumberState*) GN_ProcessingWhole
State_Create();
```

```
  me->stateList[GN_PROCESSINGWHOLEPART_STATE]->itsContext = me;
  me->stateList[GN_PROCESSINGFRACTIONALPART_STATE] = (NoNumberState*) GN_Processing
FractionState_Create();
  me->stateList[GN_PROCESSINGFRACTIONALPART_STATE]->itsContext = me;

  /* enter starting state */
  me->stateList[NONUMBER_STATE]->aSet.entryAction(me->stateList[NONUMBER_STATE]);
}

void Context_Cleanup(Context* const me) {
  NoNumberState_Destroy(me->stateList[NONUMBER_STATE]);
  GN_ProcessingWholeState_Destroy((GN_ProcessingWholeState *)me->stateList
[GN_PROCESSINGWHOLEPART_STATE]);

GN_ProcessingFractionState_Destroy((GN_ProcessingFractionState *)me->stateList
[GN_PROCESSINGFRACTIONALPART_STATE]);
  cleanUpRelations(me);
}

void Context_evDigit(Context* const me, char c) {
  Mutex_lock(me->itsMutex);
  /* this call does the entry and transition actions */
  /* and updates the currentState variable */
  if (me->stateList[me->currentState]->aSet.evDigitHandler(me->stateList[me->curren-
tState], c))
      me->stateList[me->currentState]->aSet.entryAction(me->stateList
[me->currentState]);
  Mutex_release(me->itsMutex);
}

void Context_evDot(Context* const me) {
  Mutex_lock(me->itsMutex);
  /* this call does the entry and transition actions */
  /* and updates the currentState variable */
  if (me->stateList[me->currentState]->aSet.evDotHandler(me->stateList
[me->currentState]))
    me->stateList[me->currentState]->aSet.entryAction(me->stateList
[me->currentState]);
  Mutex_release(me->itsMutex);
}

void Context_evEndOfString(Context* const me) {
  Mutex_lock(me->itsMutex);
  /* this call does the entry and transition actions */
  /* and updates the currentState variable */
  if (me->stateList[me->currentState]->aSet.evEOSHandler(me->stateList
[me->currentState]))
    me->stateList[me->currentState]->aSet.entryAction(me->stateList
[me->currentState]);
  Mutex_release(me->itsMutex);
}

void Context_evWhiteSpace(Context* const me) {
  Mutex_lock(me->itsMutex);
```

```
  /* this call does the entry and transition actions */
  /* and updates the currentState variable */
  if (me->stateList[me->currentState]->aSet.evWSHandler(me->stateList
[me->currentState]))
     me->stateList[me->currentState]->aSet.entryAction(me->stateList
[me->currentState]);
  Mutex_release(me->itsMutex);
}
struct Mutex* Context_getItsMutex(const Context* const me) {
  return (struct Mutex*)me->itsMutex;
}

void Context_setItsMutex(Context* const me, struct Mutex* p_Mutex) {
  me->itsMutex = p_Mutex;
}

Context * Context_Create(void) {
  Context* me = (Context *) malloc(sizeof(Context));
  if(me!=NULL)
          Context_Init(me);
  return me;
}
void Context_Destroy(Context* const me) {
  if(me!=NULL)
          Context_Cleanup(me);
  free(me);
}
static void cleanUpRelations(Context* const me) {
  if(me->itsMutex != NULL)
          me->itsMutex = NULL;
}
```

Code Listing 5-19: Context.c

The code for the StateActionSet is straightforward as we see in Code Listing 5-20 and Code Listing 5-21.

```
#ifndef StateActionSet_H
#define StateActionSet_H

#include "StatePattern.h"

typedef struct StateActionSet StateActionSet;
struct StateActionSet {
  ActionPtr0 entryAction;
  ActionPtr1 evDigitHandler;
  ActionPtr0 evDotHandler;
  ActionPtr0 evEOSHandler;
```

```
  ActionPtr0 evWSHandler;
  ActionPtr0 exitAction;
};

/* Constructors and destructors:*/
void StateActionSet_Init(StateActionSet* const me);
void StateActionSet_Cleanup(StateActionSet* const me);

StateActionSet * StateActionSet_Create(void);
void StateActionSet_Destroy(StateActionSet* const me);

#endif
```

Code Listing 5-20: StateActionSet.h

```
#include "StateActionSet.h"
void StateActionSet_Init(StateActionSet* const me) {
  me->entryAction = NULL;
  me->evDigitHandler = NULL;
  me->evDotHandler = NULL;
  me->evEOSHandler = NULL;
  me->evWSHandler = NULL;
  me->exitAction = NULL;
}

void StateActionSet_Cleanup(StateActionSet* const me) {
}

StateActionSet * StateActionSet_Create(void) {
  StateActionSet* me = (StateActionSet *)
malloc(sizeof(StateActionSet));
  if(me!=NULL)
        StateActionSet_Init(me);
  return me;
}

void StateActionSet_Destroy(StateActionSet* const me) {
  if(me!=NULL)
        StateActionSet_Cleanup(me);
  free(me);
}
```

Code Listing 5-21: StateActionSet.c

The NoNumberState is the first state class. Its implementation is in Code Listing 5-22 and Code Listing 5-23.

```
#ifndef NoNumberState_H
#define NoNumberState_H

#include "StatePattern.h"
#include "StateActionSet.h"
```

```
struct Context;

typedef struct NoNumberState NoNumberState;
struct NoNumberState {
  struct StateActionSet aSet;
  struct Context* itsContext;
};

/* Constructors and destructors:*/
void NoNumberState_Init(NoNumberState* const me);
void NoNumberState_Cleanup(NoNumberState* const me);

/* Operations */
void NoNumberState_NoNum2GN(NoNumberState* const me, char c);
void NoNumberState_entryAction(NoNumberState* const me);
int NoNumberState_evDigit(NoNumberState* const me, char c);
int NoNumberState_evDot(NoNumberState* const me);
int NoNumberState_evEndOfString(NoNumberState* const me);
int NoNumberState_evWhiteSpace(NoNumberState* const me);
void NoNumberState_exitAction(NoNumberState* const me);

struct Context* NoNumberState_getItsContext(const NoNumberState* const me);
void NoNumberState_setItsContext(NoNumberState* const me, struct Context* p_Context);
NoNumberState * NoNumberState_Create(void);
void NoNumberState_Destroy(NoNumberState* const me);

#endif
```

Code Listing 5-22: NoNumberState.h

```
#include "NoNumberState.h"
#include "Context.h"
#include <stdio.h>

static void cleanUpRelations(NoNumberState* const me);

void NoNumberState_Init(NoNumberState* const me) {
  {
      StateActionSet_Init(&(me->aSet));
  }
  me->itsContext = NULL;

  /* set up the aSet function pointers */
  /* remember that the initializer for the StateActionSet */
  /* sets all pointers to NULL as the default */
  me->aSet.evDigitHandler = (ActionPtr1) NoNumberState_evDigit;
  me->aSet.evDotHandler = (ActionPtr0) NoNumberState_evDot;
  me->aSet.evWSHandler = (ActionPtr0) NoNumberState_evWhiteSpace;
  me->aSet.evEOSHandler = (ActionPtr0) NoNumberState_evEndOfString;
  me->aSet.entryAction = (ActionPtr0) NoNumberState_entryAction;
  me->aSet.exitAction = (ActionPtr0) NoNumberState_exitAction;
}
```

```c
void NoNumberState_Cleanup(NoNumberState* const me) {
  cleanUpRelations(me);
}

void NoNumberState_NoNum2GN(NoNumberState* const me, char c) {
  me->itsContext->ch = c;
  me->itsContext->tensPlace = 10.0;
}

void NoNumberState_entryAction(NoNumberState* const me) {
  me->itsContext->result = 0.0;
}

int NoNumberState_evDigit(NoNumberState* const me, char c) {
  NoNumberState_exitAction(me);
  NoNumberState_NoNum2GN(me, c); /* transition action */
  /* set new state */
  me->itsContext->currentState = GN_PROCESSINGWHOLEPART_STATE;
  return 1; /* processed event */
}

int NoNumberState_evDot(NoNumberState* const me) {
  return 0; /* discarded event */
}

int NoNumberState_evEndOfString(NoNumberState* const me) {
  return 0; /* discarded event */
}

int NoNumberState_evWhiteSpace(NoNumberState* const me) {
  return 0; /* discarded event */
}

void NoNumberState_exitAction(NoNumberState* const me) {
}

struct Context* NoNumberState_getItsContext(const NoNumberState* const me) {
  return (struct Context*)me->itsContext;
}

void NoNumberState_setItsContext(NoNumberState* const me, struct Context* p_Context) {
  me->itsContext = p_Context;
}

NoNumberState * NoNumberState_Create(void) {
  NoNumberState* me = (NoNumberState *)
malloc(sizeof(NoNumberState));
  if(me!=NULL)
        NoNumberState_Init(me);
  return me;
}

void NoNumberState_Destroy(NoNumberState* const me) {
  if(me!=NULL)
        NoNumberState_Cleanup(me);
  free(me);
}
```

```
static void cleanUpRelations(NoNumberState* const me) {
  if(me->itsContext != NULL)
        me->itsContext = NULL;
}
```

Code Listing 5-23: NoNumberState.c

The next state class is GN_ProcessingWholeState. Its implementation is in Code Listing 5-24 and Code Listing 5-25.

```
#ifndef GN_ProcessingWholeState_H
#define GN_ProcessingWholeState_H

#include "StatePattern.h"
#include "StateActionSet.h"

struct Context;

typedef struct GN_ProcessingWholeState GN_ProcessingWholeState;
struct GN_ProcessingWholeState {
  struct StateActionSet aSet;
  struct Context* itsContext;
};

/* Constructors and destructors:*/
void GN_ProcessingWholeState_Init(GN_ProcessingWholeState* const me);
void GN_ProcessingWholeState_Cleanup(GN_ProcessingWholeState* const me);

/* Operations */
void GN_ProcessingWholeState_entryAction(GN_ProcessingWholeState* const me);
int GN_ProcessingWholeState_evDigit(GN_ProcessingWholeState* const me, char c);
int GN_ProcessingWholeState_evDot(GN_ProcessingWholeState* const me);
int
GN_ProcessingWholeState_evEndOfString(GN_ProcessingWholeState* const me);

int
GN_ProcessingWholeState_evWhiteSpace(GN_ProcessingWholeState* const me);
void GN_ProcessingWholeState_exitAction(GN_ProcessingWholeState* const me);
void GN_ProcessingWholeState_printResult(GN_ProcessingWholeState* const me);

struct Context* GN_ProcessingWholeState_getItsContext(const GN_ProcessingWholeState*
const me);

void
GN_ProcessingWholeState_setItsContext(GN_ProcessingWholeState* const me, struct Con-
text* p_Context);

GN_ProcessingWholeState * GN_ProcessingWholeState_Create(void);
void GN_ProcessingWholeState_Destroy(GN_ProcessingWholeState* const me);

#endif
```

Code Listing 5-24: GN_ProcessingWholeState.h

```
#include "GN_ProcessingWholeState.h"
#include "Context.h"

static void cleanUpRelations(GN_ProcessingWholeState* const me);

void GN_ProcessingWholeState_Init(GN_ProcessingWholeState* const me) {
  StateActionSet_Init(&(me->aSet));
  me->itsContext = NULL;

  /* set up the aSet function pointers */
  /* remember that the initializer for the StateActionSet */
  /* sets all pointers to NULL as the default */
  me->aSet.evDigitHandler = (ActionPtr1) GN_ProcessingWholeState_evDigit;
  me->aSet.evDotHandler = (ActionPtr0) GN_ProcessingWholeState_evDot;
  me->aSet.evWSHandler = (ActionPtr0) GN_ProcessingWholeState_evWhiteSpace;
  me->aSet.evEOSHandler = (ActionPtr0) GN_ProcessingWholeState_evEndOfString;
  me->aSet.entryAction = (ActionPtr0) GN_ProcessingWholeState_entryAction;
  me->aSet.exitAction = (ActionPtr0) GN_ProcessingWholeState_exitAction;
}

void GN_ProcessingWholeState_Cleanup(GN_ProcessingWholeState* const me) {
  cleanUpRelations(me);
}

void GN_ProcessingWholeState_entryAction(GN_ProcessingWholeState* const me) {
  me->itsContext->result = me->itsContext->result*10 + digit(me->itsContext->ch);
}

int GN_ProcessingWholeState_evDigit(GN_ProcessingWholeState* const me, char c) {
  GN_ProcessingWholeState_exitAction(me);
  me->itsContext->ch = c; /* transitiona action */
  /* note: same state is reentered in this case */
  /* so no change to itsContext->currentState */
  return 1; /* event handled */
}

int GN_ProcessingWholeState_evDot(GN_ProcessingWholeState* const me) {
  GN_ProcessingWholeState_exitAction(me);
  /* no transition action */
  me->itsContext->currentState = GN_PROCESSINGFRACTIONALPART_STATE;
  return 1; /* event handled */
}

int
GN_ProcessingWholeState_evEndOfString(GN_ProcessingWholeState* const me) {
  GN_ProcessingWholeState_exitAction(me);
  GN_ProcessingWholeState_printResult(me); /* transition action */
  me->itsContext->currentState = NONUMBER_STATE;
  return 1; /* event handled */
}

int GN_ProcessingWholeState_evWhiteSpace(GN_ProcessingWholeState* const me) {
  GN_ProcessingWholeState_exitAction(me);
  GN_ProcessingWholeState_printResult(me); /* transition action */
  me->itsContext->currentState = NONUMBER_STATE;
```

```
        return 1; /* event handled */
    }
/*## operation exitAction() */
void GN_ProcessingWholeState_exitAction(GN_ProcessingWholeState* const me) {
}

void GN_ProcessingWholeState_printResult(GN_ProcessingWholeState* const me) {
    printf("Number: %g\n", me->itsContext->result);
}

struct Context* GN_ProcessingWholeState_getItsContext(const GN_ProcessingWholeState*
const me) {
    return (struct Context*)me->itsContext;
}

void
GN_ProcessingWholeState_setItsContext(GN_ProcessingWholeState* const me, struct Con-
text* p_Context) {
    me->itsContext = p_Context;
}

GN_ProcessingWholeState * GN_ProcessingWholeState_Create(void) {
    GN_ProcessingWholeState* me = (GN_ProcessingWholeState *) malloc(sizeof
(GN_ProcessingWholeState));
    if(me!=NULL)
            GN_ProcessingWholeState_Init(me);
    return me;
}

void GN_ProcessingWholeState_Destroy(GN_ProcessingWholeState* const me) {
    if(me!=NULL)
            GN_ProcessingWholeState_Cleanup(me);
    free(me);
}

static void cleanUpRelations(GN_ProcessingWholeState* const me) {
    if(me->itsContext != NULL)
            me->itsContext = NULL;
}
```

Code Listing 5-25: GN_ProcessingWholeState.c

And finally, the code for the GN_ProcessingFractionState is given in Code Listing 5-26
and Code Listing 5-27.

```
#ifndef GN_ProcessingFractionState_H
#define GN_ProcessingFractionState_H

#include "StatePattern.h"
#include "StateActionSet.h"

struct Context;

typedef struct GN_ProcessingFractionState GN_ProcessingFractionState;
struct GN_ProcessingFractionState {
```

```
  struct StateActionSet aSet;
  struct Context* itsContext;
};

void GN_ProcessingFractionState_Init(GN_ProcessingFractionState* const me);
void
GN_ProcessingFractionState_Cleanup(GN_ProcessingFractionState* const me);

/* Operations */

void
GN_ProcessingFractionState_Frac2Frac(GN_ProcessingFractionState* const me, char c);
void
GN_ProcessingFractionState_entryAction(GN_ProcessingFractionState* const me);
int
GN_ProcessingFractionState_evDigit(GN_ProcessingFractionState* const me, char c);
int
GN_ProcessingFractionState_evDot(GN_ProcessingFractionState* const me);
int
GN_ProcessingFractionState_evEndOfString(GN_ProcessingFractionState* const me);
int
GN_ProcessingFractionState_evWhiteSpace(GN_ProcessingFractionState* const me);
void
GN_ProcessingFractionState_exitAction(GN_ProcessingFractionState* const me);
void
GN_ProcessingFractionState_printResult(GN_ProcessingFractionState* const me);

struct Context* GN_ProcessingFractionState_getItsContext(const GN_Processing
FractionState* const me);

void
GN_ProcessingFractionState_setItsContext(GN_ProcessingFractionState* const me, struct
Context* p_Context);

GN_ProcessingFractionState *
GN_ProcessingFractionState_Create(void);

void GN_ProcessingFractionState_Destroy(GN_ProcessingFractionState* const me);

#endif
```

Code Listing 5-26: GN_ProcessingFractionState.h

```
#include "GN_ProcessingFractionState.h"
#include "Context.h"

static void cleanUpRelations(GN_ProcessingFractionState* const me);

void GN_ProcessingFractionState_Init(GN_ProcessingFractionState* const me) {
  StateActionSet_Init(&(me->aSet));
  me->itsContext = NULL;

  /* set up the aSet function pointers */
  /* remember that the initializer for the StateActionSet */
```

```
    /* sets all pointers to NULL as the default */
    me->aSet.evDigitHandler = (ActionPtr1) GN_ProcessingFractionState_evDigit;
    me->aSet.evDotHandler = (ActionPtr0) GN_ProcessingFractionState_evDot;
    me->aSet.evWSHandler = (ActionPtr0) GN_ProcessingFractionState_evWhiteSpace;
    me->aSet.evEOSHandler = (ActionPtr0) GN_ProcessingFractionState_evEndOfString;
    me->aSet.entryAction = (ActionPtr0) GN_ProcessingFractionState_entryAction;
    me->aSet.exitAction = (ActionPtr0) GN_ProcessingFractionState_exitAction;
}

void
GN_ProcessingFractionState_Cleanup(GN_ProcessingFractionState* const me) {
    cleanUpRelations(me);
}

void
GN_ProcessingFractionState_Frac2Frac(GN_ProcessingFractionState* const me, char c) {
    me->itsContext->ch = c;
    me->itsContext->result += digit(me->itsContext->ch)/me->itsContext->tensPlace;
    me->itsContext->tensPlace *= 10;
}

void
GN_ProcessingFractionState_entryAction(GN_ProcessingFractionState* const me) {
}

int
GN_ProcessingFractionState_evDigit(GN_ProcessingFractionState* const me, char c) {
    GN_ProcessingFractionState_entryAction(me);
    GN_ProcessingFractionState_Frac2Frac(me, c); /* transition action */
    /* note: state doesn't change, so the same */
    /* state is reentered */
    return 1; /* event handled */
}

int
GN_ProcessingFractionState_evDot(GN_ProcessingFractionState* const me) {
    return 0; /* event discarded */
}

int
GN_ProcessingFractionState_evEndOfString(GN_ProcessingFractionState* const me) {
    GN_ProcessingFractionState_exitAction(me);
    GN_ProcessingFractionState_printResult(me); /* transition action */
    me->itsContext->currentState = NONUMBER_STATE;
    return 1; /* event handled */
}

int
GN_ProcessingFractionState_evWhiteSpace(GN_ProcessingFractionState* const me) {
    GN_ProcessingFractionState_exitAction(me);
    GN_ProcessingFractionState_printResult(me); /* transition action */
    me->itsContext->currentState = NONUMBER_STATE;
    return 1; /* event handled */
}
```

```
void
GN_ProcessingFractionState_exitAction(GN_ProcessingFractionState* const me) {
}

void
GN_ProcessingFractionState_printResult(GN_ProcessingFractionState* const me) {
   printf("Number: %g\n", me->itsContext->result);
}

struct Context* GN_ProcessingFractionState_getItsContext(const GN_ProcessingFraction-
State* const me) {
   return (struct Context*)me->itsContext;
}

void
GN_ProcessingFractionState_setItsContext(GN_ProcessingFractionState* const me, struct
Context* p_Context) {
  me->itsContext = p_Context;
}

GN_ProcessingFractionState *
GN_ProcessingFractionState_Create(void) {
  GN_ProcessingFractionState* me = (GN_ProcessingFractionState *) malloc(sizeof
(GN_ProcessingFractionState));
  if(me!=NULL)
        GN_ProcessingFractionState_Init(me);
  return me;
}

void
GN_ProcessingFractionState_Destroy(GN_ProcessingFractionState* const me) {
  if(me!=NULL)
        GN_ProcessingFractionState_Cleanup(me);
  free(me);
}

static void cleanUpRelations(GN_ProcessingFractionState* const me) {
  if(me->itsContext != NULL)
        me->itsContext = NULL;
}
```

Code Listing 5-27: GN_ProcessingFractionState.c

It is interesting to see how these elements interact in execution of a simple case. Figure 5-19 shows the interaction flow for processing the string "1.3" sent from a client of the state machine. In this figure, you can see how the event handlers of the Context invoke the event handlers of the then-current state to process the incoming events. You may find it interesting to compare this figure with the corresponding figure for the State Table Pattern, Figure 5-17.

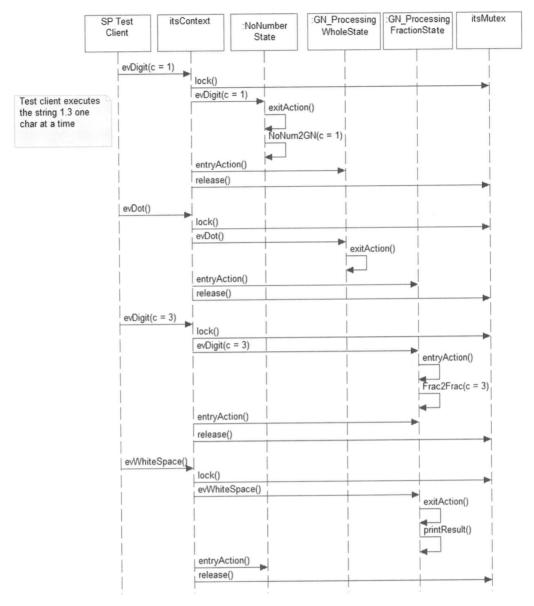

Figure 5-19: State Pattern sequence for execution of "1.3" string

5.7 AND-States

The last two patterns in the chapter will deal with the implementation of AND-states. For these patterns, we'll use the state machine from Figure 5-7 as the example state machine with AND-states and we'll add a few actions to make it more "real." Figure 5-20 shows the example state machine we will demonstrate with the pattern.

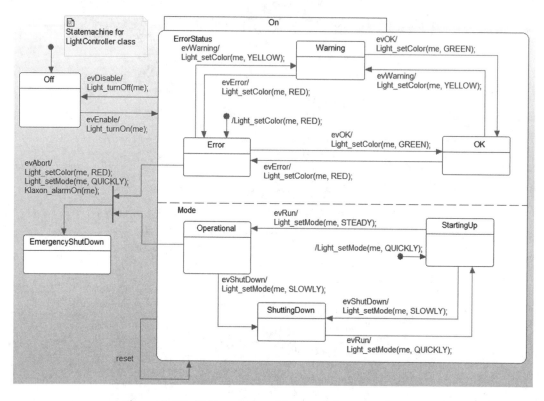

Figure 5-20: AND-state machine used for example case

The semantics of AND-states is that they are "logically concurrent." This means that in some sense they execute simultaneously, but what does that mean in a world of run-to-completion semantics? If two AND-states process different events, then AND-states are more of a (significant) notational convenience. Simultaneity only has importance when the same event is processed within two AND-states. In that case, what is the meaning of concurrency? In this context, it means that *the semantics of the system are not dependent upon which AND-state processes the event first.* It also implies that when AND-states are active, each active AND-state must receive, in some sense, its own copy of the event sent to the state machine, and is free to act on it or discard it as appropriate. What we cannot rely upon is which AND-state will act on the event first[5]. If we need to ensure that they execute in a particular order, then perhaps AND-states were not the appropriate mechanism to employ.

A naïve but overly heavyweight approach would be to spawn threads for the different state machines. While this can be made to work, in practice I have never seen AND-states implemented in this fashion, and for good reason. The logical concurrency really only requires:

---

[5] The UML has no notion of event or transition priority. Events are always handled in order of their arrival.

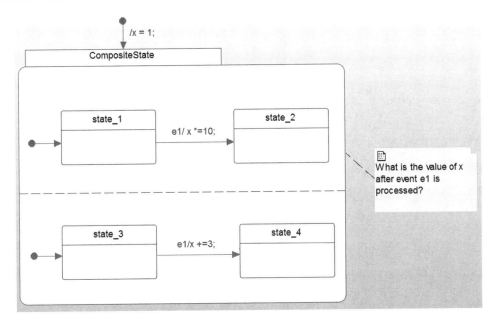

Figure 5-21: Race condition

- events are handled run-to-completion
- all AND-state event handlers must complete before the next event sent to the state machine is processed

So which AND-state processes the event first? The upper one? The one you drew first? The one that appears first in the code? *It doesn't matter.* That's the point.

Note that this does raise the specter of race conditions, as discussed in Chapter 4. For example, consider the state machine in Figure 5-21. It contains a race condition. The e1 event is handled in two AND-states; each of which has an action manipulating the variable x. If the upper AND-state processes its event first[6], the resulting value of x is 33. If the lower one processes it first, the answer is 40.

Don't do that![7]

If you need to force a specific order of event execution, then you can use a simple pattern called *event propagation*. If you want to force the upper AND-state to execute first, then create a different event for the second transition and generate that event in an action of the first transition. This pattern, applied to the previous figure, is shown in Figure 5-22, where the GEN

---

[6] Notice that x is set to the value 1 before entering the CompositeState.
[7] A Law of Douglass is that any language sufficiently powerful to be useful is also sufficiently powerful to say nonsense. My rule is to try to say only smart things.

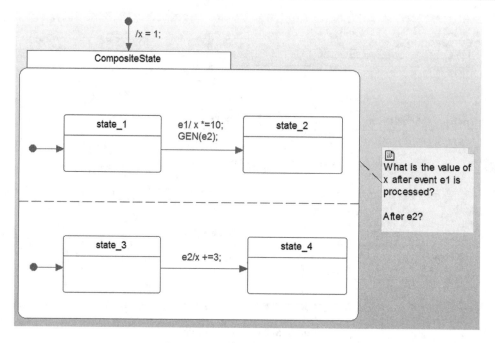

Figure 5-22: Event propagation

macro generates the event using whatever synchronous or asynchronous approach you've implemented for your state machine.

There are a number of ways commonly used to implement AND-states. The first, and least desirable, is to flatten the state space. This is problematic, as was pointed out in Figure 5-6 because of state explosion – you must compute the cross product of the state spaces and that's often very ugly and difficult to read and understand because of the sheer number of resulting states. It does work, however, and in simple cases it might be easiest.

A more common solution is to decompose the object owning the AND-states into a set of different objects that communicate. This is the Decomposed AND-State Pattern discussed below. It is also possible to elaborate one of the previous patterns we've presented into managing multiple AND-states in the event receptors. There are other solutions as well, but these are the most common.

5.8 Decomposed AND-State Pattern

This pattern extends the State Pattern to implement AND-states. It does so by creating a collaboration of different objects, each of which is allocated one of the AND-states.

5.8.1 Abstract

The Decomposed And-State Pattern takes an object that owns a state machine and decomposes it into a set of objects that interact. The main object owns the overarching state machine and other objects own individual AND-states.

5.8.2 Problem

This pattern addresses the problem of implementing state machines that contain orthogonal regions (AND-states) in a way that preserves the AND-state design.

5.8.3 Pattern Structure

Figure 5-23 shows the structure of the pattern. Instead of a singular class that implements the state machine, the state problem is decomposed into a collaboration of objects that relate to the Context.

5.8.4 Collaboration Roles

This section describes the pattern roles.

5.8.4.1 AbstractAndStateClass

This class is an abstraction of the various concrete AND-state classes. It provides an event handler for each event received by the `Context` and a link back to the `Context` for execution of actions and manipulation of variables within the `Context`. In the C implementation, this is a notional class, not usually represented directly but it does show what is inherited by the subclasses.

5.8.4.2 ActionSet

This class contains a set of pointers, one per incoming event. It is owned by the `Abstract-AndStateClass` and implemented by the `ConcreteAndStateClass`. In the initializer of the latter class, the function pointers are set to point to the functions that implement the actions within that concrete state.

5.8.4.3 AndStateList

This class maintains a list of orthogonal regions (containing all the nested states within an AND-state) for a given composite. There is an entry for every state; if the list is empty, then that state has no nested AND-states. If it is not empty, then the list entry contains a list of the orthogonal regions of AND-states; each region is represented by a `ConcreteAndStateClass` and it implements all the nested states within that region.

Consider the example state machine in Figure 5-20. `STATE_N` in this case is 1. The 0-th state is the `Off` state; this state has no nested AND-states so this list entry is empty. The 1st state is the `On` state; this state has two nested orthogonal regions, `Color` and `Rate`. Thus the

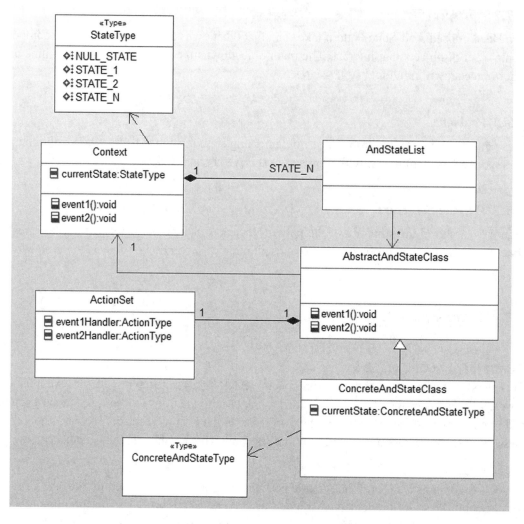

Figure 5-23: Decomposed And-State Pattern structure

`Context` list entry for State 1 is not empty. In this case, the list entry for State 1 has two lists, one for each orthogonal region. The `ConcreteAndStateClass Color` manages three states – `Red`, `Yellow`, and `Green`. The `ConcreteAndStateClass Rate` also manages three states – `Steady`, `Flashing_Slowly`, and `Flashing_Quickly`. So the lists look like this:

- `Context` state list:
  - State 0 (`Off`) – Empty
  - State 1 (`On`) list:
    - `Color` (manages `Red`, `Yellow`, `Green` states)
    - `Rate` (manages `Steady`, `Flashing_Slowly`, `Flashing_Quickly` states)

You'll notice that this gives us lists of lists of states. This is often implemented with arrays because the sizes of these lists are known at design time. If you need to further nest AND-states, you can add additional levels of lists, with appropriate behavior at each level.

5.8.4.4 ConcreteAndStateClass

This represents one of the AND-state classes. The design will contain one such class, with its own implementation of its event receptors, for each AND-state. In the basic pattern, each and-state class will itself use the Multiple Event Receptor Pattern, but other approaches are possible.

5.8.4.5 ConcreteAndStateType

Each AND-state has its unique set of nested states; this type is an enumerated list of them. One such type will exist for each `ConcreteAndStateClass`. This class is not visible to the clients of the `Context`.

5.8.4.6 Context

The `Context` is the ultimate owner of the state machine. It has the variables and functions that are manipulated by the `ConcreteAndStateClasses`. This class presents a set of event receptors to its clients (alternatively, the Single Event Receptor Pattern or other state pattern can be mixed in here as well). The basic rules of state change are enforced here:

- If the new state being entered has no nested AND-states, it executes normally as in the Multiple Event Receptor Pattern.
- If the new state being entered has nested AND-states, then its entry action is performed prior to the entry actions of the entered nested AND-states.
- If the composite state having AND-states is unchanged (that is, the event transition is local within the current composite state), then the event is delegated to its set of AND-state classes for handling, including the execution of the actions.
- If the composite state is left, then the exit actions of the composite state execute after the exit actions for each of its nested AND-states.

5.8.5 Consequences

This pattern provides extensible mechanisms for managing state machines that contain AND-states. Each class is relatively simple because of the delegation of responsibilities for state management – the `Context` manages the high level states, and the `ConcreteAndStateClasses` manage the nested AND-states.

On the other hand, this pattern does require the management of lists, of lists which will normally be managed as lists of pointers to lists of pointers. This level of indirection, while powerful, can be the source of obscure mistakes if not carefully managed.

5.8.6 Implementation Strategies

As mentioned in the previous section, the common implementation of the lists is as arrays of pointers to arrays of pointers.

5.8.7 Related Patterns

This pattern can be used in conjunction with the state pattern – in this case, each state class will be the Context for its AND-states.

5.8.8 Example

In the example shown in Figure 5-24, the LightController plays the part of the Context pattern role and the ErrorStateClass and ModeStateClass play the part of the ConcreteAndStateClass in the pattern. Supporting elements, such as enumerated types and the StateActionSet are also shown.

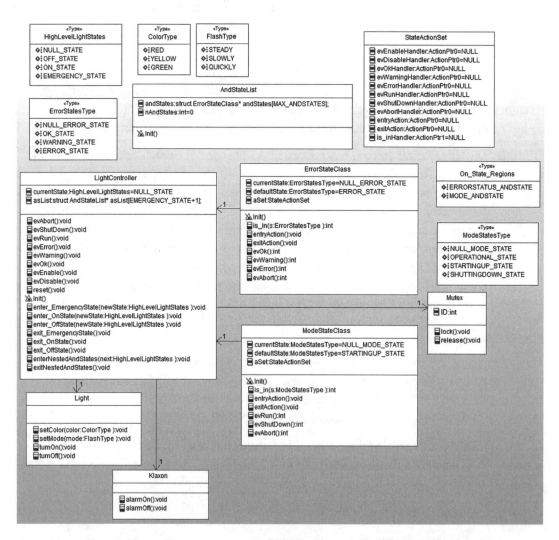

Figure 5-24: Example problem with Decomposed And-State Pattern

The code for this example is shown in the next several code listings. The LightPkg.h file in Code Listing 5-28 holds the basic types used in the other files.

```c
#ifndef LightPkg_H
#define LightPkg_H

struct AndStateList;
struct DecompBuilder;
struct DecompClient;
struct ErrorStateClass;
struct Klaxon;
struct Light;
struct LightController;
struct ModeStateClass;
struct Mutex;
struct StateActionSet;

typedef enum HighLevelLightStates {
    NULL_STATE,
    OFF_STATE,
    ON_STATE,
    EMERGENCY_STATE
} HighLevelLightStates;

typedef enum ColorType {
    RED,
    YELLOW,
    GREEN
} ColorType;

typedef enum FlashType {
    STEADY,
    SLOWLY,
    QUICKLY
} FlashType;

typedef enum ErrorStatesType {
    NULL_ERROR_STATE,
    OK_STATE,
    WARNING_STATE,
    ERROR_STATE
} ErrorStatesType;

typedef enum ModeStatesType {
    NULL_MODE_STATE,
    OPERATIONAL_STATE,
    STARTINGUP_STATE,
    SHUTTINGDOWN_STATE
} ModeStatesType;

typedef enum On_State_Regions {
    ERRORSTATUS_ANDSTATE,
    MODE_ANDSTATE
} On_State_Regions;
```

```
#define MAX_ANDSTATES 2

/* This action pointer accepts the me object pointer and no other parameters. */
typedef int (*ActionPtr0)(void*);
/* This action pointer accepts the me object pointer and no other parameters. */
typedef int (*ActionPtr1)(void*, int);

#endif
```

Code Listing 5-28: LightPkg.h

The LightController class provides the state machine context for the example. It is shown in Code Listing 5-29 and Code Listing 5-30.

```
#ifndef LightController_H
#define LightController_H

#include "LightPkg.h"
#include "AndStateList.h"
#include "ErrorStateClass.h"
#include "ModeStateClass.h"

struct Klaxon;
struct Light;
struct Mutex;

typedef struct LightController LightController;
struct LightController {
    struct AndStateList* asList[EMERGENCY_STATE+1];
    HighLevelLightStates currentState;
    struct Klaxon* itsKlaxon;
    struct Light* itsLight;
    struct Mutex* itsMutex;
};

void LightController_evEnable(LightController* const me);
void LightController_evDisable(LightController* const me);
void LightController_reset(LightController* const me);
void LightController_evOk(LightController* const me);
void LightController_evWarning(LightController* const me);
void LightController_evError(LightController* const me);
void LightController_evRun(LightController* const me);
void LightController_evShutDown(LightController* const me);
void LightController_evAbort(LightController* const me);
void LightController_Init(LightController* const me);
void LightController_Cleanup(LightController* const me);

void LightController_enter_OffState(LightController* const me, HighLevelLightStates
newState);

void LightController_enter_OnState(LightController* const me, HighLevelLightStates
newState);
```

```
void LightController_enter_EmergencyState(LightController* const me, HighLevelLight-
States newState);

void LightController_exit_OffState(LightController* const me);

void LightController_exit_OnState(LightController* const me);

void LightController_exit_EmergencyState(LightController* const me);

void LightController_enterNestedAndStates(LightController* const me, HighLevelLight-
States next);

void LightController_exitNestedAndStates(LightController* const me);

struct Klaxon* LightController_getItsKlaxon(const LightController* const me);

void LightController_setItsKlaxon(LightController* const me, struct Klaxon* p_Klaxon);

struct Light* LightController_getItsLight(const LightController* const me);

void LightController_setItsLight(LightController* const me, struct Light* p_Light);

struct Mutex* LightController_getItsMutex(const LightController* const me);

void LightController_setItsMutex(LightController* const me, struct Mutex* p_Mutex);
LightController * LightController_Create(void);

void LightController_Destroy(LightController* const me);

#endif
```

Code Listing 5-29: LightController.h

```
#include "LightController.h"
#include "Klaxon.h"
#include "Light.h"
#include "Mutex.h"

static void cleanUpRelations(LightController* const me);

void LightController_evEnable(LightController* const me) {
  HighLevelLightStates newState;

  Mutex_lock(me->itsMutex);
  switch (me->currentState) {
    case OFF_STATE:
        newState = ON_STATE;
        LightController_exit_OffState(me);

        /* transition to On state */
        Light_turnOn(me->itsLight);
        Light_setColor(me->itsLight, RED);
        Light_setMode(me->itsLight, QUICKLY);

        /* enter new state */
        LightController_enter_OnState(me, newState);
        me->currentState = newState;
```

```
            break;
      case ON_STATE:
      case EMERGENCY_STATE:
            break;
      };
   Mutex_release(me->itsMutex);
}

void LightController_evDisable(LightController* const me) {
   HighLevelLightStates newState;

   Mutex_lock(me->itsMutex);
   switch (me->currentState) {
      case ON_STATE:
            newState = OFF_STATE;
              /* exit nested states */
                  LightController_exit_OnState(me);

                  /* transition to OFF state */
                  Light_turnOff(me->itsLight);

         /* enter new state */
         LightController_enter_OffState(me, newState);
         me->currentState = newState;
         break;
   case OFF_STATE:
   case EMERGENCY_STATE:
      break;
      };
   Mutex_release(me->itsMutex);
}

void LightController_reset(LightController* const me) {
   HighLevelLightStates newState;

   Mutex_lock(me->itsMutex);
   switch (me->currentState) {
     case ON_STATE:
      /* exit and-states, then outer state */
     newState = ON_STATE;
         LightController_exit_OnState(me);

         /* note no transition actions */
         /* enter outer state, then and-states */
         LightController_enter_OnState(me, newState);
         me->currentState = newState;
         break;

   case OFF_STATE:
   case EMERGENCY_STATE:
         break;
```

```
    };
  Mutex_release(me->itsMutex);
}
void LightController_evOk(LightController* const me) {
  ErrorStateClass* sPtr;
  ActionPtr0 aPtr;
  HighLevelLightStates newState;

  Mutex_lock(me->itsMutex);
  switch (me->currentState) {
      case OFF_STATE:
        break;
      case ON_STATE:
      /* note we're not exiting On state */
      /* so just delegate to and-states */
      newState = me->currentState;
      if (me->asList[newState]) {
        sPtr = me->asList[newState]->andStates[ERRORSTATUS_ANDSTATE];
        if (sPtr) {
            aPtr = sPtr->aSet.evOkHandler;
            if (aPtr)
                aPtr(sPtr);
            };
        sPtr = me->asList[newState]->andStates[MODE_ANDSTATE];
        if (sPtr) {
         aPtr = sPtr->aSet.evOkHandler;
            if (aPtr)
                aPtr(sPtr);
            };
        };
    break;
  case EMERGENCY_STATE:
    break;
  };
  Mutex_release(me->itsMutex);
}
void LightController_evWarning(LightController* const me) {
  ErrorStateClass* sPtr;
  ActionPtr0 aPtr;
  HighLevelLightStates newState;

  Mutex_lock(me->itsMutex);
  switch (me->currentState) {
    case OFF_STATE:
      break;
    case ON_STATE:
        /* note we're not exiting On state */
        /* so just delegate to and-states */
        newState = me->currentState;
```

```
            if (me->asList[newState]) {
              sPtr = me->asList[newState]->andStates[ERRORSTATUS_ANDSTATE];
              if (sPtr) {
                aPtr = sPtr->aSet.evWarningHandler;
                if (aPtr)
                    aPtr(sPtr);
              };
            sPtr = me->asList[newState]->andStates[MODE_ANDSTATE];
              if (sPtr) {
               aPtr = sPtr->aSet.evWarningHandler;
                if (aPtr)
                    aPtr(sPtr);
                };
            };
          break;
      case EMERGENCY_STATE:
          break;
      };
    Mutex_release(me->itsMutex);
    /*#]*/
}

void LightController_evError(LightController* const me) {
    ErrorStateClass* sPtr;
    ActionPtr0 aPtr;
    HighLevelLightStates newState;

    Mutex_lock(me->itsMutex);
    switch (me->currentState) {
      case OFF_STATE:
          break;
      case ON_STATE:
          /* note we're not exiting On state */
          /* so just delegate to AND-states */
          newState = me->currentState;
          if (me->asList[newState]) {
            sPtr = me->asList[newState]->andStates[ERRORSTATUS_ANDSTATE];
            if (sPtr) {
              aPtr = sPtr->aSet.evErrorHandler;
              if (aPtr)
                  aPtr(sPtr);
              };
            sPtr = me->asList[newState]->andStates[MODE_ANDSTATE];
            if (sPtr) {
            aPtr = sPtr->aSet.evErrorHandler;
              if (aPtr)
                  aPtr(sPtr);
              };
            };
        break;
    case EMERGENCY_STATE:
```

```
        break;
    };
  Mutex_release(me->itsMutex);
}
void LightController_evRun(LightController* const me) {
  ErrorStateClass* sPtr;
  ActionPtr0 aPtr;
  HighLevelLightStates newState;

  Mutex_lock(me->itsMutex);
  switch (me->currentState) {
    case OFF_STATE:
      break;
    case ON_STATE:
      /* note we're not exiting On state */
      /* so just delegate to and-states */
      newState = me->currentState;
      if (me->asList[newState]) {
        sPtr = me->asList[newState]->andStates[ERRORSTATUS_ANDSTATE];
        if (sPtr) {
          aPtr = sPtr->aSet.evRunHandler;
          if (aPtr)
            aPtr(sPtr);
        };
        sPtr = me->asList[newState]->andStates[MODE_ANDSTATE];
        if (sPtr) {
          aPtr = sPtr->aSet.evRunHandler;
          if (aPtr)
            aPtr(sPtr);
        };
      };
      break;
    case EMERGENCY_STATE:
      break;
    };
  Mutex_release(me->itsMutex);
}
void LightController_evShutDown(LightController* const me) {
  ErrorStateClass* sPtr;
  ActionPtr0 aPtr;
  HighLevelLightStates newState;

  Mutex_lock(me->itsMutex);
  switch (me->currentState) {
    case OFF_STATE:
      break;
    case ON_STATE:
      /* note we're not exiting On state */
      /* so just delegate to and-states */
      newState = me->currentState;
```

```
            if (me->asList[newState]) {
                sPtr = me->asList[newState]->andStates[ERRORSTATUS_ANDSTATE];
                if (sPtr) {
                    aPtr = sPtr->aSet.evShutDownHandler;
                    if (aPtr)
                        aPtr(sPtr);
                };
                sPtr = me->asList[newState]->andStates[MODE_ANDSTATE];
                if (sPtr) {
                 aPtr = sPtr->aSet.evShutDownHandler;
                    if (aPtr)
                        aPtr(sPtr);
                };
            };
        break;
    case EMERGENCY_STATE:
            break;
        };
    Mutex_release(me->itsMutex);
}

void LightController_evAbort(LightController* const me) {
    ErrorStateClass* sPtr;
    ActionPtr1 is_inPtr;
    HighLevelLightStates newState;
    int precond1=0, precond2=0;

    Mutex_lock(me->itsMutex);
    switch (me->currentState) {
        case ON_STATE:
            /* are the preconditions true: */
            /* ie. Error and Operational states? */
            if (me->asList[me->currentState]) {
             sPtr = me->asList[me->currentState]->andStates[ERRORSTATUS_ANDSTATE];
             if (sPtr) {
                    is_inPtr = sPtr->aSet.is_inHandler;
                    if (is_inPtr)
                        precond1 = is_inPtr(sPtr, ERROR_STATE);
                };
             sPtr = me->asList[me->currentState]->andStates[MODE_ANDSTATE];
             if (sPtr) {
                    is_inPtr = sPtr->aSet.is_inHandler;
                    if (is_inPtr)
                        precond2 = is_inPtr(sPtr, OPERATIONAL_STATE);
                };
                };
        if (precond1 && precond2) {
         newState = EMERGENCY_STATE;
         /* exit nested states */
                LightController_exit_OnState(me);
```

```
                    /* transition to OFF state */
                    Light_setColor(me->itsLight, RED);
                    Light_setMode(me->itsLight, QUICKLY);
                    Klaxon_alarmOn(me->itsKlaxon);

                    /* enter new state */
                    LightController_enter_EmergencyState(me, newState);
                    me->currentState = newState;
                };
            break;
        case OFF_STATE:
        case EMERGENCY_STATE:
            break;
        };
    Mutex_release(me->itsMutex);
}

void LightController_Init(LightController* const me) {
    me->currentState = NULL_STATE;
    me->itsKlaxon = NULL;
    me->itsLight = NULL;
    me->itsMutex = NULL;
    int j;
    HighLevelLightStates st;
    ErrorStateClass* andStatePtr = NULL;

    for (st=NULL_STATE;st<=EMERGENCY_STATE; st++)
        me->asList[st] = NULL;

    /* create all and-state lists and states */
    /* In this case, there is only one, the ON_STATE */
    /* which has two nested orthogonal regions */
    me->asList[ON_STATE] = AndStateList_Create();
    me->asList[ON_STATE]->nAndStates = 2;
    andStatePtr = ErrorStateClass_Create();
    if (andStatePtr) {
        me->asList[ON_STATE]->andStates[ERRORSTATUS_ANDSTATE] = andStatePtr;
        andStatePtr->itsLightController = me;
        };
    andStatePtr = (ErrorStateClass*)ModeStateClass_Create();
    if (andStatePtr) {
        me->asList[ON_STATE]->andStates[MODE_ANDSTATE] = andStatePtr;
        andStatePtr->itsLightController = me;
        };

    /* enter initial state */
    me->currentState = OFF_STATE;
    LightController_enter_OffState(me, OFF_STATE);
    /* any nested and-states? */
    if (me->asList[me->currentState])
        /* enter the default state in each and-state */
```

```
      for (j=0; j<me->asList[me->currentState]->nAndStates; j++)
          me->asList[me->currentState]->andStates[j]->aSet.entryAction(me->asList[me-
>currentState]->andStates[j]);
}

void LightController_Cleanup(LightController* const me) {
  cleanUpRelations(me);
}

void LightController_enter_OffState(LightController* const me, HighLevelLightStates
newState) {
  LightController_enterNestedAndStates(me, newState);
}

void LightController_enter_OnState(LightController* const me, HighLevelLightStates
newState) {
  LightController_enterNestedAndStates(me, newState);
}

void LightController_enter_EmergencyState(LightController* const me, HighLevelLight-
States newState) {
  LightController_enterNestedAndStates(me, newState);
}

void LightController_exit_OffState(LightController* const me) {
  LightController_exitNestedAndStates(me);
}

void LightController_exit_OnState(LightController* const me) {
  LightController_exitNestedAndStates(me);
}

void LightController_exit_EmergencyState(LightController* const me) {
  LightController_exitNestedAndStates(me);
}

void LightController_enterNestedAndStates(LightController* const me, HighLevelLight-
States next) {
  ErrorStateClass* sPtr;
  ActionPtr0 aPtr;
  int j;

  if (me->asList[next]) {
    /* and-states to enter? */
    for (j=0; j<me->asList[next]->nAndStates; j++ ) {
        sPtr = me->asList[next]->andStates[j];
        if (sPtr) {
            aPtr = sPtr->aSet.entryAction;
            if (aPtr)
               aPtr(sPtr);
            sPtr->currentState = sPtr->defaultState;
            };
        };
    };
}
```

```
void LightController_exitNestedAndStates(LightController* const me) {
  ErrorStateClass* sPtr;
  ActionPtr0 aPtr;
  int j;

  if (me->asList[me->currentState]) { /* any nested states to exit? */
    for (j=0; j<me->asList[me->currentState]->nAndStates; j++ ) {
      sPtr = me->asList[me->currentState]->andStates[j];
      if (sPtr) {
        aPtr = sPtr->aSet.exitAction;
        if (aPtr)
          aPtr(sPtr);
        sPtr->currentState = NULL_ERROR_STATE;
        }; // if sPtr
      }; // for
    }; // if asList
}

struct Klaxon* LightController_getItsKlaxon(const LightController* const me) {
  return (struct Klaxon*)me->itsKlaxon;
}

void LightController_setItsKlaxon(LightController* const me, struct Klaxon* p_Klaxon) {
  me->itsKlaxon = p_Klaxon;
}

struct Light* LightController_getItsLight(const LightController* const me) {
  return (struct Light*)me->itsLight;
}

void LightController_setItsLight(LightController* const me, struct Light* p_Light) {
  me->itsLight = p_Light;
}

struct Mutex* LightController_getItsMutex(const LightController* const me) {
  return (struct Mutex*)me->itsMutex;
}

void LightController_setItsMutex(LightController* const me, struct Mutex* p_Mutex) {
  me->itsMutex = p_Mutex;
}

LightController * LightController_Create(void) {
  LightController* me = (LightController *)
malloc(sizeof(LightController));
  if(me!=NULL)
    LightController_Init(me);
  return me;
}

void LightController_Destroy(LightController* const me) {
  if(me!=NULL)
    LightController_Cleanup(me);
  free(me);
}
```

```
static void cleanUpRelations(LightController* const me) {
  if(me->itsKlaxon != NULL)
    me->itsKlaxon = NULL;
  if(me->itsLight != NULL)
    me->itsLight = NULL;
  if(me->itsMutex != NULL)
    me->itsMutex = NULL;
}
```

Code Listing 5-30: LightController.c

The AndStateList class provides a level of indirection to a ConcreteAndState class that implements the state machine for all its nested states. It primarily provides an array of pointers to ConcreteAndState classes. Its code appears in Code Listing 5-31 and Code Listing 5-32.

```
#ifndef AndStateList_H
#define AndStateList_H

#include "LightPkg.h"

struct ErrorStateClass;

typedef struct AndStateList AndStateList;
struct AndStateList {
  struct ErrorStateClass* andStates[MAX_ANDSTATES];
  int nAndStates;
};

void AndStateList_Init(AndStateList* const me);
void AndStateList_Cleanup(AndStateList* const me);
AndStateList * AndStateList_Create(void);
void AndStateList_Destroy(AndStateList* const me);

#endif
```

Code Listing 5-31: AndStateList.h

```
#include "AndStateList.h"
#include "ErrorStateClass.h"

void AndStateList_Init(AndStateList* const me) {
  me->nAndStates = 0;
  int j;
  for (j=0; j<MAX_ANDSTATES; j++)
    me->andStates[j] = NULL;
}

void AndStateList_Cleanup(AndStateList* const me) {
}

AndStateList * AndStateList_Create(void) {
```

```
   AndStateList* me = (AndStateList *)
malloc(sizeof(AndStateList));
  if(me!=NULL)
      AndStateList_Init(me);
  return me;
}

void AndStateList_Destroy(AndStateList* const me) {
  if(me!=NULL)
      AndStateList_Cleanup(me);
  free(me);
}
```

Code Listing 5-32: AndStateList.c

Similarly, the StateActionSet is a collection of function pointers, one per event handler plus a couple of the entry and exit actions and the is_in() function (which identifies if the state class is in the specified state). The reason I made it a class is to provide the initialization behavior (setting all the pointers to NULL).

```
#ifndef StateActionSet_H
#define StateActionSet_H

#include "LightPkg.h"

typedef struct StateActionSet StateActionSet;
struct StateActionSet {
  ActionPtr0 entryAction;
  ActionPtr0 evAbortHandler;
  ActionPtr0 evDisableHandler;
  ActionPtr0 evEnableHandler;
  ActionPtr0 evErrorHandler;
  ActionPtr0 evOkHandler;
  ActionPtr0 evRunHandler;
  ActionPtr0 evShutDownHandler;
  ActionPtr0 evWarningHandler;
  ActionPtr0 exitAction;
  ActionPtr1 is_inHandler;
};

void StateActionSet_Init(StateActionSet* const me);
void StateActionSet_Cleanup(StateActionSet* const me);
StateActionSet * StateActionSet_Create(void);
void StateActionSet_Destroy(StateActionSet* const me);

#endif
```

Code Listing 33: StateActionSet.h

```
#include "StateActionSet.h"

void StateActionSet_Init(StateActionSet* const me) {
  me->entryAction = NULL;
  me->evAbortHandler = NULL;
  me->evDisableHandler = NULL;
  me->evEnableHandler = NULL;
  me->evErrorHandler = NULL;
  me->evOkHandler = NULL;
  me->evRunHandler = NULL;
  me->evShutDownHandler = NULL;
  me->evWarningHandler = NULL;
  me->exitAction = NULL;
  me->is_inHandler = NULL;
}

void StateActionSet_Cleanup(StateActionSet* const me) {
}

StateActionSet * StateActionSet_Create(void) {
  StateActionSet* me = (StateActionSet *)
malloc(sizeof(StateActionSet));
  if(me!=NULL)
      StateActionSet_Init(me);
  return me;
}

void StateActionSet_Destroy(StateActionSet* const me) {
  if(me!=NULL)
      StateActionSet_Cleanup(me);
  free(me);
}
```

Code Listing 34: StateActionSet.c

The last four code lists are for the header and implementation files of the ErrorStateClass and ModeStateClass, which each implement one of the orthogonal regions of the composite state On. Note that the aSet attribute variable is in the same position for both these classes, meaning that we can invoke them using a cast pointer without getting into trouble – that is, if we have a pointer to an ErrorStateClass struct, in reality it might be a pointer to a ModeStateClass but it still works because the aSet that holds the function pointers to the functions to invoke is in the same position offset in both classes.

```
#ifndef ErrorStateClass_H
#define ErrorStateClass_H

#include "LightPkg.h"
#include "StateActionSet.h"

struct LightController;
```

```
typedef struct ErrorStateClass ErrorStateClass;
struct ErrorStateClass {
  struct StateActionSet aSet;
  ErrorStatesType currentState;
  ErrorStatesType defaultState;
  struct LightController* itsLightController;
};

void ErrorStateClass_Init(ErrorStateClass* const me);
void ErrorStateClass_Cleanup(ErrorStateClass* const me);

/* Operations */
void ErrorStateClass_entryAction(ErrorStateClass* const me);
int ErrorStateClass_evAbort(ErrorStateClass* const me);
int ErrorStateClass_evError(ErrorStateClass* const me);
int ErrorStateClass_evOk(ErrorStateClass* const me);
int ErrorStateClass_evWarning(ErrorStateClass* const me);

void ErrorStateClass_exitAction(ErrorStateClass* const me);
int ErrorStateClass_is_in(ErrorStateClass* const me, ErrorStatesType s);

struct LightController* ErrorStateClass_getItsLightController(const ErrorStateClass*
const me);

void ErrorStateClass_setItsLightController(ErrorStateClass* const me, struct LightCon-
troller* p_LightController);

ErrorStateClass * ErrorStateClass_Create(void);

void ErrorStateClass_Destroy(ErrorStateClass* const me);
#endif
```

Code Listing 5-35: ErrorStateClass.h

```
#include "ErrorStateClass.h"
#include "LightController.h"

static void cleanUpRelations(ErrorStateClass* const me);

void ErrorStateClass_Init(ErrorStateClass* const me) {
  me->currentState = NULL_ERROR_STATE;
  me->defaultState = ERROR_STATE;
  StateActionSet_Init(&(me->aSet));
  me->itsLightController = NULL;
  /*
  set up the aSet function pointers
  remember that the initializer for the StateActionSet
  sets all pointers to NULL as the default

  Note: In this implementation, pointers are only set to non-NULL
  for actual handlers (except for entry() and exit()). The caller
  can tell if the event is handled by checking if the pointer is NULL.

  The higher level On state handles events reset and evDisable.
  */
```

```
  me->aSet.evOkHandler = (ActionPtr0) ErrorStateClass_evOk;
  me->aSet.evWarningHandler = (ActionPtr0) ErrorStateClass_evWarning;
  me->aSet.evErrorHandler = (ActionPtr0) ErrorStateClass_evError;
  me->aSet.evAbortHandler = (ActionPtr0) ErrorStateClass_evAbort;
  me->aSet.entryAction = (ActionPtr0) ErrorStateClass_entryAction;
  me->aSet.exitAction = (ActionPtr0) ErrorStateClass_exitAction;
  me->aSet.is_inHandler = (ActionPtr1) ErrorStateClass_is_in
}

void ErrorStateClass_Cleanup(ErrorStateClass* const me) {
  cleanUpRelations(me);
}

void ErrorStateClass_entryAction(ErrorStateClass* const me) {
}

int ErrorStateClass_evAbort(ErrorStateClass* const me) {
  switch(me->currentState) {
    case OK_STATE:
    case WARNING_STATE:
    case ERROR_STATE:

  case NULL_ERROR_STATE:
       break;
    };
  return 1;
}

int ErrorStateClass_evError(ErrorStateClass* const me) {
  switch(me->currentState) {
    case OK_STATE:
       Light_setColor(me->itsLightController->itsLight, RED);
       me->currentState = ERROR_STATE;
       break;
    case WARNING_STATE:
       Light_setColor(me->itsLightController->itsLight, RED);
       me->currentState = ERROR_STATE;
       break;
    case ERROR_STATE:
       case NULL_ERROR_STATE:
       break;
    };
  return 1;
}

int ErrorStateClass_evOk(ErrorStateClass* const me) {
  switch(me->currentState) {
       Light_setColor(me->itsLightController->itsLight, GREEN);
       me->currentState = OK_STATE;
       break;
    case ERROR_STATE:
       Light_setColor(me->itsLightController->itsLight, GREEN);
       me->currentState = OK_STATE;
       break;
```

```
      case WARNING_STATE:
      case OK_STATE:
        case NULL_ERROR_STATE:
          break;
      };
  return 1;
  /*#]*/
}

/*## operation evWarning() */
int ErrorStateClass_evWarning(ErrorStateClass* const me) {
  /*#[ operation evWarning() */
  switch(me->currentState) {
     case OK_STATE:
        Light_setColor(me->itsLightController->itsLight, YELLOW);
        me->currentState = WARNING_STATE;
        break;
     case ERROR_STATE:
        Light_setColor(me->itsLightController->itsLight, YELLOW);
        me->currentState = WARNING_STATE;
        break;
     case WARNING_STATE:
   case NULL_ERROR_STATE:
        break;
     };
  return 1;
}

void ErrorStateClass_exitAction(ErrorStateClass* const me) {
}

int ErrorStateClass_is_in(ErrorStateClass* const me, ErrorStatesType s) {
  return s == me->currentState;
}

struct LightController* ErrorStateClass_getItsLightController(const ErrorStateClass*
const me) {
  return (struct LightController*)me->itsLightController;
}

void ErrorStateClass_setItsLightController(ErrorStateClass* const me, struct LightCon-
troller* p_LightController) {
  me->itsLightController = p_LightController;
}

ErrorStateClass * ErrorStateClass_Create(void) {
  ErrorStateClass* me = (ErrorStateClass *)
malloc(sizeof(ErrorStateClass));
  if(me!=NULL)
      ErrorStateClass_Init(me);
  return me;
}

void ErrorStateClass_Destroy(ErrorStateClass* const me) {
  if(me!=NULL)
```

```
        ErrorStateClass_Cleanup(me);
    free(me);
}

static void cleanUpRelations(ErrorStateClass* const me) {
    if(me->itsLightController != NULL)
    me->itsLightController = NULL;
}
```

Code Listing 5-36: ErrorState.c

```
#ifndef ModeStateClass_H
#define ModeStateClass_H

#include "LightPkg.h"
#include "StateActionSet.h"
struct LightController;

typedef struct ModeStateClass ModeStateClass;
struct ModeStateClass {
    struct StateActionSet aSet;
    ModeStatesType currentState;
    ModeStatesType defaultState;
    struct LightController* itsLightController;
};

void ModeStateClass_Init(ModeStateClass* const me);

void ModeStateClass_Cleanup(ModeStateClass* const me);

/* Operations */
void ModeStateClass_entryAction(ModeStateClass* const me);
int ModeStateClass_evAbort(ModeStateClass* const me);
int ModeStateClass_evRun(ModeStateClass* const me);
int ModeStateClass_evShutDown(ModeStateClass* const me);
void ModeStateClass_exitAction(ModeStateClass* const me);
int ModeStateClass_is_in(ModeStateClass* const me, ModeStatesType s);

struct LightController* ModeStateClass_getItsLightController(const ModeStateClass*
const me);

void ModeStateClass_setItsLightController(ModeStateClass* const me, struct LightCon-
troller* p_LightController);

ModeStateClass * ModeStateClass_Create(void);

void ModeStateClass_Destroy(ModeStateClass* const me);
#endif
```

Code Listing 5-37: ModeStateClass.h

```
#include "ModeStateClass.h"
#include "LightController.h"

static void cleanUpRelations(ModeStateClass* const me);

void ModeStateClass_Init(ModeStateClass* const me) {
```

```
    me->currentState = NULL_MODE_STATE;
    me->defaultState = STARTINGUP_STATE;
    {
        StateActionSet_Init(&(me->aSet));
    }
    me->itsLightController = NULL;
    /*
    set up the aSet function pointers
    remember that the initializer for the StateActionSet
    sets all pointers to NULL as the default

    Note: In this implementation, pointers are only set to non-NULL
    for actual handlers (except for entry() and exit()). The caller
    can tell if the event is handled by checking if the pointer is NULL.

    The higher level On state handles events reset and evDisable.
    */
    me->aSet.evRunHandler = (ActionPtr0) ModeStateClass_evRun;
    me->aSet.evShutDownHandler = (ActionPtr0) ModeStateClass_evShutDown;
    me->aSet.evAbortHandler = (ActionPtr0) ModeStateClass_evAbort;
    me->aSet.entryAction = (ActionPtr0) ModeStateClass_entryAction;
    me->aSet.exitAction = (ActionPtr0) ModeStateClass_exitAction;
    me->aSet.is_inHandler = (ActionPtr1) ModeStateClass_is_in;
}
void ModeStateClass_Cleanup(ModeStateClass* const me) {
    cleanUpRelations(me);
}

void ModeStateClass_entryAction(ModeStateClass* const me) {
}

int ModeStateClass_evAbort(ModeStateClass* const me) {
    return 1;
}
int ModeStateClass_evRun(ModeStateClass* const me) {
    switch (me->currentState) {
        case STARTINGUP_STATE:
            Light_setMode(me->itsLightController->itsLight, STEADY);
            me->currentState = OPERATIONAL_STATE;
            break;
        case SHUTTINGDOWN_STATE:
            Light_setMode(me->itsLightController->itsLight, QUICKLY);
            me->currentState = STARTINGUP_STATE;
            break;
        case OPERATIONAL_STATE:
        case NULL_MODE_STATE:
            break;
    };
    return 1;
}

int ModeStateClass_evShutDown(ModeStateClass* const me) {
```

```
    switch (me->currentState) {
      case OPERATIONAL_STATE:
          Light_setMode(me->itsLightController->itsLight, SLOWLY);
          me->currentState = SHUTTINGDOWN_STATE;
          break;
      case STARTINGUP_STATE:
          Light_setMode(me->itsLightController->itsLight, SLOWLY);
          me->currentState = SHUTTINGDOWN_STATE;
          break;
      case SHUTTINGDOWN_STATE:
          break;
      case NULL_MODE_STATE:
          break;
      };
   return 1;
}

void ModeStateClass_exitAction(ModeStateClass* const me) {
}

int ModeStateClass_is_in(ModeStateClass* const me, ModeStatesType s) {
   return s == me->currentState;
}

struct LightController* ModeStateClass_getItsLightController(const ModeStateClass*
const me) {
   return (struct LightController*)me->itsLightController;
}

void ModeStateClass_setItsLightController(ModeStateClass* const me, struct LightCon-
troller* p_LightController) {
   me->itsLightController = p_LightController;
}

ModeStateClass * ModeStateClass_Create(void) {
   ModeStateClass* me = (ModeStateClass *)
malloc(sizeof(ModeStateClass));
   if(me!=NULL)
      ModeStateClass_Init(me);
   return me;
}

void ModeStateClass_Destroy(ModeStateClass* const me) {
   if(me!=NULL)
   ModeStateClass_Cleanup(me);
   free(me);
}

static void cleanUpRelations(ModeStateClass* const me) {
   if(me->itsLightController != NULL)
   me->itsLightController = NULL;
}
```

Code Listing 5-38: ModeStateClass.c

Figure 5-25 shows a sample run of the model shown in Figure 5-24 in which the `LightControllerClient` sends the events `evEnable`, `evError`, `evRun`, and then `evAbort`. Note the use of the `is_in()` function to make sure that the preconditions of the join are met before the event is taken.

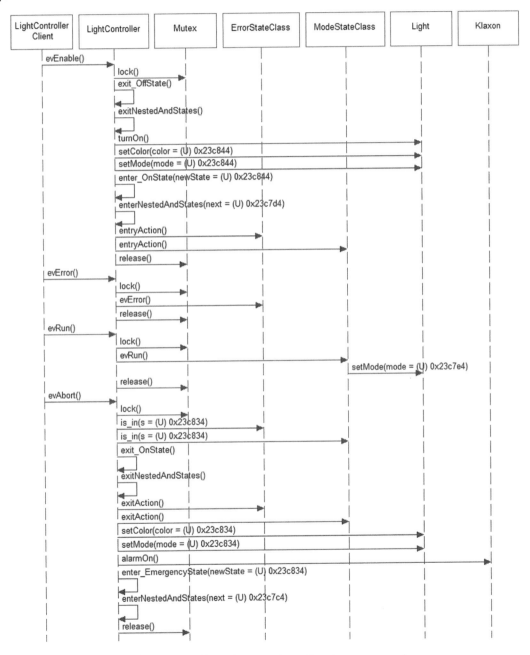

Figure 5-25: Sample execution of and-state example

5.9 OK, What Have We Learned?

This chapter presents a number of patterns for the implementation of state machines. Each of these implementation patterns offers benefits and costs, so which you use depends upon what you are trying to optimize. The Single Event Receptor Pattern uses a single event receptor and implements the state behavior internally as a large switch-case statement. It requires the creation of a type associated with the event to pass to the receptor. In contrast, the Multiple Event Receptor Pattern uses separate event handlers for each event so that the event type is not explicitly represented. This makes the interface to the client slightly richer. The State Table Pattern scales up to large state spaces and provides performance that is independent of the size of the state space – good for large state spaces. However, it takes more set-up time than the other approaches and may waste memory for sparse state spaces. The State Pattern creates separate objects for each state. The implementation is less complex than for the State Table Pattern and distributes the complexity of the state machine across several objects. The Decomposed And-State Pattern provides a straightforward way of implementing AND-states but requires the use of polymorphism.

Safety and Reliability Patterns

Design Patterns for Embedded Systems in C
DOI:10.1016/B978-1-85617-707-8.00006-6
Copyright © 2011 Elsevier Inc.

Patterns in this chapter

One's Complement Pattern – Adds a bitwise inverted copy of primitive data elements to identify when data is corrupted *in vivo*

CRC Pattern – Adds a cyclic redundancy check to identify when bits of the data have been corrupted *in vivo*

Smart Data Pattern – Adds behavior to the data to ensure that the data's preconditions and constraints are adhered to

Channel Pattern – Arranges the processing of sensory data as a series of transformational steps to provide a unit of large-scale redundancy

Protected Single Channel Pattern – Elaborates the channel pattern by adding data and processing validation checks at various points within the single channel

Dual Channel Pattern – Creates multiple channels to aid in the identification of errors and failures and, optionally, allows continuation of service in the presence of such as fault

---

6.1 A Little Bit About Safety and Reliability

Reliability is a measure of the "up-time" or "availability" of a system – specifically, it is the probability that a computation will successfully complete before the system fails. Put another way, it is a stochastic measure of the probability that the system can deliver a service. It is normally estimated with Mean Time Between Failure (MTBF) or a probability of successful service delivery (e.g., 0.9999). Reliability analysis is normally managed within Fault Means and Effect Analysis (FMEA) such as shown in Figure 6.1.

Safety is distinct from reliability. A safe system is one that does not incur too much risk to persons or equipment. A *hazard* is an event or condition that leads to an *accident* or *mishap* – a loss of some kind, such as injury or death. The risk associated with a hazard is the product of the severity of the hazard and its likelihood.

Hazards arise in five fundamental ways:

- Release of energy
- Release of toxins
- Interference of life support functions
- Supplying misleading information to safety personnel or control systems
- Failure to alarm when hazardous conditions arise

Description	Function	Failure Rate / million hrs	Failure Mode	Failure Effects	Potential Cases	Corrective Actions	Criticality
O2 Sensor	Senses O2 conc in the blood	110	LED fails	System cannot determine O2 conc	• Power disconnect • LED emitter fails	• use gold contacts • Use high rated LED	High
		1750	Connection to system fails	System cannot determine O2 conc	• RS232 wire disconnect • EMI	• Use high rated connectors • Use message CRCs	High
Display of O2 conc	Displays patient data to physician	6000	Wrong value displayed	Misdiagnosis and mistreatment of acute conditions	• Internal data corruption • Computation error	• Data stored redundantly • Computation done with two separate algorithms	High
...							
...							
...							
...							

Figure 6-1: FMEA example

6.1.1 Safety and Reliability Related Faults

Accidents can occur either because the system contains errors or because some aspect of the system fails. An error is a systematic fault and may be due to incorrect requirements, poor design or poor implementation. An error is always present in the system even though it may not be visible. When an error is visible, it is said to be *manifest*. Failures are different in that they occur at some point in time and result in a failed state. That is, the system worked properly before but something changed, such as a metal rod breaking in the drive train of your car (a *persistent failure*) or a bit was flipped in your car's engine control computer but is repaired as the system runs (a *volatile failure*). A fault is either an error or a failure.

I make this distinction about the different kinds of faults because the kinds of mitigation behavior for the two kinds is different. Specifically, failures can be addressed by *homogeneous redundancy* – multiple replicas of the processing channel (a set of software elements). If you assume the system is safe enough if it remains safe in the case of a single point failure, then having an identical backup channel can address such a failure. However, if the fault is an error, then all identical copies of the channel will exhibit the same fault. In this case, you must have *heterogeneous redundancy* (also known as *n-way programming*) – multiple channels that provide the service using different code and/or designs-in order to continue functioning safely.

Faults have relations to various elements in your system. Behavior around faults may be specified in requirements. An element can be the source of the fault. Another element may be the identifier of the fault. Yet another element may mitigate or extenuate the fault.

I believe that there is no such thing as "safe software" because it is the composite system – the electronics, mechanical, and software aspects working together in their operational environment – that is safe or unsafe. Hardware faults can be mitigated with software and vice versa. The system as a whole is safe or unsafe. Similarly, the reliability of a system is a function of all its constituent parts, since other parts can be designed to continue to provide services normally provided by failed parts. There is extensive literature on how to assess safety and reliability and how to construct safe and reliable systems, but a detailed discussion of those topics is beyond the scope of this book. More information can be had in my *Real-Time UML Workshop for Embedded Systems* or my *Real-Time Agility* books[1] as well as many other sources.

6.1.2 Achieving Safety and Reliability

In general, all safety-critical systems and high-reliability systems must contain and properly manage redundancy to achieve their safety and reliability requirements. How these elements are managed is different depending on the particular mix of safety and reliability concerns. This redundancy can be "redundancy-in-the-large" (architectural redundancy) or "redundancy-in-the-small" (detailed design redundancy). In this chapter we will consider a few patterns of both varieties. More safety-related architectural patterns can be found in my book *Real-Time Design Patterns*[2].

In the safety and reliability patterns in this chapter, there are three kinds of related elements. *Fault Manifestors* are elements that can, if they contain a fault, lead to a safety or reliability concern. *Fault Identifiers* are elements that can identify when a failure has occurred or when an error has become manifest. *Fault Extenuators* are elements that extenuate (reduce) a fault by either making the fault-related hazard less severe or by making the manifestation of the fault at the system level less likely. The patterns in this chapter differ in how these elements are organized and managed. Feedforward error correction patterns address the concern by providing enough redundancy to identify and deal with the fault and continue processing. Feedback error correction patterns address the faults by repeating one or more computational steps. Fault-safe state patterns address safety by going to a fault-safe state, but this reduces the reliability of the system[3].

And now, let's get to the patterns. First we'll start with some detailed design-level patterns, which some people call *idioms*.

---

[1] Douglass, B.P., 2009. *Real-Time UML Workshop for Embedded Systems*. Elsevier. Douglass, B.P., 2009. *Real-Time Agility*. Addison-Wesley.

[2] Douglass, B.P., 2002. *Real-Time Design Patterns: Robust Scalable Architecture for Real-Time Systems*. Addison-Wesley.

[3] In general, safety and reliability are opposing concerns when there is a fault-safe state and they are cooperating concerns when there is not.

6.2 One's Complement Pattern

The One's Complement Pattern is useful to detect when memory is corrupted by outside influences or hardware faults.

6.2.1 Abstract

The One's Complement Pattern provides a detailed design pattern for the identification of single or multiple memory bit corruption. This can occur due to EMI (Electromagnetic interference), heat, hardware faults, software faults, or other external causes. The pattern works by storing critical data twice – once in normal form and once in one's complement (bit-wise inverted) form. When the data are read, the one's complement form is reinverted and then compared to the value of the normal form. If the values match, then that value is returned, otherwise error processing ensues.

6.2.2 Problem

This pattern addresses the problem that variables may be corrupted by a variety of causes such as environmental factors (e.g., EMI, heat, radiation), hardware faults (e.g., power fluctuation, memory cell faults, address line shorts), or software faults (other software erroneously modifying memory). This pattern addresses the problem of identifying data corruption for small sets of critical data values.

6.2.3 Pattern Structure

The basic pattern structure is shown in Figure 6-2. Since this is a detailed design pattern, its scope is primarily within a single class.

6.2.4 Collaboration Roles

This section describes the roles for this pattern.

6.2.4.1 DataClient
The `DataClient` is an element that either sets or retrieves the value of the datum.

6.2.4.2 OnesCompProtectedDataElement
The `OnesCompProtectedDataElement` is the class that combines the storage of the data with services that can detect bit-level data corruption. This class stores the values of the data twice; once in regular form and once as a bit-wise complement. When data are retrieved, inverted data are reinverted and compared to the value stored normally. If the values match, that value is returned; if not, the class provides a function to handle the error. The actual behavior of the error handler is system- and context-specific.

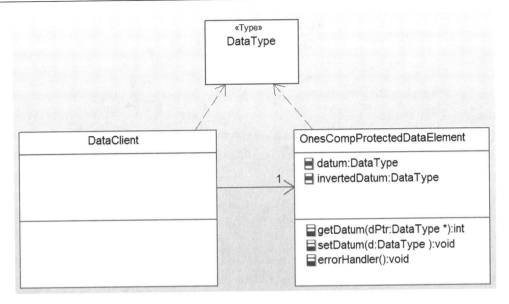

Figure 6-2: One's Complement Pattern

6.2.5 Consequences

This pattern uses twice the memory for storage of critical data; in addition, there is some performance hit on both data storage and retrieval. On the plus side, the pattern can identify errors due to EMI or due to stuck or shorted memory or address bits.

6.2.6 Implementation Strategies

This pattern is very easy to implement in C. For primitive types, the ~ operator computes the bit inversion. For structured types, you must iterate over the primitive parts of the data (see the example below).

6.2.7 Related Patterns

This pattern provides a very reliable way to identify faults that affect a single memory location. For very large data structures, the memory impact of complete data replication may be too large; in such cases, the Cyclic Redundancy Check (CRC) Pattern provides a reasonably reliable alternative.

6.2.8 Example

Figure 6-3 shows a simple application of this pattern. OwnShipAttitude stores the attitude (orientation, as specified in its attributes roll, pitch, and yaw) using the One's Complement Pattern. When the getAttitude() function is called, it reinverts the inverted copy and

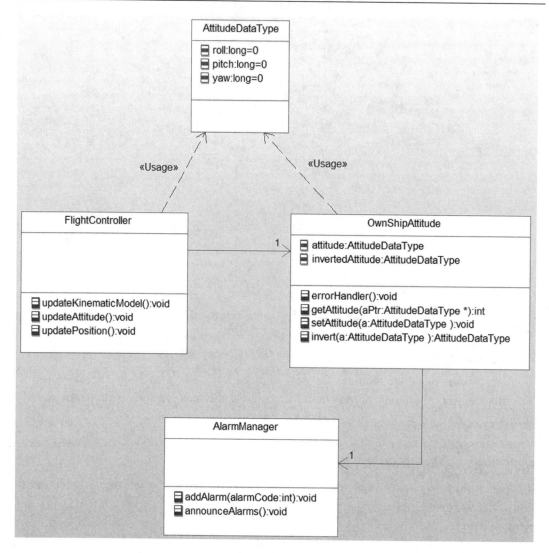

Figure 6-3: One's Complement example

compares it to normally stored values. If there is a difference, then the errorHandler() function is invoked which, in this case, calls the addAlarm() function of the AlarmHandler. If the values are the same, the value is returned to the client.

The code for the OwnShipAttitude (fulfilling the role of the OnesCompProtectedDataElement in the pattern) is shown in Code Listing 6-1 and Code Listing 6-2. Note that the error code passed to the alarm manager ATTITUDE_MEMORY_FAULT is simply a #define-d value.

```
#ifndef OwnShipAttitude_H
#define OwnShipAttitude_H

#include "AttitudeDataType.h"
struct AlarmManager;

typedef struct OwnShipAttitude OwnShipAttitude;
struct OwnShipAttitude {
  struct AttitudeDataType attitude;
  struct AttitudeDataType invertedAttitude;
  struct AlarmManager* itsAlarmManager;
};

void OwnShipAttitude_Init(OwnShipAttitude* const me);

void OwnShipAttitude_Cleanup(OwnShipAttitude* const me);

/* Operations */
void OwnShipAttitude_errorHandler(OwnShipAttitude* const me);

int OwnShipAttitude_getAttitude(OwnShipAttitude* const me, AttitudeDataType * aPtr);

AttitudeDataType OwnShipAttitude_invert(OwnShipAttitude* const me, AttitudeDataType a);

void OwnShipAttitude_setAttitude(OwnShipAttitude* const me, AttitudeDataType a);

struct AlarmManager* OwnShipAttitude_getItsAlarmManager(const OwnShipAttitude* const me);

void OwnShipAttitude_setItsAlarmManager(OwnShipAttitude* const
me, struct AlarmManager* p_AlarmManager);

OwnShipAttitude * OwnShipAttitude_Create(void);

void OwnShipAttitude_Destroy(OwnShipAttitude* const me);
#endif
```

Code Listing 6-1: OwnShipAttitude.h

```
#include "OwnShipAttitude.h"
#include "AlarmManager.h"

static void cleanUpRelations(OwnShipAttitude* const me);

void OwnShipAttitude_Init(OwnShipAttitude* const me) {
    AttitudeDataType_Init(&(me->attitude));
    AttitudeDataType_Init(&(me->invertedAttitude));
    me->itsAlarmManager = NULL;
}
void OwnShipAttitude_Cleanup(OwnShipAttitude* const me) {
```

```
    cleanUpRelations(me);
}

void OwnShipAttitude_errorHandler(OwnShipAttitude* const me) {
    AlarmManager_addAlarm(me->itsAlarmManager,
ATTITUDE_MEMORY_FAULT);
}

int OwnShipAttitude_getAttitude(OwnShipAttitude* const me,
AttitudeDataType * aPtr) {
    AttitudeDataType ia = OwnShipAttitude_invert(me, me-  >invertedAttitude);

    if (me->attitude.roll == ia.roll && me->attitude.yaw == ia.yaw &&
        me->attitude.pitch == ia.pitch ) {
        *aPtr = me->attitude;
        return 1;
        }
    else {
        OwnShipAttitude_errorHandler(me);
        return 0;
        };
}

AttitudeDataType OwnShipAttitude_invert(OwnShipAttitude* const
me, AttitudeDataType a) {
    a.roll = ~a.roll;
    a.yaw = ~a.yaw;
    a.pitch = ~a.pitch;
    return a;
}

void OwnShipAttitude_setAttitude(OwnShipAttitude* const me,
AttitudeDataType a) {
    me->attitude = a;
    me->invertedAttitude = OwnShipAttitude_invert(me, a);
}

struct AlarmManager* OwnShipAttitude_getItsAlarmManager(const
OwnShipAttitude* const me) {
    return (struct AlarmManager*)me->itsAlarmManager;
}

void OwnShipAttitude_setItsAlarmManager(OwnShipAttitude* const
me, struct AlarmManager* p_AlarmManager) {
    me->itsAlarmManager = p_AlarmManager;
}

OwnShipAttitude * OwnShipAttitude_Create(void) {
    OwnShipAttitude* me = (OwnShipAttitude *)
malloc(sizeof(OwnShipAttitude));
```

```
    if(me!=NULL)
      OwnShipAttitude_Init(me);
    return me;
  }

  void OwnShipAttitude_Destroy(OwnShipAttitude* const me) {
    if(me!=NULL)
      OwnShipAttitude_Cleanup(me);
    free(me);
  }

  static void cleanUpRelations(OwnShipAttitude* const me) {
    if(me->itsAlarmManager != NULL)
        me->itsAlarmManager = NULL;
  }
```

Code Listing 6-2: OwnShipAttitude.c

6.3 CRC Pattern

The Cyclic Redundancy Check (CRC) Pattern calculates a fixed-size error-detection code based on a cyclic polynomial that can detect corruption in data sets far larger than the size of the code.

6.3.1 Abstract

The CRC Pattern computes a fixed-length binary code, called a CRC value, on your data to detect whether or not they have been corrupted. This code is stored in addition to the data values and is set when data are updated and checked when data are read.

A detailed explanation of the mathematics is beyond the scope of this book[4] but CRCs are both common and useful in practice. CRCs are characterized by the bit-length of the polynomials used to compute them. While algorithmic computation can be complex and time consuming, table-driven algorithms are quite time efficient. CRCs provide good detection of single and multiple bit errors for arbitrary length data fields and so are good for large data structures.

6.3.2 Problem

This pattern addresses the problem that variables may be corrupted by a variety of causes such as environmental factors (e.g., EMI, heat, radiation), hardware faults (e.g., power fluctuation, memory cell faults, address line shorts), or software faults (other software erroneously modifying memory). This pattern addresses the problem of data corruption in large data sets.

---

[4] See for example, "A Painless Guide to CRC Error Detection Algorithms" at http://www.ross.net/crc/download/ crc_v3.txt for more information.

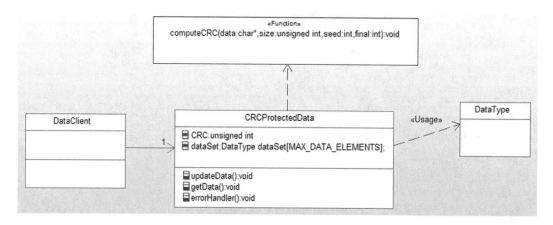

Figure 6-4: CRC Pattern

6.3.3 Pattern Structure

The CRC Pattern structure is shown in Figure 6-4. The CRCProtectedData class uses a generic computeCRC() function to calculate the CRC over its data. While data are shown as an array, it can be any contiguous block of data.

6.3.4 Collaboration Roles

This section describes the roles for this pattern.

6.3.4.1 computeCRCData Function

This is a general function for the computation of a CRC of a specific polynomial bit-length. Common bit lengths include 12, 16, and 32. In practice, this is usually implemented with a table-driven algorithm for computational efficiency.

6.3.4.2 CRCProtectedData

This is the class that owns and protects the data. When data are stored, the CRC is calculated. When data are read, the CRC is recalculated and compared to the stored CRC; if the CRCs differ, then the errorHandler() is called, otherwise the retrieved data are returned to the client.

6.3.4.3 DataClient

The DataClient is the element that stores and/or retrieves data from the CRCProtectedData class.

6.3.4.4 DataType

The is the base data type of the data structure owned by the CRCProtectedDataClass. This might be a primitive type, such as int or double, or it may be a complex data type such as an array or struct.

6.3.5 Consequences

With a little bit of memory used by the CRC data table and a relatively small amount of computation time to compute the polynomial, this pattern provides good detection of single (and small number of) bit errors. It is used often in communications messaging because such connections tend to be extremely unreliable. In addition, it can be used for in-memory error detection for harsh EMI environments or for mission-critical data. It is excellent for large data sets with relatively few bit errors.

6.3.6 Implementation Strategies

As mentioned, the table-driven algorithm is almost always used because it provides good performance with only a minor memory space penalty. If memory is tight, the polynomial can be computed algorithmically but this has a significant impact on run-time performance.

6.3.7 Related Patterns

The One's Complement Pattern replicates the data in a bit-inverted form, and so is useful for small data sets. This pattern is better for large data sets because it has a smaller memory impact.

6.3.8 Example

Figure 6-5 shows a medical application of this pattern. In this case, several clients set or get patient data, which is defined by the `PatientDataType`. The `PatientData` class provides functions to set and get the data, protecting them via calls to the `CRCCalculator`. A single CRC protects the entire patient data set.

The code listing in Code Listing 6-3 is adapted from free software written by Jack Klein, and released under the GNU General Public License Agreement[5]. It computes a CRC-16 but other polynomial bit-lengths are available for use.

```
/* crcccitt.c - a demonstration of look up table based CRC
 *              computation using the non-reversed CCITT_CRC
 *              polynomial 0x1021 (truncated)
 *
 * Copyright (C) 2000 Jack Klein
 *              Macmillan Computer Publishing
 *
```

---

[5] For information concerning usage of this code, contact the Free Software Foundation, Inc., 675 Mass Ave, Cambridge, MA 02139, USA

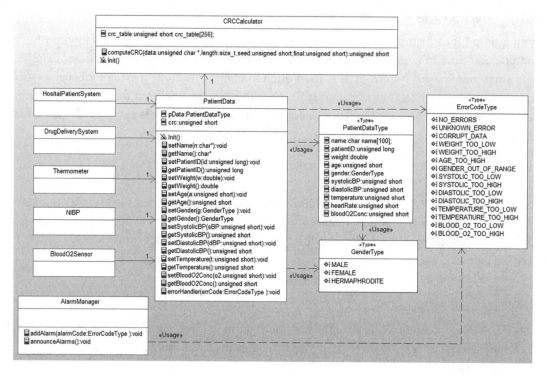

Figure 6-5: CRC example

```
* This program is free software; you can redistribute it
* and/or modify it under the terms of the GNU General
* Public License as published by the Free Software
* Foundation; either version 2 of the License, or
* (at your option) any later version.
*
* This program is distributed in the hope that it will
* be useful, but WITHOUT ANY WARRANTY; without even the
* implied warranty of MERCHANTABILITY or FITNESS FOR A
* PARTICULAR PURPOSE. See the GNU General Public License
* for more details.
*
* You should have received a copy of the GNU General
* Public License along with this program; if not, write
* to the Free Software Foundation, Inc., 675 Mass Ave,
* Cambridge, MA 02139, USA.
*
* Jack Klein may be contacted by email at:
*   The_C_Guru@yahoo.com
*
```

```
* Modified slightly by Bruce Powel Douglass, 2010
*/


/*# # dependency stdio */
#include <stdio.h>
/*# # dependency stdlib */
#include <stdlib.h>
/*# # dependency string */
#include <string.h>

static unsigned short crc_table[256] = {
  0x0000, 0x1021, 0x2042, 0x3063, 0x4084, 0x50a5,
  0x60c6, 0x70e7, 0x8108, 0x9129, 0xa14a, 0xb16b,
  0xc18c, 0xd1ad, 0xe1ce, 0xf1ef, 0x1231, 0x0210,
  0x3273, 0x2252, 0x52b5, 0x4294, 0x72f7, 0x62d6,
  0x9339, 0x8318, 0xb37b, 0xa35a, 0xd3bd, 0xc39c,
  0xf3ff, 0xe3de, 0x2462, 0x3443, 0x0420, 0x1401,
  0x64e6, 0x74c7, 0x44a4, 0x5485, 0xa56a, 0xb54b,
  0x8528, 0x9509, 0xe5ee, 0xf5cf, 0xc5ac, 0xd58d,
  0x3653, 0x2672, 0x1611, 0x0630, 0x76d7, 0x66f6,
  0x5695, 0x46b4, 0xb75b, 0xa77a, 0x9719, 0x8738,
  0xf7df, 0xe7fe, 0xd79d, 0xc7bc, 0x48c4, 0x58e5,
  0x6886, 0x78a7, 0x0840, 0x1861, 0x2802, 0x3823,
  0xc9cc, 0xd9ed, 0xe98e, 0xf9af, 0x8948, 0x9969,
  0xa90a, 0xb92b, 0x5af5, 0x4ad4, 0x7ab7, 0x6a96,
  0x1a71, 0x0a50, 0x3a33, 0x2a12, 0xdbfd, 0xcbdc,
  0xfbbf, 0xeb9e, 0x9b79, 0x8b58, 0xbb3b, 0xab1a,
  0x6ca6, 0x7c87, 0x4ce4, 0x5cc5, 0x2c22, 0x3c03,
  0x0c60, 0x1c41, 0xedae, 0xfd8f, 0xcdec, 0xddcd,
  0xad2a, 0xbd0b, 0x8d68, 0x9d49, 0x7e97, 0x6eb6,
  0x5ed5, 0x4ef4, 0x3e13, 0x2e32, 0x1e51, 0x0e70,
  0xff9f, 0xefbe, 0xdfdd, 0xcffc, 0xbf1b, 0xaf3a,
  0x9f59, 0x8f78, 0x9188, 0x81a9, 0xb1ca, 0xa1eb,
  0xd10c, 0xc12d, 0xf14e, 0xe16f, 0x1080, 0x00a1,
  0x30c2, 0x20e3, 0x5004, 0x4025, 0x7046, 0x6067,
  0x83b9, 0x9398, 0xa3fb, 0xb3da, 0xc33d, 0xd31c,
  0xe37f, 0xf35e, 0x02b1, 0x1290, 0x22f3, 0x32d2,
  0x4235, 0x5214, 0x6277, 0x7256, 0xb5ea, 0xa5cb,
  0x95a8, 0x8589, 0xf56e, 0xe54f, 0xd52c, 0xc50d,
  0x34e2, 0x24c3, 0x14a0, 0x0481, 0x7466, 0x6447,
  0x5424, 0x4405, 0xa7db, 0xb7fa, 0x8799, 0x97b8,
  0xe75f, 0xf77e, 0xc71d, 0xd73c, 0x26d3, 0x36f2,
  0x0691, 0x16b0, 0x6657, 0x7676, 0x4615, 0x5634,
  0xd94c, 0xc96d, 0xf90e, 0xe92f, 0x99c8, 0x89e9,
  0xb98a, 0xa9ab, 0x5844, 0x4865, 0x7806, 0x6827,
  0x18c0, 0x08e1, 0x3882, 0x28a3, 0xcb7d, 0xdb5c,
  0xeb3f, 0xfb1e, 0x8bf9, 0x9bd8, 0xabbb, 0xbb9a,
  0x4a75, 0x5a54, 0x6a37, 0x7a16, 0x0af1, 0x1ad0,
  0x2ab3, 0x3a92, 0xfd2e, 0xed0f, 0xdd6c, 0xcd4d,
  0xbdaa, 0xad8b, 0x9de8, 0x8dc9, 0x7c26, 0x6c07,
```

```
    0x5c64, 0x4c45, 0x3ca2, 0x2c83, 0x1ce0, 0x0cc1,
    0xef1f, 0xff3e, 0xcf5d, 0xdf7c, 0xaf9b, 0xbfba,
    0x8fd9, 0x9ff8, 0x6e17, 0x7e36, 0x4e55, 0x5e74,
    0x2e93, 0x3eb2, 0x0ed1, 0x1ef0
};

unsigned short computeCRC(unsigned char * data, size_t length, unsigned short seed,
unsigned short final) {
    size_t count;
    unsigned int crc = seed;
    unsigned int temp;

    for (count = 0; count < length; ++count)
    {
      temp = (*data++ ^ (crc >> 8)) & 0xff;
      crc = crc_table[temp] ^ (crc << 8);
    }
    return (unsigned short)(crc ^ final);
}
```

Code Listing 6-3: CRCCalculator.c

The data types used in the example appear in Code Listing 6-4. Note that `ErrorCodeType`
includes other errors, such as data element out of range, as is common in such applications.

```
#ifndef CRCExample_H
#define CRCExample_H

struct AlarmManager;
struct BloodO2Sensor;
struct CRCBuilderClass;
struct CRCClient2;
struct DrugDeliverySystem;
struct HositalPatientSystem;
struct NIBP;
struct PatientData;
struct Thermometer;

typedef enum ErrorCodeType {
  NO_ERRORS,
  UNKNOWN_ERROR,
  CORRUPT_DATA,
  WEIGHT_TOO_LOW,
  WEIGHT_TOO_HIGH,
  AGE_TOO_HIGH,
  GENDER_OUT_OF_RANGE,
  SYSTOLIC_TOO_LOW,
```

```
    SYSTOLIC_TOO_HIGH,
    DIASTOLIC_TOO_LOW,
    DIASTOLIC_TOO_HIGH,
    TEMPERATURE_TOO_LOW,
    TEMPERATIURE_TOO_HIGH,
    BLOOD_02_TOO_LOW,
    BLOOD_02_TOO_HIGH
} ErrorCodeType;

typedef enum GenderType {
    MALE,
    FEMALE,
    HERMAPHRODITE
} GenderType;

typedef struct PatientDataType {
    unsigned short age;
    unsigned short blood02Conc;
    unsigned short diastolicBP;
    GenderType gender;
    unsigned short heartRate;
    char name[100];
    unsigned long patientID;
    unsigned short systolicBP;
    unsigned short temperature;
    double weight;
} PatientDataType;

#endif
```

Code Listing 6-4: CRCExample.c

Of course the main action is in the `PatientData` class code, shown in Code Listing 6-5 and Code Listing 6-6.

```
#ifndef PatientData_H
#define PatientData_H

#include "CRCExample.h"
struct AlarmManager;

typedef struct PatientData PatientData;
struct PatientData {
    PatientDataType pData;
    unsigned short crc;
    struct AlarmManager* itsAlarmManager;
};
```

```
void PatientData_Init(PatientData* const me);
void PatientData_Cleanup(PatientData* const me);

/* Operations */

void PatientData_errorHandler(PatientData* const me,
ErrorCodeType errCode);

unsigned short PatientData_getAge(PatientData* const me);

unsigned short PatientData_getBloodO2Conc(PatientData* const me);

unsigned short PatientData_getDiastolicBP(PatientData* const me);

GenderType PatientData_getGender(PatientData* const me);

char* PatientData_getName(PatientData* const me);

unsigned long PatientData_getPatientID(PatientData* const me);

unsigned short PatientData_getSystolicBP(PatientData* const me);

unsigned short PatientData_getTemperature(PatientData* const me);

double PatientData_getWeight(PatientData* const me);

void PatientData_setAge(PatientData* const me, unsigned short a);

void PatientData_setBloodO2Conc(PatientData* const me, unsigned short o2);

void PatientData_setDiastolicBP(PatientData* const me, unsigned short dBP);

void PatientData_setGender(PatientData* const me, GenderType g);

void PatientData_setName(PatientData* const me, char* n);

void PatientData_setPatientID(PatientData* const me, unsigned long id);

void PatientData_setSystolicBP(PatientData* const me, unsigned short sBP);

void PatientData_setTemperature(PatientData* const me, unsigned short t);

void PatientData_setWeight(PatientData* const me, double w);

int PatientData_checkData(PatientData* const me);

struct AlarmManager* PatientData_getItsAlarmManager(const PatientData* const me);
```

```
void PatientData_setItsAlarmManager(PatientData* const me, struct
AlarmManager* p_AlarmManager);

PatientData * PatientData_Create(void);

void PatientData_Destroy(PatientData* const me);
#endif
```

Code Listing 6-5: PatientData.h

```
#include "PatientData.h"
#include "AlarmManager.h"
#include <string.h>
#include <stdio.h>

static void cleanUpRelations(PatientData* const me);

void PatientData_Init(PatientData* const me) {
  me->itsAlarmManager = NULL;
  strcpy(me->pData.name, " ");
  me->pData.patientID = 0;
  me->pData.weight = 0;
  me->pData.age = 0;
  me->pData.gender = HERMAPHRODITE;
  me->pData.systolicBP = 0;
  me->pData.diastolicBP = 0;
  me->pData.temperature = 0;
  me->pData.heartRate = 0;
  me->pData.bloodO2Conc = 0;
  me->crc = computeCRC((unsigned char *)&me->pData, sizeof(me->pData), 0xffff, 0);
  }
}

void PatientData_Cleanup(PatientData* const me) {
  cleanUpRelations(me);
}

int PatientData_checkData(PatientData* const me) {
  unsigned short result = computeCRC((unsigned char *)&me->pData,
sizeof(me->pData), 0xffff, 0);
  printf("computed CRC = %d, stored CRC = %d\n", result, me->crc);
  return result == me->crc;
}

void PatientData_errorHandler(PatientData* const me,
ErrorCodeType errCode) {
  AlarmManager_addAlarm(me->itsAlarmManager, errCode);
}
```

```
unsigned short PatientData_getAge(PatientData* const me) {
  if (PatientData_checkData(me))
     return me->pData.age;
  else {
     PatientData_errorHandler(me, CORRUPT_DATA);
     return 0;
     };
}

unsigned short PatientData_getBloodO2Conc(PatientData* const me)
{
  if (PatientData_checkData(me))
     return me->pData.bloodO2Conc;
  else {
     PatientData_errorHandler(me, CORRUPT_DATA);
     return 0;
     };
}

unsigned short PatientData_getDiastolicBP(PatientData* const me)
{
  if (PatientData_checkData(me))
     return me->pData.diastolicBP;
  else {
     PatientData_errorHandler(me, CORRUPT_DATA);
     return 0;
     };
}

GenderType PatientData_getGender(PatientData* const me) {
  if (PatientData_checkData(me))
     return me->pData.gender;
  else {
     PatientData_errorHandler(me, CORRUPT_DATA);
     return 0;
     };
}

char* PatientData_getName(PatientData* const me) {
  if (PatientData_checkData(me))
     return me->pData.name;
  else {
     PatientData_errorHandler(me, CORRUPT_DATA);
     return 0;
     };
}

unsigned long PatientData_getPatientID(PatientData* const me) {
  if (PatientData_checkData(me))
     return me->pData.patientID;
```

```
    else {
        PatientData_errorHandler(me, CORRUPT_DATA);
        return 0;
        };
}

unsigned short PatientData_getSystolicBP(PatientData* const me) {
  if (PatientData_checkData(me))
      return me->pData.systolicBP;
  else {
      PatientData_errorHandler(me, CORRUPT_DATA);
      return 0;
      };
}


unsigned short PatientData_getTemperature(PatientData* const me)
{
  if (PatientData_checkData(me))
      return me->pData.temperature;
  else {
      PatientData_errorHandler(me, CORRUPT_DATA);
      return 0;
      };
}

double PatientData_getWeight(PatientData* const me) {
  if (PatientData_checkData(me))
      return me->pData.weight;
  else {
      PatientData_errorHandler(me, CORRUPT_DATA);
      return 0;
      };
}

void PatientData_setAge(PatientData* const me, unsigned short a) {
  if (PatientData_checkData(me)) {
      me->pData.age = a;
      me->crc = computeCRC((unsigned char *)&me->pData, sizeof(me->pData), 0xffff, 0);
      if (!PatientData_checkData(me))
          PatientData_errorHandler(me, CORRUPT_DATA);
      }
  else {
      printf("Set failed\n");
      PatientData_errorHandler(me, CORRUPT_DATA);
      };
}

void PatientData_setBloodO2Conc(PatientData* const me, unsigned
short o2) {
  if (PatientData_checkData(me)) {
```

```
       me->pData.bloodO2Conc = o2;
       me->crc = computeCRC((unsigned char *)&me->pData, sizeof(me->pData), 0xffff, 0);
       if (!PatientData_checkData(me))
           PatientData_errorHandler(me, CORRUPT_DATA);
       }
   else {
       printf("Set failed\n");
       PatientData_errorHandler(me, CORRUPT_DATA);
       };
}


void PatientData_setDiastolicBP(PatientData* const me, unsigned short dBP) {
   if (PatientData_checkData(me)) {
       me->pData.diastolicBP = dBP;
       me->crc = computeCRC((unsigned char *)&me->pData, sizeof(me->pData), 0xffff, 0);
       if (!PatientData_checkData(me))
           PatientData_errorHandler(me, CORRUPT_DATA);
       }
   else {
       printf("Set failed\n");
       PatientData_errorHandler(me, CORRUPT_DATA);
       };
}

void PatientData_setGender(PatientData* const me, GenderType g) {
   if (PatientData_checkData(me)) {
       me->pData.gender = g;
       me->crc = computeCRC((unsigned char *)&me->pData, sizeof(me->pData), 0xffff, 0);
       if (!PatientData_checkData(me))
           PatientData_errorHandler(me, CORRUPT_DATA);
       }
   else {
       printf("Set failed\n");
       PatientData_errorHandler(me, CORRUPT_DATA);
       };
}


void PatientData_setName(PatientData* const me, char* n) {
   if (PatientData_checkData(me)) {
       strcpy(me->pData.name, n);
       me->crc = computeCRC((unsigned char *)&me->pData, sizeof(me->pData), 0xffff, 0);
       if (!PatientData_checkData(me))
           PatientData_errorHandler(me, CORRUPT_DATA);
       }
   else {
       printf("Set failed\n");
       PatientData_errorHandler(me, CORRUPT_DATA);
       };
}

void PatientData_setPatientID(PatientData* const me, unsigned long id) {
   if (PatientData_checkData(me)) {
```

```
            me->pData.patientID = id;
            me->crc = computeCRC((unsigned char *)&me->pData, sizeof(me->pData), 0xffff, 0);
            if (!PatientData_checkData(me))
                PatientData_errorHandler(me, CORRUPT_DATA);

        }
    else {
        printf("Set failed\n");
        PatientData_errorHandler(me, CORRUPT_DATA);
        };
}
void PatientData_setSystolicBP(PatientData* const me, unsigned short sBP) {
    if (PatientData_checkData(me)) {
        me->pData.systolicBP = sBP;
        me->crc = computeCRC((unsigned char *)&me->pData, sizeof(me->pData), 0xffff, 0);
        if (!PatientData_checkData(me))
            PatientData_errorHandler(me, CORRUPT_DATA);

        }
    else {
        printf("Set failed\n");
        PatientData_errorHandler(me, CORRUPT_DATA);
        };
}
void PatientData_setTemperature(PatientData* const me, unsigned short t) {
    if (PatientData_checkData(me)) {
        me->pData.temperature = t;
        me->crc = computeCRC((unsigned char *)&me->pData, sizeof(me->pData), 0xffff, 0);
        if (!PatientData_checkData(me))
            PatientData_errorHandler(me, CORRUPT_DATA);

        }
    else {
        printf("Set failed\n");
        PatientData_errorHandler(me, CORRUPT_DATA);
        };
}
void PatientData_setWeight(PatientData* const me, double w) {
    if (PatientData_checkData(me)) {
        me->pData.weight = w;
        me->crc = computeCRC((unsigned char *)&me->pData, sizeof(me->pData), 0xffff, 0);
        if (!PatientData_checkData(me))
            PatientData_errorHandler(me, CORRUPT_DATA);

        }
    else {
        printf("Set failed\n");
        PatientData_errorHandler(me, CORRUPT_DATA);
        };
}
struct AlarmManager* PatientData_getItsAlarmManager(const PatientData* const me) {
    return (struct AlarmManager*)me->itsAlarmManager;
}
```

```
void  PatientData_setItsAlarmManager(PatientData*  const  me,  struct  AlarmManager*
p_AlarmManager) {
  me->itsAlarmManager = p_AlarmManager;
}
PatientData * PatientData_Create(void) {
  PatientData* me = (PatientData *) malloc(sizeof(PatientData));
  if(me!=NULL)
    PatientData_Init(me);
  return me;
}

void PatientData_Destroy(PatientData* const me) {
  if(me!=NULL)
    PatientData_Cleanup(me);
  free(me);
}
static void cleanUpRelations(PatientData* const me) {
  if(me->itsAlarmManager != NULL)
    me->itsAlarmManager = NULL;
}
```

Code Listing 6-6: PatientData.c

6.4 Smart Data Pattern

One of the biggest problems I see with embedded C code in actual application is that functions all
have preconditions for their proper execution but these functions don't explicitly check that those
preconditons are actually satisfied. This falls under the heading of what is commonly known as
"defensive design" – a development paradigm in which we design with proactive run-time defenses
in place to detect problems. The Smart Data Pattern codifies that paradigm for scalar data elements.

6.4.1 Abstract

One of the reasons that languages such as Ada are considered "safer[6]" than C is that they have
run-time checking. These checks include array index range checking, definition of subtypes
and subranges (such as defining a subtype Color to be not only limited to a set of colors but
generate errors if that range is exceeded), and parameter range checks. Indeed although the C++
language is not inherently "safe" by this standard, because data structures may be bound
together with operations that provide run-time checking, it can be made "safe" by intention if
not by default. This pattern, although landing more within the accepted range of "idiom" rather
than "pattern," addresses those concerns by creating C classes that explicitly perform these
checks.

---

[6] Using the term loosely and not in the precise sense defined in this chapter.

6.4.2 Problem

The most common idiom for error checking of data ranges in C is for functions that have no other return value: return a 0 value for function success and a -1 value for failure and set `errno` to an error code.

The problem is, of course, that the most common way to "handle" the return values is to ignore them, resulting in difficult-to-debug systems. Some people even put error checking into the system during development but remove such checks in the final release. I would argue that the time when you *really* want to know that the altitude determination program in your aircraft isn't producing the right value is when you are in the flying aircraft.

The problem this pattern addresses is to build functions and data types that essentially check themselves and provide error detection means that cannot be easily ignored.

6.4.3 Pattern Structure

As mentioned above, this pattern falls into the idiomatic (tiny pattern) range rather than a collaboration pattern. The key concepts of the pattern are to 1) build self-checking types whenever possible; 2) check incoming parameter values for appropriate range checking; and 3) check consistency and reasonableness among one or a set of parameters. Figure 6-6 shows the structure of the pattern.

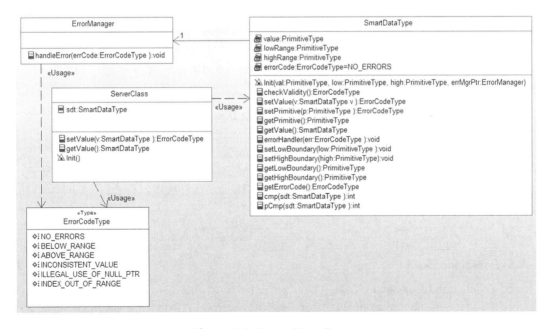

Figure 6-6: Smart Data Pattern

6.4.4 Collaboration Roles

This section describes the roles for this pattern.

6.4.4.1 ErrorCodeType

The ErrorCodeType is an enumeration of possible error codes. This can be, and is often, done with ints but enums clarify the intention and usage of the codes more clearly.

6.4.4.2 ErrorManager

The ErrorManager class handles the errors identified within the SubrangeType; the actions are system-, context-, and error-specific.

6.4.4.3 ServerClass

This role in the pattern is an element that uses the SmartDataType.

6.4.4.4 SmartDataType

One of the disadvantages of data structures is that while they all have preconditions and rules for proper usage, they have no intrinsic behavior to enforce the conditions and rules. This class binds together the data (of type PrimitiveType) with operations that ensure they stay in range.

This class provides a variety of functions. The SmartDataType_Init() function takes five parameters – the me pointer to point to the instance data, the initial value being set, the low and high values specifying the valid data subrange, and the address of the ErrorManager. If the entire range of the primitive type is valid, then you can simply use the primitive type or use the low and high values to the min and max for the type (such as SHRT_MIN and SHRT_MAX, or INT_MIN and INT_MAX declared in <limits.h>, for example). To create a new instance on the heap, simply use SmartDataType_Create(), which takes three parameters – the value, the low boundary, and the high boundary and returns a pointer to the newly created instance.

SmartDataTypes can be assigned to each other with the SmartDataType_setValue() and SmartDataType_getValue() functions and compared with the boundary with the SmartDataType_cmp() function and without the boundaries with the SmartDataType_pCmp() function. The held data can be independently set with the SmartDataType_setPrimitive() and SmartDataType_getPrimitive() functions. The low and high boundaries can also be set and read independently.

6.4.5 Consequences

The downside for using smart data types is the performance overhead for executing the operations. The upside is that data are self-protecting and provide automatic checking when data are set. It is also possible for the programmers to avoid using the functions and access the values directly if they are so inclined, defeating the purpose of the smart data type.

6.4.6 Implementation Strategies

The key implementation strategy is to create a struct with a set of defined operations, and then only use those operations to access the values. The pattern can be extended to include an enumerated type of the units of the value when dealing with numbers with units, such as currency (e.g., euro and dollar), mass (such as pounds, kilograms), and distance (such as inches, feet, yards, fathoms, miles, and even meters). This is a useful extension when dealing with data from different clients that may require conversion. The conversion functions can be built into the class so that you both set and get the values in a specified unit.

6.4.7 Related Patterns

The idea of building smart types is extended pointers with the Smart Pointer Pattern.

6.4.8 Example

Figure 6-7 shows a simple medical example of using smart data types. The `ServerClass` role from the pattern is fulfilled by the `PatientDataClass` in the figure. It has three *SmartInt* variables (weight, age, and heart rate) and two `SmartColor` variables (foreground color and background color). I included the use of an enumerated type `ColorType` to show how the pattern can be used for types other than built-in types.

Code Listing 6-7 gives the basic use types for the example.

```
#ifndef SmartDataExample_H
#define SmartDataExample_H

struct AlarmManager;
struct PatientDataClass;
struct SmartColor;
struct SmartDataClient;
struct SmartInt;

typedef enum ErrorCodeType {
    NO_ERRORS,
    BELOW_RANGE,
    ABOVE_RANGE,
    INCONSISTENT_VALUE,
    ILLEGAL_USE_OF_NULL_PTR,
    INDEX_OUT_OF_RANGE
} ErrorCodeType;
```

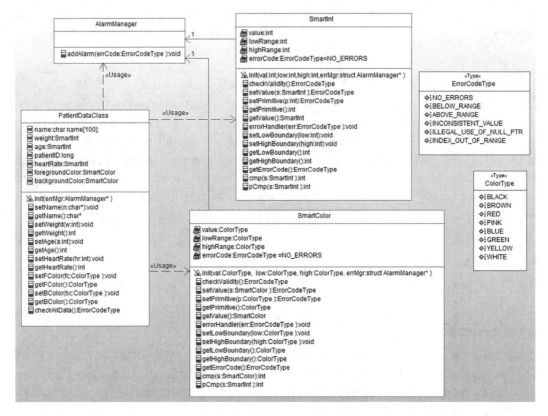

Figure 6-7: Smart Data example

```
typedef enum ColorType {
  BLACK,
  BROWN,
  RED,
  PINK,
  BLUE,
  GREEN,
  YELLOW,
  WHITE
}
ColorType;

#endif
```

Code Listing 6-7: SmartDataExample.h

The SmartInt data type is used for three variables in the PatientDataClass. It provides the basic functions identified in the SmartData role of the pattern, where int is the primitive type. Its code is shown in Code Listing 6-8 and Code Listing 6-9.

```
#ifndef SmartInt_H
#define SmartInt_H

#include "SmartDataExample.h"
struct AlarmManager;

typedef struct SmartInt SmartInt;
struct SmartInt {
  ErrorCodeType errorCode;
  int highRange;
  int lowRange;
  int value;
  struct AlarmManager* itsAlarmManager;
};

/* Constructors and destructors:*/
void SmartInt_Init(SmartInt* const me, int val, int low, int
high, struct AlarmManager* errMgr);

void SmartInt_Cleanup(SmartInt* const me);

/* Operations */
ErrorCodeType SmartInt_checkValidity(SmartInt* const me);

int SmartInt_cmp(SmartInt* const me, SmartInt s);

void SmartInt_errorHandler(SmartInt* const me, ErrorCodeType err);

ErrorCodeType SmartInt_getErrorCode(SmartInt* const me);

int SmartInt_getHighBoundary(SmartInt* const me);

int SmartInt_getLowBoundary(SmartInt* const me);

int SmartInt_getPrimitive(SmartInt* const me);

SmartInt SmartInt_getValue(SmartInt* const me);

int SmartInt_pCmp(SmartInt* const me, SmartInt s);

void SmartInt_setHighBoundary(SmartInt* const me, int high);

void SmartInt_setLowBoundary(SmartInt* const me, int low);

ErrorCodeType SmartInt_setPrimitive(SmartInt* const me, int p);
ErrorCodeType SmartInt_setValue(SmartInt* const me, SmartInt s);
struct AlarmManager* SmartInt_getItsAlarmManager(const SmartInt* const me);
```

```
void SmartInt_setItsAlarmManager(SmartInt* const me, struct
AlarmManager* p_AlarmManager);

SmartInt * SmartInt_Create(int val, int low, int high, struct AlarmManager* errMgr);

void SmartInt_Destroy(SmartInt* const me);
#endif
```

Code Listing 6-8: SmartInt.h

```
#include "SmartInt.h"
#include "AlarmManager.h"

static void cleanUpRelations(SmartInt* const me);

void SmartInt_Init(SmartInt* const me, int val, int low, int
high, struct AlarmManager* errMgr) {
  me->errorCode = NO_ERRORS;
  me->itsAlarmManager = NULL;

  me->lowRange = low;
  me->highRange = high;
  me->value = val;
  me->itsAlarmManager = errMgr;
}

void SmartInt_Cleanup(SmartInt* const me) {
  cleanUpRelations(me);
}

ErrorCodeType SmartInt_checkValidity(SmartInt* const me) {
  if (me->value < me->lowRange)
      return BELOW_RANGE;
  else if (me->value > me->highRange)
      return ABOVE_RANGE;
  else
      return NO_ERRORS;
}

int SmartInt_cmp(SmartInt* const me, SmartInt s) {
  return memcmp(me, &s, sizeof(s));
}

void SmartInt_errorHandler(SmartInt* const me, ErrorCodeType err) {
  AlarmManager_addAlarm(me->itsAlarmManager, err);
}

ErrorCodeType SmartInt_getErrorCode(SmartInt* const me) {
  return me->errorCode;
}
```

```
int SmartInt_getHighBoundary(SmartInt* const me) {
  return me->highRange;
}
int SmartInt_getLowBoundary(SmartInt* const me) {
  return me->lowRange;
}

int SmartInt_getPrimitive(SmartInt* const me) {
  return me->value;
}

SmartInt SmartInt_getValue(SmartInt* const me) {
  /*#[ operation getValue() */
  return *me;
  /*#]*/
}

int SmartInt_pCmp(SmartInt* const me, SmartInt s) {
  int a = SmartInt_getPrimitive(&s);
  return memcmp(&me->value, &a, sizeof(me->value));
}

void SmartInt_setHighBoundary(SmartInt* const me, int high) {
  me->highRange = high;
}

void SmartInt_setLowBoundary(SmartInt* const me, int low) {
  me->lowRange = low;
}

ErrorCodeType SmartInt_setPrimitive(SmartInt* const me, int p) {
  me->errorCode = NO_ERRORS;
  if (p < me->lowRange) {
      me->errorCode = BELOW_RANGE;
      AlarmManager_addAlarm(me->itsAlarmManager, me->errorCode);
      }
  else if (p > me->highRange) {
      me->errorCode = ABOVE_RANGE;
      AlarmManager_addAlarm(me->itsAlarmManager, me->errorCode);
      }
  else
      me->value = p;

  return me->errorCode;
}

ErrorCodeType SmartInt_setValue(SmartInt* const me, SmartInt s) {
  me->errorCode = SmartInt_checkValidity(&s);
  if (me->errorCode == NO_ERRORS)
      *me = s;
  else
      SmartInt_errorHandler(me, me->errorCode);
```

```
    return me->errorCode;
}
struct AlarmManager* SmartInt_getItsAlarmManager(const SmartInt* const me) {
    return (struct AlarmManager*)me->itsAlarmManager;
}

void SmartInt_setItsAlarmManager(SmartInt* const me, struct
AlarmManager* p_AlarmManager) {
    me->itsAlarmManager = p_AlarmManager;
}

SmartInt * SmartInt_Create(int val, int low, int high, struct AlarmManager* errMgr) {
    SmartInt* me = (SmartInt *) malloc(sizeof(SmartInt));
    if(me!=NULL)
        SmartInt_Init(me, val, low, high, errMgr);
    return me;
}

void SmartInt_Destroy(SmartInt* const me) {
    if(me!=NULL)
        SmartInt_Cleanup(me);
    free(me);
}

static void cleanUpRelations(SmartInt* const me) {
    if(me->itsAlarmManager != NULL)
        me->itsAlarmManager = NULL;
}
```

Code Listing 6-9: SmartInt.c

Similarly, the code for the SmartColor class is given in Code Listing 6-10 and Code
Listing 6-11.

```
#ifndef SmartColor_H
#define SmartColor_H

#include "SmartDataExample.h"
#include "SmartInt.h"
struct AlarmManager;

typedef struct SmartColor SmartColor;
struct SmartColor {
    ErrorCodeType errorCode;
    ColorType highRange;
    ColorType lowRange;
    ColorType value;
```

```
  struct AlarmManager* itsAlarmManager;
};
/* Constructors and destructors:*/

void SmartColor_Init(SmartColor* const me, ColorType val, ColorType low,
ColorType high, struct AlarmManager* errMgr);

void SmartColor_Cleanup(SmartColor* const me);


/* Operations */
ErrorCodeType SmartColor_checkValidity(SmartColor* const me);

int SmartColor_cmp(SmartColor* const me, SmartColor s);

void SmartColor_errorHandler(SmartColor* const me, ErrorCodeType err);

ErrorCodeType SmartColor_getErrorCode(SmartColor* const me);

ColorType SmartColor_getHighBoundary(SmartColor* const me);

ColorType SmartColor_getLowBoundary(SmartColor* const me);

ColorType SmartColor_getPrimitive(SmartColor* const me);

SmartColor SmartColor_getValue(SmartColor* const me);

int SmartColor_pCmp(SmartColor* const me, SmartInt s);

void SmartColor_setHighBoundary(SmartColor* const me, ColorType high);

void SmartColor_setLowBoundary(SmartColor* const me, ColorType low);

ErrorCodeType SmartColor_setPrimitive(SmartColor* const me, ColorType p);

ErrorCodeType SmartColor_setValue(SmartColor* const me, SmartColor s);

struct AlarmManager* SmartColor_getItsAlarmManager(const SmartColor* const me);

void SmartColor_setItsAlarmManager(SmartColor* const me,
struct AlarmManager* p_AlarmManager);

SmartColor * SmartColor_Create(ColorType val, ColorType low, ColorType high, struct
AlarmManager* errMgr);

void SmartColor_Destroy(SmartColor* const me);
#endif
```

Code Listing 6-10: SmartColor.h

```
#include "SmartColor.h"
#include "AlarmManager.h"
static void cleanUpRelations(SmartColor* const me);

void SmartColor_Init(SmartColor* const me, ColorType val, ColorType low, ColorType high,
struct AlarmManager* errMgr) {
  me->errorCode = NO_ERRORS;
  me->itsAlarmManager = NULL;
```

```
    me->lowRange = low;
    me->highRange = high;
    me->value = val;
    me->itsAlarmManager = errMgr;
}

void SmartColor_Cleanup(SmartColor* const me) {
    cleanUpRelations(me);
}

ErrorCodeType SmartColor_checkValidity(SmartColor* const me) {
    if (me->value < me->lowRange)
        return BELOW_RANGE;
    else if (me->value > me->highRange)
        return ABOVE_RANGE;
    else
        return NO_ERRORS;
}

int SmartColor_cmp(SmartColor* const me, SmartColor s) {
    return memcmp(me, &s, sizeof(s));
}

void SmartColor_errorHandler(SmartColor* const me, ErrorCodeType err) {
    AlarmManager_addAlarm(me->itsAlarmManager, err);
}

ErrorCodeType SmartColor_getErrorCode(SmartColor* const me) {
    return me->errorCode;
}

ColorType SmartColor_getHighBoundary(SmartColor* const me) {
    return me->highRange;
}

ColorType SmartColor_getLowBoundary(SmartColor* const me) {
    return me->lowRange;
}

ColorType SmartColor_getPrimitive(SmartColor* const me) {
    return me->value;
}

SmartColor SmartColor_getValue(SmartColor* const me) {
    return *me;
}

int SmartColor_pCmp(SmartColor* const me, SmartInt s) {
    ColorType a = SmartInt_getPrimitive(&s);
    return memcmp(&me->value, &a, sizeof(me->value));
}

void SmartColor_setHighBoundary(SmartColor* const me, ColorType high) {
    me->highRange = high;
}
```

```
void SmartColor_setLowBoundary(SmartColor* const me, ColorType low) {
  me->lowRange = low;
}
ErrorCodeType SmartColor_setPrimitive(SmartColor* const me, ColorType p) {
  me->errorCode = NO_ERRORS;
  if (p < me->lowRange) {
      me->errorCode = BELOW_RANGE;
      AlarmManager_addAlarm(me->itsAlarmManager, me->errorCode);
      }
  else if (p > me->highRange) {
      me->errorCode = ABOVE_RANGE;
      AlarmManager_addAlarm(me->itsAlarmManager, me->errorCode);
      }
  else
      me->value = p;

  return me->errorCode;
}

ErrorCodeType SmartColor_setValue(SmartColor* const me, SmartColor s) {
  me->errorCode = SmartColor_checkValidity(&s);
  if (me->errorCode == NO_ERRORS)
      *me = s;
  else
      SmartColor_errorHandler(me, me->errorCode);
  return me->errorCode;
}

struct AlarmManager* SmartColor_getItsAlarmManager(const SmartColor* const me) {
  return (struct AlarmManager*)me->itsAlarmManager;
}

void SmartColor_setItsAlarmManager(SmartColor* const me, struct AlarmManager* p_Alarm-
Manager) {
  me->itsAlarmManager = p_AlarmManager;
}

SmartColor * SmartColor_Create(ColorType val, ColorType low, ColorType high, struct
AlarmManager* errMgr) {
  SmartColor* me = (SmartColor *) malloc(sizeof(SmartColor));
  if(me!=NULL)
    SmartColor_Init(me, val, low, high, errMgr);
  return me;
}

void SmartColor_Destroy(SmartColor* const me) {
  if(me!=NULL)
    SmartColor_Cleanup(me);
  free(me);
}

static void cleanUpRelations(SmartColor* const me) {
```

```
    if(me->itsAlarmManager != NULL)
      me->itsAlarmManager = NULL;
  }
```

Code Listing 6-11: SmartColor.c

Finally, the `PatientDataClass` uses these types to store values. In this case, it provides the access functions for the primitive types to its clients but uses smart data types for itself.

```
#ifndef PatientDataClass_H
#define PatientDataClass_H

#include "SmartDataExample.h"
#include "AlarmManager.h"
#include "SmartColor.h"
#include "SmartInt.h"

typedef struct PatientDataClass PatientDataClass;
struct PatientDataClass {
  SmartInt age;
  SmartColor backgroundColor;
  SmartColor foregroundColor;
  SmartInt heartRate;
  char name[100];
  long patientID;
  SmartInt weight;
};

/* Constructors and destructors:*/
void PatientDataClass_Init(PatientDataClass* const me, AlarmManager* errMgr);

void PatientDataClass_Cleanup(PatientDataClass* const me);

/* Operations */
ErrorCodeType PatientDataClass_checkAllData(PatientDataClass* const me);

int PatientDataClass_getAge(PatientDataClass* const me);
ColorType PatientDataClass_getBColor(PatientDataClass* const me);
ColorType PatientDataClass_getFColor(PatientDataClass* const me);
int PatientDataClass_getHeartRate(PatientDataClass* const me);
char* PatientDataClass_getName(PatientDataClass* const me);
int PatientDataClass_getWeight(PatientDataClass* const me);
void PatientDataClass_setAge(PatientDataClass* const me, int a);
void PatientDataClass_setBColor(PatientDataClass* const me, ColorType bc);
void PatientDataClass_setFColor(PatientDataClass* const me, ColorType fc);
void PatientDataClass_setHeartRate(PatientDataClass* const me, int hr);
```

```
void PatientDataClass_setName(PatientDataClass* const me, char* n);

void PatientDataClass_setWeight(PatientDataClass* const me, int w);

PatientDataClass * PatientDataClass_Create(AlarmManager* errMgr);

void PatientDataClass_Destroy(PatientDataClass* const me);
#endif
```

Code Listing 6-12: PatientDataClass.h

```
#include "PatientDataClass.h"
void PatientDataClass_Init(PatientDataClass* const me,
AlarmManager* errMgr) {
  strcpy(me->name, " ");
  me->patientID = 0;
  /* initialize smart variables */
  SmartInt_Init(&me->weight, 0, 0, 500, errMgr);
  SmartInt_Init(&me->age, 0, 0, 130, errMgr);
  SmartInt_Init(&me->heartRate, 0, 0, 400, errMgr);
  SmartColor_Init(&me->foregroundColor, WHITE, BLACK, WHITE, errMgr);
  SmartColor_Init(&me->backgroundColor, BLACK, BLACK, WHITE, errMgr);
}

void PatientDataClass_Cleanup(PatientDataClass* const me) {
}

ErrorCodeType PatientDataClass_checkAllData(PatientDataClass* const me) {
  ErrorCodeType res;
  res = SmartInt_checkValidity(&me->weight);
  if (res != NO_ERRORS)
      return res;
  res = SmartInt_checkValidity(&me->age);
  if (res != NO_ERRORS)
      return res;
  res = SmartInt_checkValidity(&me->heartRate);
  if (res != NO_ERRORS)
      return res;
  res = SmartColor_checkValidity(&me->foregroundColor);
  if (res != NO_ERRORS)
      return res;
  res = SmartColor_checkValidity(&me->backgroundColor);
  if (res != NO_ERRORS)
      return res;
}

int PatientDataClass_getAge(PatientDataClass* const me) {
  return SmartInt_getPrimitive(&me->age);
}

ColorType PatientDataClass_getBColor(PatientDataClass* const me) {
  return SmartColor_getPrimitive(&me->backgroundColor);
}
```

```
ColorType PatientDataClass_getFColor(PatientDataClass* const me) {
   return SmartColor_getPrimitive(&me->foregroundColor);
}

int PatientDataClass_getHeartRate(PatientDataClass* const me) {
   return SmartInt_getPrimitive(&me->heartRate);
}

char* PatientDataClass_getName(PatientDataClass* const me) {
   return me->name;
}

int PatientDataClass_getWeight(PatientDataClass* const me) {
   return SmartInt_getPrimitive(&me->weight);
}

void PatientDataClass_setAge(PatientDataClass* const me, int a) {
   SmartInt_setPrimitive(&me->age, a);
}

void PatientDataClass_setBColor(PatientDataClass* const me, ColorType bc) {
   SmartColor_setPrimitive(&me->backgroundColor, bc);
}

void PatientDataClass_setFColor(PatientDataClass* const me, ColorType fc) {
   SmartColor_setPrimitive(&me->foregroundColor, fc);
}

void PatientDataClass_setHeartRate(PatientDataClass* const me, int hr) {
   SmartInt_setPrimitive(&me->heartRate, hr);
}

void PatientDataClass_setName(PatientDataClass* const me, char* n) {
   strcpy(me->name, n);
}

void PatientDataClass_setWeight(PatientDataClass* const me, int w) {
   SmartInt_setPrimitive(&me->weight, w);
}

PatientDataClass * PatientDataClass_Create(AlarmManager* errMgr) {
   PatientDataClass* me = (PatientDataClass *)
malloc(sizeof(PatientDataClass));
   if(me!=NULL)
      PatientDataClass_Init(me, errMgr);
   return me;
}
void PatientDataClass_Destroy(PatientDataClass* const me) {
   if(me!=NULL)
      PatientDataClass_Cleanup(me);
   free(me);
}
```

Code Listing 6-13: PatientDataClass.c

6.5 Channel Pattern

The Channel Pattern is a larger-scale pattern than the previous patterns we've discussed. The Channel Pattern is used to support medium- to large-scale redundancy to help identify when run-time faults occur and possibly (depending on how it's used) continue to provide services in the presence of such a fault.

6.5.1 Abstract

A *channel* is an architectural structure that contains software (and possibly hardware as well) that performs end-to-end processing; that is, it processes data from raw acquisition, through a series of data processing steps, to physical actuation in the real world. The advantage of a channel is that it provides an independent, self-contained unit of functionality that can be replicated in different ways to address different safety and reliability concerns.

6.5.2 Problem

This pattern provides a basic element of large-scale redundancy that can be used in different ways to address safety and reliability concerns.

6.5.3 Pattern Structure

The basic structure of the pattern is shown in Figure 6-8. The generalization shown in the figure can be notional only for implementations without true inheritance in C. The basic idea is that the software objects are arranged through a series of links so that data are acquired and processed through a series of data transformational steps until it results in an output that controls one or more actuators. The data types may change (for example, it may be acquired as a `double` but result in an `int` output to the actuator) or they may remain the same throughout the processing.

6.5.4 Collaboration Roles

This section describes the roles for this pattern.

6.5.4.1 AbstractDataTransform

This class is a placeholder for one of several possible `ConcreteDataTransforms`. In an object-oriented language or implementation, true generalization can be used but in C, it is more common to use this as a notional concept to give the intent rather than drive the implementation. It has two optional[7] associations. One is to itself – that is, the next transformation in the list. The other is to the output `ActuationDeviceDriver` – this is in lieu of the former

---

[7] Common implementation is a pointer, set to `NULL` if the link isn't active.

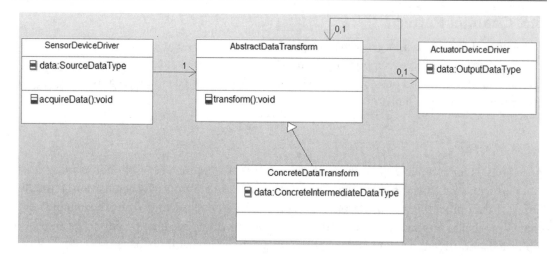

Figure 6-8: Channel Pattern

association. Put another way, the `AbstractDataTransform` has a link to the next processing step in the chain or to the output device driver, but not both.

6.5.4.2 ActuatorDeviceDriver

This class uses the computed output values to drive one or more actuators, such as a motor, light, or heating unit.

6.5.4.3 ConcreteDataTransform

This class performs a mathematical transformation on the incoming data value (from either the sensor or a previous `ConcreteDataTransform`) and produces an output (sent to the next `ConcreteDataTransform` in the sequence or the `ActuatorDeviceDriver`). In simple cases, there may be only a single `ConcreteDataTransform` class, but in more complex situations there may be dozens.

6.5.4.4 SensorDeviceDriver

This class acquires physical data from one or more sensors to put into the channel for processing.

6.5.5 Consequences

The channel in this simple form provides only a little value. Its true value lies in the use of multiple instances of the channel. There are a number of different ways to combine the channels, as well as the data transformations, to address safety and reliability concerns.

In general, the benefits are providing a clear means for identifying faults and, depending on the pattern, providing redundant channels to continue to provide services or to go to a "fault-safe state." The drawbacks of using the pattern are the extra memory (both code and data), processing time, and hardware required, depending on the specific usage of the channel.

6.5.6 Implementation Strategies

As mentioned, the pattern may be implemented with true generalization but it is more common in C to think of the `AbstractDataTransformation` as a placeholder for some particular `ConcreteDataTransform` class; the links are from different types of `ConcereteDataTransform` classes that each perform some special step in the overall data transformation function.

Data are passed through using one of two primary strategies. The simplest is to acquire a datum and pass it through the series of transformations and out to the actuator before acquiring the next. In higher bandwidth applications multiple instances of the data may be present in the channel at the same time, one data instance per transform, with data passed from one transform to the next at the same time as all the other data instances.

6.5.7 Related Patterns

This pattern provides medium- to large-scale redundancy. It is the basis for many other patterns including the Protected Single-Channel Pattern and Homogeneous Redundancy Pattern discussed later in this chapter. Other patterns include Triple Modular Redundancy (TMR) Pattern, in which three channels run in parallel and vote on the outcome, and the Sanity Check Pattern in which a primary channel is checked for reasonableness by a lighter-weight secondary channel. A number of such patterns are discussed in my *Real-Time Design Patterns: Robust Scalable Architecture for Real-Time Systems* (Addison-Wesley, 2002).

6.5.8 Example

I will provide a relatively simple example here of the use of the pattern. The classes shown in Figure 6-9 implement a simple electromyography biofeedback device. It works by measuring the voltage across two electrodes placed on the skin over a muscle. As the muscle tension changes, the voltage potential between the electrodes changes. This is gathered (as an `int`) by the `EMGSensorDeviceDriver`. Then it is converted into a frequency by the first data transformation class; this is done by taking a set of samples and computing the fundamental frequency of those data. This results in a computed frequency that is then filtered with a moving average filter; this is a filter that smoothes out rapid frequency changes by averaging the change over several (frequency) data samples measured sequentially in time. Lastly, the filtered frequency is converted into a pleasing color (specified via RGB values) and this color is sent to the light device driver.

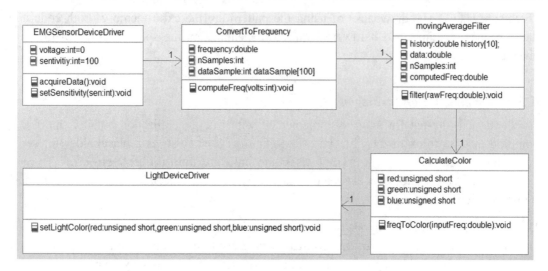

Figure 6-9: Channel Pattern example

Figure 6-10 shows the same pattern but with the channel objects as a part of a larger assembly[8] in which the elements of the channel are parts. The boxes inside the EMG Channel are part objects, and this is indicated with an underlined name.

The code is very straightforward for the pattern. For initialization, the key is that the EMG channel instantiates the objects and then initializes the links (see the initRelations() function in Code Listing 6-15) between them. For operation, the EMG Channel provides the services of the start of the chain (that is, the EMG Device Driver); when acquireData() is called on the EMG Channel, it calls EMGSensorDeviceDriver_acquireData() which gets the data and passes them on to the ConvertToFrequency_computeFreq() function. This function does the conversion and then calls the MovingAverageFilter_filter() function which smooths out high frequency artifacts and then passes the output of that onto CalculateColor_freqToColor(). This function computes the RGB values associated with the frequency and calls LightDeviceDriver_setLightColor() to turn the light to the correct color. In this case, the processing is all synchronous. It is possible to pass the data among the ConcreteDataTransformations asynchronously as well, by putting queues between them and having them pend on data input to those queues.

The next two code listings, Code Listing 6-14 and Code Listing 6-15 give the code for the EMGChannel class header and implementation, respectively.

---

[8] In UML, such an assembly is known as a *structured class*.

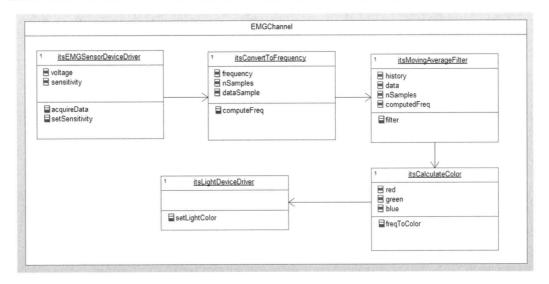

Figure 6-10: Channel Pattern (object view)

```c
#ifndef EMGChannel_H
#define EMGChannel_H

#include "CalculateColor.h"
#include "ConvertToFrequency.h"
#include "EMGSensorDeviceDriver.h"
#include "LightDeviceDriver.h"
#include "movingAverageFilter.h"

typedef struct EMGChannel EMGChannel;
struct EMGChannel {
  struct CalculateColor itsCalculateColor;
  struct ConvertToFrequency itsConvertToFrequency;
  struct EMGSensorDeviceDriver itsEMGSensorDeviceDriver;
  struct LightDeviceDriver itsLightDeviceDriver;
  struct movingAverageFilter itsMovingAverageFilter;
};

/* Constructors and destructors:*/
void EMGChannel_Init(EMGChannel* const me);
void EMGChannel_Cleanup(EMGChannel* const me);

/* Operations */
void EMGChannel_acquireData(EMGChannel* const me);
double EMGChannel_getFrequency(EMGChannel* const me);
long EMGChannel_getLightColor(EMGChannel* const me);
int EMGChannel_getVoltage(EMGChannel* const me);
void EMGChannel_setSensitivity(EMGChannel* const me, int sen);
```

```
struct CalculateColor* EMGChannel_getItsCalculateColor(const EMGChannel* const me);

struct ConvertToFrequency* EMGChannel_getItsConvertToFrequency(const
EMGChannel* const me);

struct EMGSensorDeviceDriver* EMGChannel_getItsEMGSensorDeviceDriver(const
EMGChannel* const me);

struct LightDeviceDriver* EMGChannel_getItsLightDeviceDriver(const
EMGChannel* const me);

struct movingAverageFilter* EMGChannel_getItsMovingAverageFilter(const
EMGChannel* const me);

EMGChannel * EMGChannel_Create(void);

void EMGChannel_Destroy(EMGChannel* const me);
#endif
```

Code Listing 6-14: EMGChannel.h

```
#include "EMGChannel.h"

static void initRelations(EMGChannel* const me);

static void cleanUpRelations(EMGChannel* const me);

void EMGChannel_Init(EMGChannel* const me) {
  initRelations(me);
}

void EMGChannel_Cleanup(EMGChannel* const me) {
  cleanUpRelations(me);
}

void EMGChannel_acquireData(EMGChannel* const me) {
  /* delegate to the appropriate part */
  EMGSensorDeviceDriver_acquireData(&me->itsEMGSensorDeviceDriver);
}

double EMGChannel_getFrequency(EMGChannel* const me) {
  return me->itsMovingAverageFilter.computedFreq;
}

long EMGChannel_getLightColor(EMGChannel* const me) {
  return me->itsCalculateColor.red<<16 + me->itsCalculateColor.green<<8 + me->its-
CalculateColor.blue;
}
```

```
int EMGChannel_getVoltage(EMGChannel* const me) {
  return me->itsEMGSensorDeviceDriver.voltage;
}

void EMGChannel_setSensitivity(EMGChannel* const me, int sen) {
  EMGSensorDeviceDriver_setSensitivity(&me->itsEMGSensorDeviceDriver, sen);
}

struct CalculateColor* EMGChannel_getItsCalculateColor(const EMGChannel* const me) {
  return (struct CalculateColor*)&(me->itsCalculateColor);
}

struct ConvertToFrequency* EMGChannel_getItsConvertToFrequency(const EMGChannel*
const me) {
  return (struct ConvertToFrequency*)&(me->itsConvertToFrequency);
}

struct EMGSensorDeviceDriver* EMGChannel_getItsEMGSensorDeviceDriver(const EMGChan-
nel* const me) {
  return (struct EMGSensorDeviceDriver*)&(me->itsEMGSensorDeviceDriver);
}

struct LightDeviceDriver* EMGChannel_getItsLightDeviceDriver(const EMGChannel* const
me) {
  return (struct LightDeviceDriver*)&(me->itsLightDeviceDriver);
}

struct movingAverageFilter*
EMGChannel_getItsMovingAverageFilter(const EMGChannel* const me) {
  return (struct movingAverageFilter*)&(me->itsMovingAverageFilter);
}

EMGChannel * EMGChannel_Create(void) {
  EMGChannel* me = (EMGChannel *) malloc(sizeof(EMGChannel));
  if(me!=NULL)
    EMGChannel_Init(me);
  return me;
}

void EMGChannel_Destroy(EMGChannel* const me) {
  if(me!=NULL)
    EMGChannel_Cleanup(me);
  free(me);
}

static void initRelations(EMGChannel* const me) {
  CalculateColor_Init(&(me->itsCalculateColor));
  ConvertToFrequency_Init(&(me->itsConvertToFrequency));
  EMGSensorDeviceDriver_Init(&(me->itsEMGSensorDeviceDriver));
```

```
   LightDeviceDriver_Init(&(me->itsLightDeviceDriver));
   movingAverageFilter_Init(&(me->itsMovingAverageFilter));
   EMGSensorDeviceDriver_setItsConvertToFrequency(&(me-
>itsEMGSensorDeviceDriver),&(me->itsConvertToFrequency));
   ConvertToFrequency_setItsMovingAverageFilter(&(me->itsConvertToFrequency),&(me-
>itsMovingAverageFilter));
   movingAverageFilter_setItsCalculateColor(&(me->itsMovingAverageFilter),&(me-
>itsCalculateColor));
   CalculateColor_setItsLightDeviceDriver(&(me->itsCalculateColor),&(me-
>itsLightDeviceDriver));
}

static void cleanUpRelations(EMGChannel* const me) {
   movingAverageFilter_Cleanup(&(me->itsMovingAverageFilter));
   LightDeviceDriver_Cleanup(&(me->itsLightDeviceDriver));
   EMGSensorDeviceDriver_Cleanup(&(me-
>itsEMGSensorDeviceDriver));
   ConvertToFrequency_Cleanup(&(me->itsConvertToFrequency));
   CalculateColor_Cleanup(&(me->itsCalculateColor));
}
```

Code Listing 6-15: EMGChannel.c

6.6 Protected Single Channel Pattern

The Protected Single Channel Pattern is a simple elaboration of the Channel pattern in which data checks are added at one or more ConcreteDataTransformations. It provides light-weight redundancy but typically cannot continue to provide service if a fault is discovered.

6.6.1 Abstract

The Protected Single Channel Pattern uses a single channel to handle the processing of sensory data up to and including actuation based on that data. Safety and reliability are enhanced through the addition of checks at key points in the channel, which may require some additional hardware. Because there is only a single channel, this pattern will not be able to continue to function in the presence of persistent faults, but it detects and may be able to handle transient faults.

6.6.2 Problem

Redundancy is expensive, not only in terms of development effort but also in terms of replicated hardware. In some cases, fully redundant channels may not be necessary. For example, if there is a fault-safe state (a state known to always be safe), the system must only be able to detect a fault and enter that state. This lowers the reliability of the system, but it may still meet the user needs.

6.6.3 Pattern Structure

The pattern, shown in Figure 6-11, is a simple elaboration of the Channel Pattern (see Figure 6-8). All we've added here are some checks on some of the transformation steps. In fact, these checks need not be implemented in separate classes, but can be implemented internal to the `ConcreteTransformations` themselves. These checks can be any kind of checks, such as:

- Range checking (are data in the required range?)
- Reasonableness checks (are data in the expected range for the given situation?)
- Consistency checks (is the value consistent with other data being computed at the same time?)
- Backwards transformational checks (can we recompute the input values from the output values?)
- Data validity checks, such as we've seen before with the One's Complement and CRC Patterns

and so on.

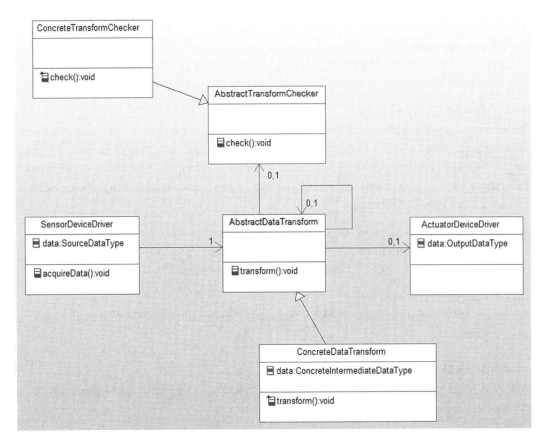

Figure 6-11: Protected Single Channel Pattern

6.6.4 Collaboration Roles

This section describes the roles for this pattern.

6.6.4.1 AbstractDataTransform

This class is a placeholder for one of several possible `ConcreteDataTransforms`. In an object-oriented language or implementation, true generalization can be used but in C, it is more common to use this as a notional concept to give the intent rather than drive the implementation. It has two optional[9] associations. One is to itself – that is, the next transformation in the list. The other is to the output `ActuationDeviceDriver` – this is in lieu of the former association. Put another way, the `AbstractDataTransform` has a link to the next processing step in the chain or to the output device driver, but not both.

This class has also an optional link to `AbstractTransformChecker` for points in the transformation chain where checking is appropriate.

6.6.4.2 AbstractTransformChecker

This is a placeholder for an element that can check on the validity of the processing or data in their current state.

6.6.4.3 ActuatorDeviceDriver

This class uses the computed output values to drive one or more actuators, such as a motor, light, or heating unit.

6.6.4.4 ConcreteDataTransform

This class performs a mathematical transformation on the incoming data value (from either the sensor or a previous `ConcreteDataTransform`) and produces an output (sent to the next `ConcreteDataTransform` in the sequence or the `ActuatorDeviceDriver`). In simple cases, there may be only a single `ConcreteDataTransform` class, but in more complex situations there may be dozens.

Certain of these classes form checkpoints – points in the transformation chain at which the validality of the transformations will be checked. For those cases, the optional link to the specific `ConcreteTransformChecker` will be implemented.

6.6.4.5 ConcreteTransformChecker

This is a specific transformation checking class that checks the transformation at a single point in the chain.

6.6.4.6 SensorDeviceDriver

This class acquires physical data from one or more sensors to put into the channel for processing.

---

[9] Common implementation is a pointer, set to `NULL` if the link isn't active.

6.6.5 Implementation Strategies

In general, the benefits are providing a clear means for identifying faults and, depending on the pattern, providing redundant channels to continue to provide services or to go to a "fault-safe state." The drawbacks of using the pattern are the extra memory (both code and data), processing time, and hardware required, depending on the specific usage of the channel.

As mentioned, the pattern may be implemented with true generalization, but it is more common in C to think of the `AbstractDataTransformation` as a placeholder for some particular `ConcreteDataTransform` class; the links are from different subtypes of `ConcereteDataTransform` classes that each perform some special step in the overall data transformation function.

Data are passed through using one of two primary strategies. The simplest is to acquire a datum and pass it through the series of transformations and out to the actuator before acquiring the next. In higher bandwidth applications multiple instances of the data may be present in the channel at the same time, one data instance per transform, with the data passed from one transform to the next at the same time as all the other data instances.

In addition, the data checks may be put "in place" in some of the transformation steps rather than implement them as a separate class. That is, in fact, the specialization of this pattern from the more generic Channel Pattern.

6.6.6 Related Patterns

This pattern provides medium- to large-scale redundancy. It is one of many patterns based on the Channel Pattern discussed earlier in this chapter. A number of such patterns are discussed in my *Real-Time Design Patterns: Robust Scalable Architecture for Real-Time Systems* (Addison-Wesley, 2002).

The transformation checks can themselves be based on patterns such as the One's Complement and CRC Patterns discussed earlier.

6.6.7 Example

In Figure 6-12, I purposely chose a situation that demanded a bit more creative application of the pattern. This is a heating system with sensor sets for multiple rooms, each having sensors for fan speed, temperature, and thermostat settings. Because the example isn't laid out exactly as the pattern, I indicated which pattern role each class plays by adding a stereotype[10] to each.

---

[10] A stereotype is "a special kind of" marking that appears inside «» marks. In this case, the stereotypes are the names of the pattern roles.

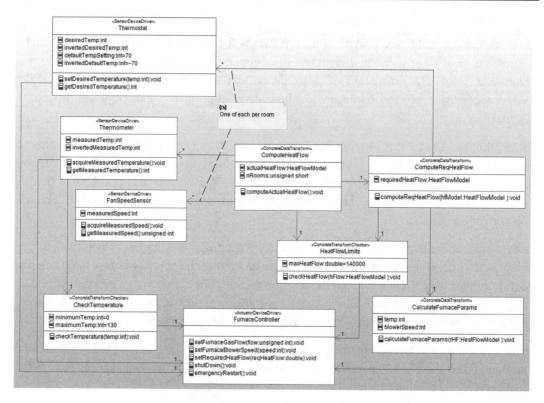

Figure 6-12: Protected Single Channel example

I further modified the pattern by making the first `ConcreteDataTransform` pull the data from its `SensorDeviceDrivers` (the thermometer and room fan) associated with each room. The next transform, `ComputeRequiredHeatFlow`, not only gets the actual heat flow from the previous transform but reads each room's thermostat to get the desired temperature and computes the desired output heat flow from the furnace. The difference between actual and desired temperatures results in a desired heat flow. This is passed on to the `CalculateFurnaceParams`, which computes the necessary furnace gas flow, furnace blower speeds, and commands the `FurnaceController`.

In terms of safety, there are several checks in this protected single channel. The Thermometers are checked for actual temperature, and if the temperature is out of bounds, this leads to either emergency restart of the furnace (if below 0° F) or shutting the furnace down (if over 130° F)[11]. The actual heat flow and (computed) desired heat flow are checked to ensure that they don't exceed 14,000 BTUs. If this limit is exceeded, the furnace is shut down.

---

[11]   That's -18° C and 54° C for you metric-speaking types.

In addition, both the `Thermometer` and `Thermostat` use the One's Complement Pattern (described earlier in this chapter) to detect memory corruption. In the case of the `Thermometer`, the behavior on corruption is to shut the furnace down, while in the case of the `Thermostat`, the behavior is to use the default temperature setting of 70° F. If the default is corrupted, then it calls `FurnaceController_shutDown()`.

I don't want to give all the code here, but we can look at a few relevant classes. The `Thermostat` class implements a One's Complement Pattern to ensure its data isn't corrupted. Its code is shown in Code Listing 6-16 and Code Listing 6-17.

```
#ifndef Thermostat_H
#define Thermostat_H

struct FurnaceController;

typedef struct Thermostat Thermostat;
struct Thermostat {
  int defaultTempSetting;
  int desiredTemp;
  int invertedDefaultTemp;
  int invertedDesiredTemp;
  struct FurnaceController* itsFurnaceController;
};

/* Constructors and destructors:*/
void Thermostat_Init(Thermostat* const me);
void Thermostat_Cleanup(Thermostat* const me);

/* Operations */
int Thermostat_getDesiredTemperature(Thermostat* const me);

void Thermostat_setDesiredTemperature(Thermostat* const me, int temp);

struct FurnaceController* Thermostat_getItsFurnaceController(const
Thermostat* const me);

void Thermostat_setItsFurnaceController(Thermostat* const me,
struct FurnaceController* p_FurnaceController);

Thermostat * Thermostat_Create(void);

void Thermostat_Destroy(Thermostat* const me);
#endif
```

Code Listing 6-16: Thermostat.h

```c
#include "Thermostat.h"
#include "FurnaceController.h"

static void cleanUpRelations(Thermostat* const me);

void Thermostat_Init(Thermostat* const me) {
  me->defaultTempSetting = 70;
  me->invertedDefaultTemp = ~70;
  me->itsFurnaceController = NULL;
}

void Thermostat_Cleanup(Thermostat* const me) {
  cleanUpRelations(me);
}

int Thermostat_getDesiredTemperature(Thermostat* const me) {
  if (me->desiredTemp == ~me->invertedDesiredTemp)
      return me->desiredTemp;
  else
      if (me->defaultTempSetting == ~me->invertedDefaultTemp) {
          Thermostat_setDesiredTemperature(me, me->defaultTempSetting);
          return me->defaultTempSetting;
          }
      else {
          FurnaceController_shutDown(me->itsFurnaceController);
          return me->defaultTempSetting;
          };
}

void Thermostat_setDesiredTemperature(Thermostat* const me, int temp) {
  if (me->desiredTemp == ~me->invertedDesiredTemp) {
      me->desiredTemp = temp;
      me->invertedDesiredTemp = ~temp;
      if (me->desiredTemp == ~me->invertedDesiredTemp)
          return;
      }
  else
      FurnaceController_shutDown(me->itsFurnaceController);
}

struct FurnaceController* Thermostat_getItsFurnaceController(const
Thermostat* const me) {
  return (struct FurnaceController*)me->itsFurnaceController;
}

void Thermostat_setItsFurnaceController(Thermostat* const me, struct FurnaceControl-
ler* p_FurnaceController) {
  me->itsFurnaceController = p_FurnaceController;
}
```

```
Thermostat * Thermostat_Create(void) {
  Thermostat* me = (Thermostat *) malloc(sizeof(Thermostat));
  if(me!=NULL)
    Thermostat_Init(me);
  return me;
}

void Thermostat_Destroy(Thermostat* const me) {
  if(me!=NULL)
    Thermostat_Cleanup(me);
  free(me);
}

static void cleanUpRelations(Thermostat* const me) {
  if(me->itsFurnaceController != NULL)
    me->itsFurnaceController = NULL;
}
```

Code Listing 6-17: Thermostat.c

The CheckTemperature class provides a reasonableness check on the temperature. Code
Listing 6-18 and Code Listing 6-19 give the header and implementation.

```
#ifndef CheckTemperature_H
#define CheckTemperature_H

struct FurnaceController;

typedef struct CheckTemperature CheckTemperature;
struct CheckTemperature {
  int maximumTemp;
  int minimumTemp;
  struct FurnaceController* itsFurnaceController;
};

void CheckTemperature_Init(CheckTemperature* const me);

void CheckTemperature_Cleanup(CheckTemperature* const me);

void CheckTemperature_checkTemperature(CheckTemperature* const me, int temp);

struct FurnaceController*
CheckTemperature_getItsFurnaceController(const CheckTemperature* const me);

void CheckTemperature_setItsFurnaceController(CheckTemperature* const
me, struct FurnaceController* p_FurnaceController);

CheckTemperature * CheckTemperature_Create(void);
```

```
void CheckTemperature_Destroy(CheckTemperature* const me);
#endif
```

Code Listing 6-18: CheckTemperature.h

```
#include "CheckTemperature.h"
#include "FurnaceController.h"

static void cleanUpRelations(CheckTemperature* const me);

void CheckTemperature_Init(CheckTemperature* const me) {
  me->maximumTemp = 130;
  me->minimumTemp = 0;
  me->itsFurnaceController = NULL;
}


void CheckTemperature_Cleanup(CheckTemperature* const me) {
  cleanUpRelations(me);
}


void CheckTemperature_checkTemperature(CheckTemperature* const me, int temp) {
  if (temp < me->minimumTemp)
      FurnaceController_emergencyRestart(me-
>itsFurnaceController);
  if (temp > me->maximumTemp)
      FurnaceController_shutDown(me->itsFurnaceController);
}


struct FurnaceController*
CheckTemperature_getItsFurnaceController(const CheckTemperature* const me) {
  return (struct FurnaceController*)me->itsFurnaceController;
}


void CheckTemperature_setItsFurnaceController(CheckTemperature*
const me, struct FurnaceController* p_FurnaceController) {
  me->itsFurnaceController = p_FurnaceController;
}


CheckTemperature * CheckTemperature_Create(void) {
  CheckTemperature* me = (CheckTemperature *)
malloc(sizeof(CheckTemperature));
  if(me!=NULL)
    CheckTemperature_Init(me);
  return me;
}


void CheckTemperature_Destroy(CheckTemperature* const me) {
  if(me!=NULL)
    CheckTemperature_Cleanup(me);
  free(me);
}
```

```
static void cleanUpRelations(CheckTemperature* const me) {
  if(me->itsFurnaceController != NULL)
          me->itsFurnaceController = NULL;
}
```

Code Listing 6-19: CheckTemperature.c

The `CheckTemperature` class is used by the `Thermometer`. The last code listings, **Code Listing 6-20** and **Code Listing 6-21**, give the header and implementation for this class.

```
#ifndef Thermometer_H
#define Thermometer_H

struct CheckTemperature;
struct FurnaceController;

typedef struct Thermometer Thermometer;
struct Thermometer {
  int invertedMeasuredTemp;
  int measuredTemp;
  struct CheckTemperature* itsCheckTemperature;
  struct FurnaceController* itsFurnaceController;
};

/* Constructors and destructors:*/

void Thermometer_Init(Thermometer* const me);

void Thermometer_Cleanup(Thermometer* const me);

/* Operations */
void Thermometer_acquireMeasuredTemperature(Thermometer* const me);

int Thermometer_getMeasuredTemperature(Thermometer* const me);

struct CheckTemperature* Thermometer_getItsCheckTemperature(const
Thermometer* const me);

void Thermometer_setItsCheckTemperature(Thermometer* const me,
struct CheckTemperature* p_CheckTemperature);

struct FurnaceController* Thermometer_getItsFurnaceController(const
Thermometer* const me);

void Thermometer_setItsFurnaceController(Thermometer* const me,
struct FurnaceController* p_FurnaceController);
```

```
Thermometer * Thermometer_Create(void);

void Thermometer_Destroy(Thermometer* const me);

struct CheckTemperature*
Thermometer_getItsCheckTemperature_1(const Thermometer* const me);

void Thermometer_setItsCheckTemperature_1(Thermometer* const me,
struct CheckTemperature* p_CheckTemperature);

#endif
```

Code Listing 6-20: Thermometer.h

```
#include "Thermometer.h"
#include "CheckTemperature.h"
#include "FurnaceController.h"

static void cleanUpRelations(Thermometer* const me);

void Thermometer_Init(Thermometer* const me) {
  me->itsCheckTemperature = NULL;
  me->itsFurnaceController = NULL;
}

void Thermometer_Cleanup(Thermometer* const me) {
  cleanUpRelations(me);
}

void Thermometer_acquireMeasuredTemperature(Thermometer* const me) {
  /* access the hardware here */
}

int Thermometer_getMeasuredTemperature(Thermometer* const me) {
  if (me->measuredTemp == ~me->invertedMeasuredTemp) {
      CheckTemperature_checkTemperature(me->itsCheckTemperature,
me->measuredTemp);
      return me->measuredTemp;
      }
  else {
      FurnaceController_shutDown(me->itsFurnaceController);
      return 0;
      }
}

struct CheckTemperature* Thermometer_getItsCheckTemperature(const
Thermometer* const me) {
  return (struct CheckTemperature*)me->itsCheckTemperature;
}

void Thermometer_setItsCheckTemperature(Thermometer* const me,
struct CheckTemperature* p_CheckTemperature) {
  me->itsCheckTemperature = p_CheckTemperature;
}
```

```
struct FurnaceController* Thermometer_getItsFurnaceController(const
Thermometer* const me) {
  return (struct FurnaceController*)me->itsFurnaceController;
}

void Thermometer_setItsFurnaceController(Thermometer* const me,
struct FurnaceController* p_FurnaceController) {
  me->itsFurnaceController = p_FurnaceController;
}

Thermometer * Thermometer_Create(void) {
  Thermometer* me = (Thermometer *)
  malloc(sizeof(Thermometer));
  if(me!=NULL)
    Thermometer_Init(me);
  return me;
}

void Thermometer_Destroy(Thermometer* const me) {
  if(me!=NULL)
    Thermometer_Cleanup(me);
  free(me);
}

static void cleanUpRelations(Thermometer* const me) {
  if(me->itsCheckTemperature != NULL)
    me->itsCheckTemperature = NULL;
  if(me->itsFurnaceController != NULL)
          me->itsFurnaceController = NULL;
}
```

Code Listing 6-21: Thermometer.c

6.7 Dual Channel Pattern

The Dual Channel Pattern is primarily a pattern to improve reliability by offering multiple channels, thereby addressing redundancy concerns at the architectural level. If the channels are identical (called the Homogeneous Redundancy Channel), the pattern can address random faults (failures) but not systematic faults (errors). If the channels use a different design or implementation, the pattern is called the Heterogeneous Redundancy Pattern (also known as Diverse Design Pattern) and can address both random and systematic faults.

6.7.1 Abstract

The Dual Channel Pattern provides architectural redundancy to address safety and reliability concerns. It does this by replicating channels and embedding logic that manages them and determines when the channels will be "active."

6.7.2 Problem

This pattern provides protection against single-point faults (either failures or both failures and errors, depending on the specific subpattern selected). Depending on which pattern, the system may detect a fault in one channel by comparing it to the other and then transition to the fault-safe state OR it may use other means to detect the fault in one channel and switch to the other when a fault occurs.

6.7.3 Pattern Structure

Figure 6-13 shows the basic structure for the pattern.

Figure 6-14 shows the detailed internal structure within the pattern. It is almost exactly what appears in the Protected Single Channel Pattern, with the exception of the association from the `ConcreteTransformChecker` to the current and alternative channel. Note that the `itsChannel` association points to the *current* channel that owns the `ConcreteTransformChecker` instance while the `itsOtherChannel` points to the alternative channel. This allows the checker to disable the current channel and enable the alternative channel.

6.7.4 Collaboration Roles

This section describes the roles for this pattern. Note that the roles that are internal to the pattern can be seen in the previous section on the Protected Single Channel Pattern.

6.7.4.1 AbstractDataTransform

This class is a placeholder for one of several possible `ConcreteDataTransforms`. In an object-oriented language or implementation, true generalization can be used but in C, it is more common to use this as a notional concept to give the intent rather than drive the implementation. It has two optional[12] associations. One is to itself – that is, the next transformation in the list. The other is to the output `ActuationDeviceDriver` – this is in lieu of the former association. Put another way, the `AbstractDataTransform` has a link to the next processing step in the chain or to the output device driver, but not both.

This class also an optional link to `AbstractTransformChecker` for points in the transformation chain where checking is appropriate.

6.7.4.2 AbstractTransformChecker

This is a placeholder for an element that can check on the validity of the processing or data in its current state.

---

[12] Common implementation is a pointer, set to NULL if the link isn't active.

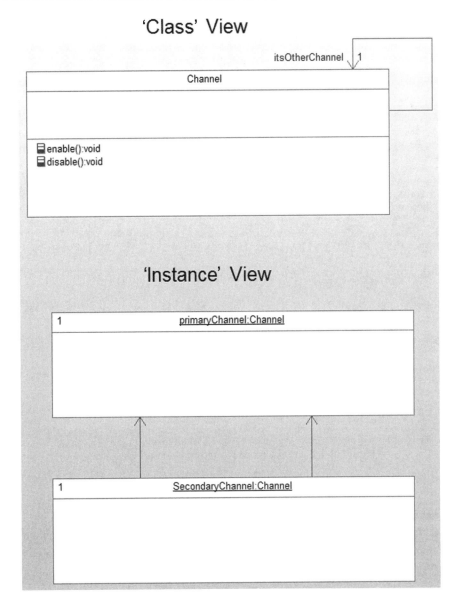

Figure 6-13: Dual Channel Pattern

6.7.4.3 ActuatorDeviceDriver

This class uses the computed output values to drive one or more actuator, such as a motor, light, or heating unit.

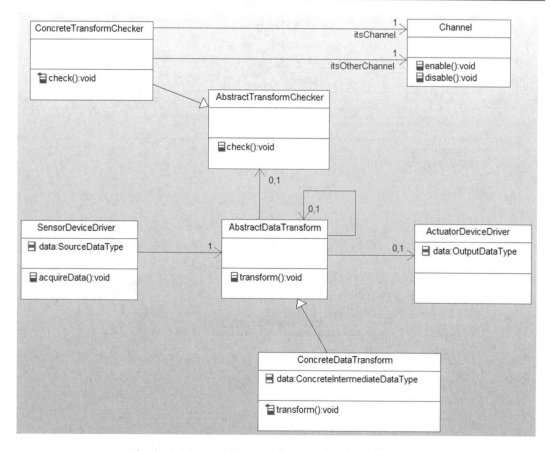

Figure 6-14: Dual Channel Pattern detailed class view

6.7.4.4 Channel

This class is the container for the instances of the others. It is important in this context because it provides a "unit of redundancy." It provides two services of itself – `enable()` and `disable()`.

6.7.4.5 ConcreteDataTransform

This class performs a mathematical transformation on the incoming data value (from either the sensor or a previous `ConcreteDataTransform`) and produces an output (sent to the next `ConcreteDataTransform` in the sequence or the `ActuatorDeviceDriver`). In simple cases, there may be only a single `ConcreteDataTransform` class, but in more complex situations there may be dozens.

Certain of these classes form checkpoints – points in the transformation chain at which the validality of the transformations will be checked. For those cases, the optional link to the specific `ConcreteTransformChecker` will be implemented.

6.7.4.6 *ConcreteTransformChecker*

This is a specific transformation checking class that checks the transformation at a single point in the chain.

6.7.4.7 *SensorDeviceDriver*

This class acquires physical data from one or more sensors to put into the channel for processing.

6.7.5 Consequences

This pattern addresses safety- or reliability-related faults by replicating the channel; this usually requires replication of some amount of hardware as well, giving it a rather high recurring (production) cost. If the channels are identical, then all replicants contain the same errors and so will manifest the error under the same circumstances.

6.7.6 Implementation Strategies

The large-scale implementation strategies are commonly identified as different subpatterns of this one. These are discussed in the next section. In addition, the management of the two channels can be implemented differently. Both channels can be run simultaneously and checked against each other – if the outputs differ above some threshold, the system transitions to a fault-safe state. Alternatively, one channel can run until a fault is detected and then the other channel can be enabled, allowing continuation of services in the presence of the fault.

6.7.7 Related Patterns

There are a number of variants on this pattern. Each gives a different cost/benefit ratio.

6.7.7.1 *Homogeneous Redundancy Pattern*

Description
This pattern variant uses identical designs and implementations for the different channels. It can effectively address single-point faults – provided that the faults are isolated within a single channel. The system can continue functionality in the presence of a failure within a channel.

Consequences
This pattern variant has a relatively high recurring cost (cost per shipped item) but a relatively low design cost (since each channel must only be designed once). Typically sensors and actuators are replicated to provide fault isolation, and often processors, memory, and other hardware must be replicated.

6.7.7.2 Heterogeneous Redundancy Pattern

Description

This pattern variant uses dual channels of different designs or different implementations to address both random and systematic faults. The system can continue to deliver services in the presence of a fault, provided that the fault is properly isolated within a single channel. Typically, sensors and actuators are replicated to provide fault isolation, and often processors, memory, and other hardware must be replicated as well as providing multiple (different) implementations of the same software functionality.

Consequences

This pattern variant has a relatively high recurring cost (cost per shipped item) and a relatively high design cost (since each channel must be designed twice). Typically, sensors and actuators are replicated to provide fault isolation, and often processors, memory, and other hardware must be replicated in addition to the software functionality. Note that this pattern variant addresses both random and systematic faults, although Nancy Leveson notes that there is usually some correlation between errors even in diverse implementations[13].

6.7.7.3 Triple Modular Redundancy (TMR) Pattern

Description

This pattern variant uses three channels of usually identical designs to address both faults. The theory of use is that if there is a single-point fault then one of the channels will disagree with the other two and the outlier is discarded. The voting mechanism (the `ChannelArbiter`) is designed to "fail on" so that at least one channel (it doesn't matter which) will result in an output if the voting mechanism fails. The system can continue to deliver services in the presence of a fault, provided that the fault is properly isolated within a single channel. Typically, sensors and actuators are replicated to provide fault isolation, and often processors, memory, and other hardware must be replicated as well as providing multiple (different) implementations of the same software functionality.

Consequences

This pattern has a very high recurring cost because the channel must be replicated three times. If all the channels have the same design (most common), then the design cost is relatively low; the only additional cost is the `ChannelArbiter` used to reject the outlier output. If the channels are of different designs, then the cost of this pattern variant is very high since each channel must be designed three times. This is a very common pattern in avionics.

---

[13] See Leveson, N., 1995. *Safeware: System Safety and Computers.* Addison-Wesley.

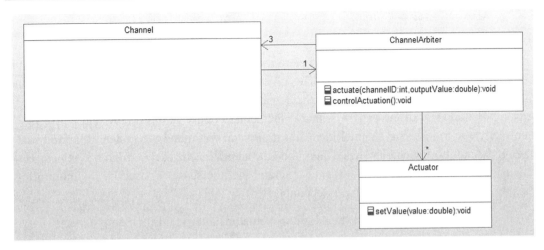

Figure 6-15: Triple Modular Redundancy Pattern

6.7.7.4 Sanity-Check Pattern

Description
This pattern variant uses two heterogeneous channels; one is the primary actuation channel and the second is a lightweight channel that checks on the output using lower fidelity computations and hardware. If the primary channel has a fault that can be detected by the lower fidelity checking channel, then the system enters its fault-safe state.

Consequences
This pattern has a low recurring cost and a medium design cost since it requires additional design work but lower fidelity redundancy. It cannot continue in the presence of a single fault.

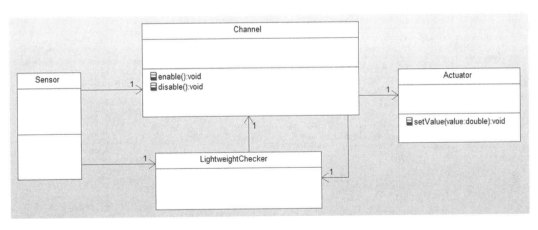

Figure 6-16: Sanity Check Pattern

6.7.7.5 Monitor-Actuator Pattern

Description

This pattern variant uses two extremely heterogeneous channels. The first, just as in the Sanity Check Pattern, is the actuator channel. This channel provides the system service. The second channel monitors the physical result of the actuation channel using one or more independent sensors. The concept of the pattern is that if the actuator channel has a fault and performs incorrectly, the monitoring channel identifies it and can command the system to a fault-safe state. If the monitoring channel has a fault, then the actuation channel is still behaving correctly. This pattern assumes there is a fault-safe state and that the sensor(s) used by the monitoring channel are not used by the actuation channel.

The primary differences between the Monitor-Actuator Pattern and the Sanity Check Pattern are:

1. The Sanity Check Pattern is a low-fidelity check on the output of the computation of the actuation channel; this does not require an independent sensor.
2. The Monitor-Actuator is a high-fidelity check on the physical result of the actuator channel and therefore requires an independent sensor.

Consequences

This is a medium-weight pattern. It requires some additional recurring cost because of the additional sensor(s) and processing capability for the monitoring channel. The extra monitoring channel requires a modest amount of additional design work as well. As mentioned, this channel provides single-point fault safety for systems in which there is a fault-safe state.

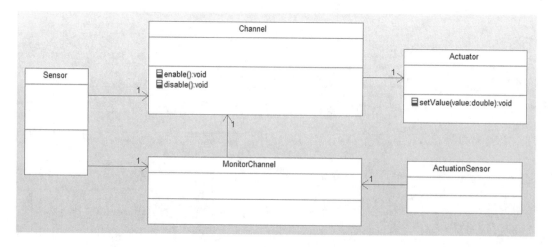

Figure 6-17: Monitor-Actuator Pattern

6.7.8 Example

The example shown in Figure 6-18 is a heterogeneous dual channel pattern for a train speed control system. One channel uses a light shining on the inside of a train wheel. By knowing the circumference of the wheel and by measuring the frequency of the mark appearance (measured by the optical sensor), one can compute the speed of the train. These data are then filtered, validated, and used to compute a desired engine output. This engine output is checked for reasonableness and, if acceptable, then sent to the engine. This is the `OpticalWheelSpeedChannel` shown in Figure 6-19.

The other channel uses a GPS to determine train speed. This channel receives ephemeris data from some set of (at least three) satellites and using the time sent (and comparing it to the time within the system), it computes the distance to the satellite. Through a process called 3D triangulation, the train's position can be calculated. This location is validated and then used to compute the speed by looking at the rate of change of positions. This speed is then validated and, if valid, is used to compute the desired engine output to maintain current speed. This is the GPSSpeedChannel shown in Figure 6-20.

In either channel, if data are identified as invalid, that channel does not command the engine speed. Because there is another channel to take over the task of determining desired engine output, the system can continue to operate even if a single point fault results in a channel shutting down. Further, because this is an extremely heterogeneous design, an error in one channel will not be replicated in the other. This design provides protection against both random and systematic faults.

There is nothing profound about the code that is produced to implement this example beyond what has been described earlier in this chapter. The interesting part is how the channels are

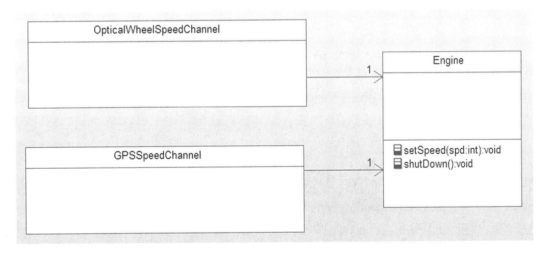

Figure 6-18: Dual Channel example

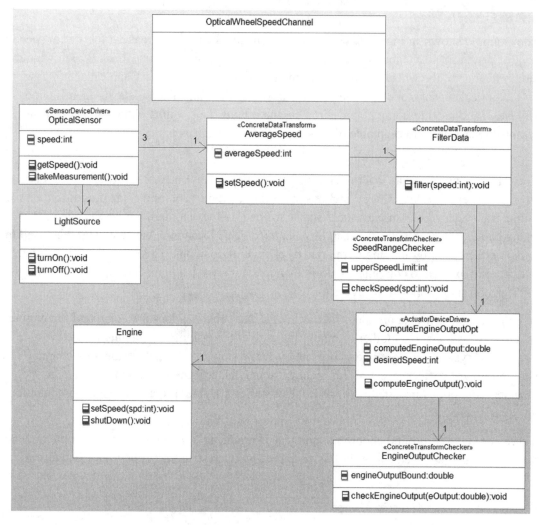

Figure 6-19: Details of Optical Wheel Speed Channel

connected to provide the ability of the system to detect a fault and continue safe operation in the presence of the fault.

6.8 Summary

This chapter has presented a number of different patterns that can be used to improve the safety and reliability of a design executing in its operational context. The first few patterns were small in scope (often referred to as "design idioms" instead of "design patterns"). The One's Complement Pattern replicated primitive data to identify run-time corruption – a problem common to harsh environments including medical operating rooms, industrial factory settings,

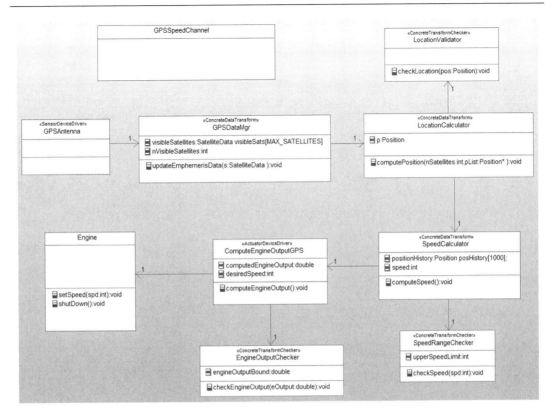

Figure 6-20: Details of GPS Speed Channel

and space. The CRC Pattern is useful in protecting larger volumes of data. The Smart Data Pattern adds behavior to data (that is, creates classes) to ensure that the data's preconditions and constraints are met.

The other patterns in this chapter have a larger scope. The fundamental pattern for the rest of the chapter is the Channel Pattern, in which the processing of sensory data into actionable commands is set up as a series of data transform steps. This pattern is elaborated into the Protected Single Channel Pattern in which checks are added at various steps to validate the processing within the channel. The Dual Channel Pattern adds two channels so that even if one fails, the other can continue providing the required service.

UML Notation

The UML is a rich and expressive language. The purpose of this appendix is to help you understand the most important notations of the language. For more detailed understanding of the richness of the UML language, the reader is referred to any of a number of books exploring the UML language in detail, such as my own *Real-Time UML: Advances in the UML for Real-Time Systems* (Addison-Wesley, 2004).

In the context of this book, UML was used to graphically depict both the structure and the behavior of the patterns. Three of the set of diagrams of the UML used in this book are discussed here: the Class Diagram, the Sequence Diagram, and the State Diagram. The UML provides a number of other diagrammatic types (not discussed here) and even for the diagrams discussed, there are other lesser-used features that are not discussed in this appendix.

1.1 Class Diagram

The Class Diagram is for showing system structure. It does this by showing Classes (specifications of software elements), Objects (instances of classes), and relations of various kinds. The UML is quite flexible about the amount of detail shown on a diagram.

1.1.1 Semantic Elements

This section describes the various elements on the Class Diagram.

1.1.1.1 Nodal Elements

1.1.1.1.1 Class
A class is a representation of an element that combines data elements (called attributes) with functions that act on them (called operations). Classes are normally implemented as structs containing the attributes.

1.1.1.1.2 Instance
An instance is a creation of a new variable of the type of the struct that defines the class.

1.1.1.1.3 Interface
An interface is a specification of a set of services (operations and/or event receptions) that a class must implement. An interface has no implementation and no attributes; it cannot be

Design Patterns for Embedded Systems in C
DOI:10.1016/B978-1-85617-707-8.00007-8

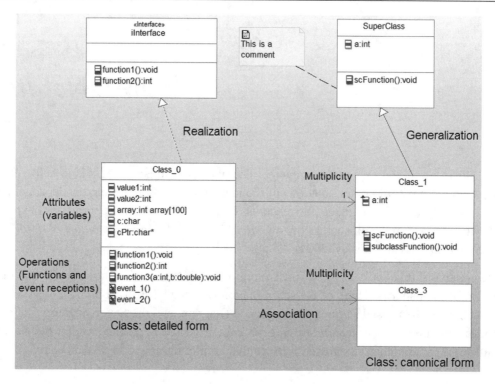

Figure A-1 Basic Class Diagram

instantiated. An interface is depicted with the same kind of box as a class, so it is visually distinguished with the «Interface» stereotype.

1.1.1.1.4 Attribute
A variable held within a class.

1.1.1.1.5 Operation
A function held within a class that typically manipulates the class attributes. In this book, the common implementation style of an operation is to use the notation <Class Name>_<Operation Name>, such as SensorClass_getData(). The first parameter passed to the class is a pointer to the instance of the class.

1.1.1.1.6 Event Reception
A special kind of operation that is invoked asynchronously and causes event transitions to take place on the state machine associated with the class.

1.1.1.1.7 Comment
A graphical way to representing textual comments. It may be anchored to one or more elements on the diagram or may be unanchored.

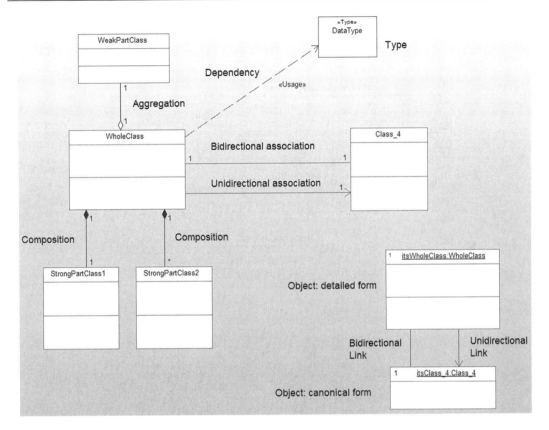

Figure A-2 More Class Diagram notations

1.1.1.2 Relations

1.1.1.2.1 Association

The specification of a navigable connection among classes. It may be unidirectional or bidirectional. Each end of the association is called an association role end and has a *multiplicity*. The multiplicity is the number of instances of the target class that participates in the interaction. "1" is the most common multiplicity and "*" means "zero or more".

An association is most commonly implemented as a pointer or collection of pointers.

1.1.1.2.2 Link

A link is an instance of an association; that is, a link exists when the pointer is populated with the address of the instance; if the pointer is NULL, then the link doesn't exist.

1.1.1.2.3 Generalization

Generalization is an is-a-specialized-kind-of relation with the arrow pointing to the more general class. The super- (more general) class's features, including attributes, operations, and relations, are all inherited by the sub- (more specific) class.

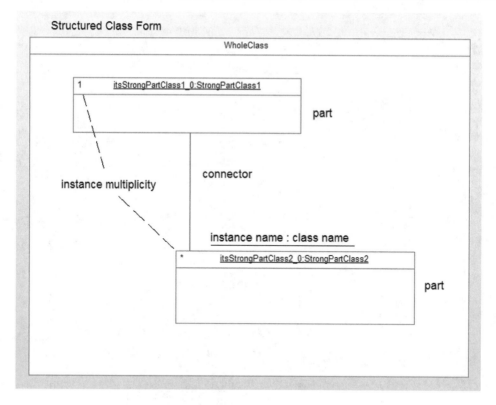

Figure A-3 Structured class notation

1.1.1.2.4 Realization

Realization is the relation between an interface and a class that implements the services specified by an interface.

1.1.1.2.5 Aggregation

Aggregation is a weak form of a whole-part association between two classes where the diamond indicates the "whole" end. It is weak in the sense that this relation does not imply creation and destruction responsibilities on the part of the whole.

1.1.1.2.6 Composition

Composition is a strong form of a whole-part association between two classes; again, the (now filled) diamond is at the end of the whole. It is strong in the sense that the whole is responsible for both the creation and destruction of the part instance.

1.1.1.2.7 Dependency

This is a general relationship not otherwise denoted. It is common to add a stereotype (such as «Usage») to indicate what kind of dependency you want. The «Usage» dependency is common

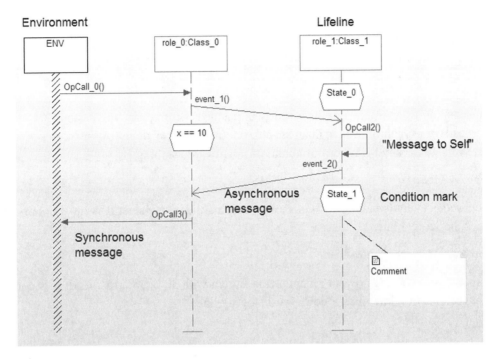

Figure A-4 Sequence Diagram

and indicates that the header of the target class or element should be included by the source class during compilation.

1.1.1.2.8 Type

A type is just that – a data type without the richness of a class to add behavior. It can be an enumerated type (enum), structured type (struct), array, typedef, or even a #define.

1.2 Sequence Diagram

A Sequence Diagram shows a particular sequence of messages exchanged between a set of instances. The instances are shown as vertical lines known as lifelines. Messages are shown as horizontal or downward slanted lines. Time flows (more or less) down the page. In addition, condition marks can be used to show state or the current value of attributes during the sequence.

1.2.1 Semantic Elements

Lifeline

A lifeline represents the usage of an instance. At the top of the lifeline, it is common to both name the instance and identify the type of class of the instance.

Environment Lifeline

This is a special lifeline that stands for "everything else." It provides a notational shorthand for elements omitted from the diagram.

Message

A representation of an interaction between two instances. It may be synchronous (as in a function call) or asynchronous (as in an asynchronous event exchange). A message is shown as a directed line that may be either horizontal or downwards sloping.

Synchronous Message

A synchronous event (function call) is shown with a solid arrowhead. It is most commonly draw as a horizontal line.

Asynchronous Message

An asynchronous message (event reception) is shown with an open arrowhead. It is common (but not required) to draw it as a downward sloping line.

Condition Mark

This is a hexagon that contains a textual condition that is true at that point in the sequence, such as the value of a variable or the current state in a state machine.

Comment

It may be anchored to one or more elements on the diagram or may be unanchored.

1.3 State Diagram

A State Diagram is a directed graph consisting of states (rounded rectangles) and transitions (arrowed lines). It defines the order of action execution for its owner class.

1.3.1 Semantic Elements

1.3.1.1 State
A state is a condition of an instance of a class. A state may contain entry actions, exit actions, and internal transitions. It is drawn as a rounded rectangle.

1.3.1.2 Nested State
States may contain lower level, nested states. For example, while the DOOR_IS_OPEN state is true, it might be in DOOR_OPENING, DOOR_CLOSING, or DOOR_WAITING_OPEN nested state. Nested states are drawn by graphically including the nested state inside a composite state.

1.3.1.3 Composite State

A state that contains nested states. These may be OR-states (as in Figure A-5 and Figure A-6) or AND-states (as in Figure A-7).

1.3.1.4 OR-State

A set of states at a given level of abstraction; the semantics of state machines are that at a given level of abstraction, the state machine must be in exactly one of its OR-States. In Figure A-5, state_0 and state_1 are OR-states; similarly, NestedState1 and NestedState2 are OR-states.

1.3.1.5 AND-State

AND-states are orthogonal regions within a composite state with the semantics that the instance must be in exactly one AND-state in each orthogonal region. AND-states are indicated with a dashed line in the composite state. While it is theoretically possible to implement AND-states in separate threads, this is almost never done. If an event is shared between AND-states, each AND-state must be allowed to independently act on it or reject it, but the order of execution of the transitions triggered by the same event is not known in principle. This leads to the possibility of race conditions.

If you look at Figure A-7, the transition from state_0 to state_1 is triggered by the event1 event. In this case, the instance enters both states UpperState1 and LowerState3; the instance is in both states at the same time. AND-states are most commonly used to model independent properties of an instance.

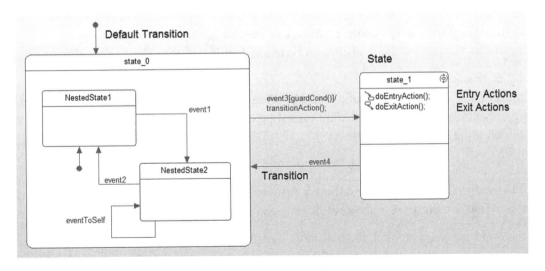

Figure A-5 Basic State Diagram

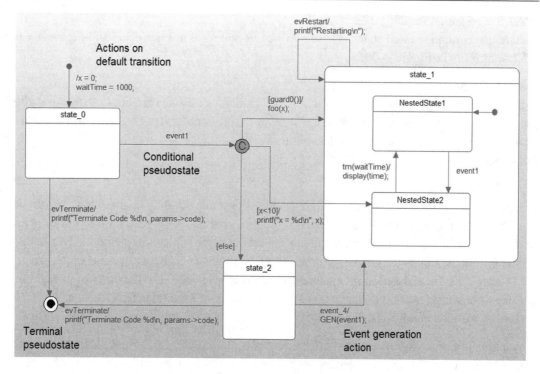

Figure A-6 Advanced State Diagram

Note also in the figure the transition from `state_1` to `state_0` triggered by the event `evGoBack`. In this case, it doesn't matter which combination of nested states is currently true; as long as the instance is in `state_1`, the transition is valid. However, in the event from nested state `LowerState3` to `state_2`, triggered by event `event10`, which nested state in `ANDStateRegion1` is true doesn't matter, but the instance must be currently in `LowerState3` state for that transition to be taken. If it is not, then the event is discarded.

1.3.1.6 Transition

A transition is a representation of a response to an event while the instance is in a specified state. It typically has an event signature consisting of an event trigger, guard, and list of actions, specified using the following syntax:

```
event trigger '[' guard condition ']' '/' action list
```

All of the fields in the event are optional. A transition is shown on a state machine as an arrowed line coming from a predecessor state and terminating on a subsequent state.

If the event-trigger is omitted, the transition "fires" as soon as the predecessor state is entered, that is, as soon as the entry actions of the predecessor state complete. See Figure A-7 for an example.

1.3.1.7 Action

An action is a primitive behavior performed under some circumstance in a state machine. This is most often one of the following:

- A function call, for example `foo(x)`
- A direct primitive action, such as `x += 10`
- The generation of an event sent to the current instance or another instance known to the current instance; a common way to indicate this is with a `GEN(event)` macro
- A list of actions

Actions are separated from the rest of the event signature with a slash ('/') character.

1.3.1.8 Entry Action

An entry action is an action performed whenever a state is entered regardless of which transition is taken.

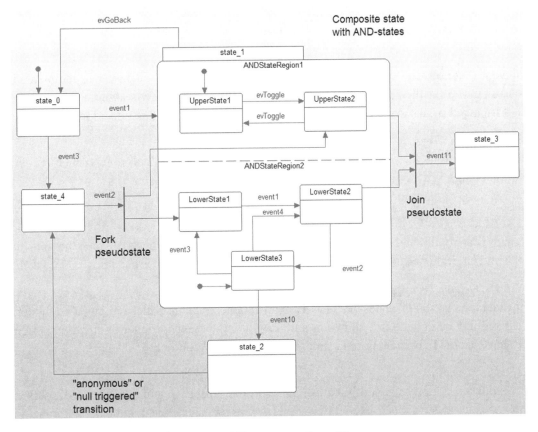

Figure A-7 AND-states on State Diagram

1.3.1.9 Exit Action

An exit action is an action performed whenever a state is left, regardless of which transition is taken.

1.3.1.10 Transition Action

A transition action is an action performed only when a specific transition is taken.

1.3.1.11 Internal Transition

An internal transition is when an action is performed when an event occurs but a state transition is not taken (that is, the exit and entry actions are not performed).

1.3.1.12 Event

An event is a specification of an event of interest that is received by the instance. It may be received synchronously or asynchronously. It may carry parameters, just like a function call. The syntax used here is that the event that carries data put those data into a `struct` called `params`. This struct may be dereferenced to access the data. For example, in Figure A-6, the `evTerminate` event carries a data element code. The actions in the state machine access the parameter by dereferencing the params `struct`.

Events may trigger a transition or an internal transition; if they do not, events are quietly discarded with no actions taken.

An event is consumed by the state machine; that is, once the state machine has completed its processing step for the current event, it is destroyed. Even if the newly assumed state has an event triggered by the same event, the event must occur again for this transition to be taken.

1.3.1.13 Event Trigger

The event trigger is the specification of an event. This may be asynchronous (which requires event queuing), synchronous, a timeout (usually indicated with a `tm(duration)` or `after (duration)` syntax), or when a state variable is changed.

1.3.1.14 Guard

A guard is an (optional) Boolean expression. If the guard evaluates to `TRUE`, then the transition is taken; if it evaluates to `FALSE`, then the event is discarded and the transition is not taken.

1.3.1.15 Default Transition

The default transition indicates that at a given level of abstraction, which, among a set of states, is entered first by default. It is shown as a transition with no event trigger or guard (although it may have actions) terminating on a state, but not beginning on one.

1.3.1.16 Pseudostate

A pseudostate is a perhaps poorly-named annotation of special semantics. The UML defines many pseudostates. In this limited description here, pseudostates include the default transition, conditional, fork, join, and terminal pseudostates.

1.3.1.17 Conditional Pseudostate

The conditional pseudostate indicates multiple potential outgoing transitions initiated by a single event with the proviso that only one such path will actually be taken. The different paths are selected on the basis of guards; that is, each transition segment will have a guard that will be evaluated to TRUE or FALSE. If multiple transition segment guards evaluate to TRUE, then any one of the true paths can be taken. If no guards evaluate to true then the event is discarded and no transition is taken. A special guard "else" is provided; if present, this transition path will be taken only if all the other transition segments are rejected.

1.3.1.18 Fork

When entering a composite state with nested AND-states, a fork can be used to specify nondefault states to enter, as in Figure A-7. The event event2 triggers the transition from state_4 to state_1, but specifically enters nested AND-states UpperState2 and LowerState1, bypassing the default states of UpperState1 and LowerState3.

1.3.1.19 Join

When leaving a composite state with nested AND-states, a join can be used to state the precondition that the transition should only be taken if the specific set of states (one per AND-state) is true when the event occurs. See Figure A-7.

1.3.1.20 Terminal Pseudostate

The terminal pseudostate is a special indicator that means that the instance no longer accepts events, usually because the instance is about to be destroyed.

Index